AF323662

Response Modeling Methodology

Series on Quality, Reliability and Engineering Statistics Vol. 8

Response Modeling Methodology

Empirical Modeling for Engineering and Science

Haim Shore

Ben-Gurion University of the Negev, Israel

World Scientific

NEW JERSEY · LONDON · SINGAPORE · BEIJING · SHANGHAI · HONG KONG · TAIPEI · CHENNAI

Published by

World Scientific Publishing Co. Pte. Ltd.

5 Toh Tuck Link, Singapore 596224

USA office: 27 Warren Street, Suite 401-402, Hackensack, NJ 07601

UK office: 57 Shelton Street, Covent Garden, London WC2H 9HE

British Library Cataloguing-in-Publication Data
A catalogue record for this book is available from the British Library.

Series on Quality, Reliability and Engineering Statistics — Vol. 8
RESPONSE MODELING METHODOLOGY (RMM)
Empirical Modeling for Engineering and Science

ISBN 981-256-102-1

Printed in Singapore by World Scientific Printers (S) Pte Ltd

Dedicated to my parents, Hava and Daniel, whose lives
have heroically defied adversity

Preface

This book is about "Empirical modeling of a response variation". Each term in this phrase requires an explanation, which will be provided. Suffice here to say that in "empirical modeling" we attempt to capture the variation observed in a response (the variable which is the subject of our modeling effort) so that as few as possible theoretical pre-conceived assumptions would be needed in the modeling effort.

While in scientific, and most often also in engineering and statistical modeling endeavors, some theory serves to derive the final model (though not necessarily the values of its parameters), in empirical modeling we let the data determine most of the model's properties. Various current major modeling approaches, like linear regression or generalized linear modeling (GLM), are examples of general methodologies for "empirical modeling". Their shortcomings, which will be discussed, have motivated the development of the new empirical modeling approach.

"Response Modeling Methodology" (RMM) is a new general system of empirical modeling, which provides the needed alternative to current methodologies. Unlike existing approaches, which engage either in empirical modeling of systematic variation (relational modeling) or in empirical modeling of random variation (modeling the distribution of a response in a steady-state condition), RMM provides platforms for both types of modeling. This is one feature, among some, which set RMM apart from currently available methodologies. The unique features of RMM will be discussed and elaborated on throughout this book.

Numerous publications that have been published in refereed journals since the original inception of RMM in the first years of this decade (a references list is provided at the end of this preface) attest to the wide scope and general effectiveness of the new approach.

In this book, RMM is introduced for the first time in a comprehensive manner, allied fitting and estimation procedures are developed, and comparisons are carried out between current major methodologies for empirical modeling, based on ten requirements that any such methodology should satisfy.

The last two parts of the book demonstrate how RMM can be applied to model systematic variation and random variation. Various areas are addressed, like chemical engineering, reliability engineering, forecasting, statistical process control and inventory analysis, to name a few.

The statistical knowledge required to understand the theory behind the new approach, or to implement it to given data, is basic. An undergraduate or a graduate student, who has been exposed to some introductory courses in statistics theory and statistical analysis, and is familiar with current basic model-fitting methodologies, will feel comfortable at reading the text.

For specific subject areas that require some preliminary knowledge, a brief introductory background is provided and the reader is referred to relevant references. Thus, when general control-chart schemes are developed for processes that are not necessarily normal, the theory behind Statistical Process Control (SPC) would not be expounded in detail. However, reference will be made to related textbooks which can assist in learning subject-area concepts, needed to understand the developed control schemes.

Additionally, this book is accessible to the large community of engineers, operations-researchers and scientists who are engaged in data-driven modeling. In particular, for practitioners of quality and reliability engineering, the new RMM constitutes a viable alternative to the current "arsenal" of modeling methodologies, used to empirically model either systematic variation in normal/non-normal environments or to empirically model random variation.

The structure of the book is now outlined briefly. We start with an introductory chapter, where basic terms and principles underlying empirical modeling are discussed. The pursuing chapters are divided into four parts:

Part I. Current Models and Modeling Methodologies
This part acquaints the reader with some current mainstream models in engineering and the sciences so that s/he will later have a good sense of how compatible are these with models developed via RMM. Models of

both systematic variation and random variation are addressed. A second purpose of this part is to provide a review of current major approaches to empirical modeling of variation (both systematic and random). This is accomplished by a general, mostly non-technical, exposition of the principles underlying current approaches. A third objective of Part I is to lay down the requirements that any empirical modeling approach should fulfill. Once these requirements are expounded, current approaches (excluding RMM) are evaluated for compliance with the requirements. RMM will similarly be evaluated, in Part II of the book, after the RMM approach is developed.

Reading Part I, the reader will be prepared for the development of the new approach in Part II.

Part II. Response Modeling Methodology – Developing and Evaluating the General Approach

This part describes the new modeling methodology and its fitting and estimation procedures. Some of the scientific and engineering relational models that have been introduced in Part I are shown to be special cases of the RMM model (though they all are). Likewise, we show for some current models of random variation (statistical distributions, response transformations and distributional approximations) that they are indeed special cases of the RMM model.

Reading Part II, the reader will understand how RMM may be used for empirical modeling of variation, so that applications introduced in Parts III and IV would be within easy grasp for a reader familiar with the basics of the subject area addressed.

Part III. Modeling Systematic Variation - Applications

This is the first of two parts that demonstrate implementation of the RMM approach. Part III addresses applications to relational modeling. It comprises five chapters. The first, Chapter 14, uses some published data-sets to derive RMM models and estimate their parameters. These are then compared to solutions obtained by current approaches. In the pursuing Chapters 15-18, four specific subject areas are addressed: Hardware reliability-engineering, software reliability-growth models, chemical-engineering and forecasting.

Part IV. Modeling Random Variation - Applications

Like its predecessor, this part of the book develops new RMM-based models for various areas in engineering and operations-management. It starts with Chapter 19, where we demonstrate the goodness-of-fit achievable when the RMM error distribution is used to model (replace) some current commonly encountered statistical distributions. Inverse normalizing transformations (INTs) and piece-wise linear approximations, both derivatives of RMM that may be useful in modeling random variation, are developed and demonstrated in Chapters 20 and 21, respectively. We then proceed to demonstrate application of RMM and its derivatives to Statistical Process Control (SPC), when traditional Shewhart charts can not be implemented (Chapter 22), and to modeling and optimization in the area of inventory analysis (Chapter 23).

Some review questions are provided in the last chapter.

Throughout the book, $\log(Y)$ means $\log_e(Y)$ $[\ln(Y)]$. Relevant references are given separately for each chapter, rendering the book more user-friendly. RMM-related materials (like updated references list, data-sets and algorithms) may be found at the author's personal site, reachable from: http://www.ie.bgu.ac.il/iem/.

The author wishes to thank Miss Chelsea Chin of World Scientific Publishing, and her team, for their valuable assistance throughout the arduous editing process. Their share in obtaining the required quality is very much appreciated.

We hope that this book will provide a fresh approach to empirical modeling in engineering, operations-management and applied science, where current tools may be too complicated to implement, require too many theoretical assumptions to apply, or do not relate respectfully to the story that the data are telling.

HAIM SHORE
Beer-Sheva
January, 2005

References

Comment: The references below are an updated list of publications, which address RMM and its derivatives

[1] Shore, H. (2000a). General control charts for variables. *International Journal of Production Research*, 38(8), 1875-1897.

[2] Shore, H. (2000b). General control charts for attributes. *IIE Transactions (Quality and Reliability Engineering)*, 32 (12), 634-637.

[3] Shore, H. (2000c). Three approaches to analyze quality data originating in non-normal populations. *Quality Engineering*, 13 (2), 277-291.

[4] Shore, H. (2001a). Modeling a non-normal response for quality improvement. *International Journal of Production Research*, 39 (17), 4049-4063.

[5] Shore, H. (2001b). Inverse normalizing transformations and a normalizing transformation. In *"Advances in Methodological and Applied Aspects of Probability and Statistics"*, Book, N. Balakrishnan (Ed.), Gordon and Breach Science Publishers. Canada. V. 2, 131-146,

[6] Shore, H. (2002a). Modeling a response with self-generated and externally-generated sources of variation. *Quality Engineering*, 14(4), 563-578.

[7] Shore, H. (2002b). Response Modeling Methodology (RMM) - Exploring the implied error distribution. *Communications in Statistics (Theory and Methods)*, 31(12), 2225-2249.

[8] Shore, H., Brauner, N., and Shacham, M. (2002). Modeling physical and thermodynamic properties via inverse normalizing transformations. *Industrial and Engineering Chemistry Research*, 41, 651-656.

[9] Shore, H. (2003). Response Modeling Methodology (RMM) - A new approach to model a chemo-response for a monotone convex/concave relationship. *Computers and Chemical Engineering*, 27(5), 715-726.

[10] Shore, H. (2004a). Response Modeling Methodology (RMM) - Validating evidence from engineering and the sciences. *Quality and Reliability Engineering International*, 20, 61-79.

[11] Shore, H. (2004b). The Random Fatigue-Limit Model as a special case of the RMM model - A comment on Pascual (2003a). *Communications in Statistics (Simulation and Computation)*, 33(2), 537-539.

[12] Shore, H. (2004c) Response Modeling Methodology (RMM) - Current distributions, transformations, and approximations as special cases of the RMM error distribution. *Communications in Statistics (Theory and Methods)*, 33(7), 1491-1510.

[13] Shore, H. (2004d). Response Modeling Methodology (RMM)- Maximum likelihood estimation procedures. *Computational Statistics and Data Analysis*. In press.

[14] Shore, H. (2005). Accurate RMM-based approximations for the CDF of the normal distribution. *Communications in Statistics (Theory and Methods)*, 34(3). In press.

Contents

12 Special Cases of the RMM Model — 203

13 Evaluating RMM for Compliance — 219

PART III. MODELING SYSTEMATIC VARIATION - APPLICATIONS — 227

14 Comparative Solutions for Relational Models — 229

PART IV. MODELING RANDOM VARIATION - APPLICATIONS   315

19 RMM Distributional Approximations   317

20 Inverse Normalizing Transformations   335

CONTENTS

Chapter 1

Introduction

This monograph is about "empirical modeling of a response variation". In this introductory chapter we elaborate on this concept and explain its various components.

Scientific and engineering modeling aims to capture the variation of a "response". The latter refers to any characteristic, or variable, where variation has been observed, and this variation is the subject of a scientific or engineering research effort. As the word implies, a "Response" is the "reaction" of an investigated characteristic to the effects exercised on it by external factors. The scientific or engineering investigation attempts to model this relationship via some mathematical expression (or a set of mathematical expressions).

Consider fatigue-life of a product that experiences stress. Fatigue-life may serve as a response and research effort is exercised to learn how it "responds" to stress. This "learning effort" result in our ability to characterize fatigue-life behavior and subsequently model the variation observed in fatigue-life in terms of variation in the observed (or applied) stress. Similarly, an electronics-engineer would never be able to design a product without the prior research effort that culminated in modeling a response of interest in terms of design factors that affect it. For example, the voltage of an electric circuit. expressed in terms of the electric resistance or capacity of components embedded in the circuit.

"Variation" is another term that needs clarification. There are two components to this term: "Systematic variation" and "Random variation". Most responses that we may wish to model exhibit both sorts of variation. Let us address the former first.

"Systematic variation" refers to observed variation in the response that reflects non-stability in the response underlying statistical distribution. For example, the mean may vary or the variance may not be constant. A "model of systematic variation" explains variation in the parameters of the statistical distribution in terms of variation in factors that are known to be related to the response. Referring to the fatigue-life example, the response variation may be modeled by a Weibull distribution. However, if stress is an influencing factor, we may expect that the mean of this distribution will not be constant. In this case, the mean may be modeled in terms of the applied stress, and again a model of systematic variation is the end result of the modeling endeavor.

Systematic variation, however, is not confined to responses with a characteristic statistical distribution. In fact, even when no random fluctuations are observed in the response, we may still observe "systematic variation". In that case, it is very likely that a mathematical (deterministic) model could be developed, which will predict, with no random error, how the response would vary when we change the value of the factor(s) known to affect it. A good simple example is the well-known equation that relates distance traveled (S) to time traveled (t): $S=Vt$. This is a deterministic model that explains systematic variation of the response, S, in terms of t. In most empirical modeling approaches, we implicitly assume that the modeled response exhibits both random and systematic types of variation. We will elaborate on it further shortly.

"Random variation" relates to random fluctuations that are observed in the response. These fluctuations form the statistical distribution (or error distribution) of the response. If the statistical distribution has parameters that are constant, then the response is said to experience only random variation. Most often, the response random fluctuations may be captured by some known statistical distribution, for example, the normal distribution. If the parameters of the distribution themselves are not constant, but vary in a systematic manner (probably due to the effects exercised by some external influencing factors), then we would say that the response experiences both "Systematic variation" and "Random variation".

As we have alluded to earlier, most current methodologies for *empirical* modeling assume that the response experiences both types of variation. For example, linear regression analysis assumes that the response has random variation captured by a normal distribution. However, this modeling methodology is intended primarily for modeling systematic variation. This is achieved by assuming that the mean of the normal distribution is not constant but varies due to some external factors. In linear regression, this transmitted variation is modeled by a linear combination of the related effects. Thus, while the normal distribution provides, in linear regression, a model for the "Random variation" component in the observed response variation, the non-constant mean provides a model for the component of "Systematic variation".

In this book, we extend the definition of a "Response" to include not only variables that experience systematic variation (namely, "respond" to variation in related factors), but also to variables that experience only random variation (no *systematic* variation is transmitted via external affecting factors). To demonstrate what is implied by this extension, suppose that we investigate a process variable, defined for a process that is assumed to be in a state of statistical stability ("in control"). This implies that the variable "experiences" only random variation, which may be captured (represented or modeled) by some known statistical distribution. Traditionally, the common assumption is that this random variation is characteristic to the variable, and it originates in a large set of random causes none of which plays a dominant role in determining the value of the response. As we will realize later in the book, this conception is not upheld by RMM. The RMM model assumes that either or both random variation *and* systematic variation may be transmitted to the response. Thus, a variable that experiences only random variation may be conceived as "responding" to random variation transmitted to it from without, no less than a response whose systematic variation is a reflection ("a responding") to systematic variation transmitted to it by external factors. The extension, made in this book, of the term "Response" to include variables that are subjected to only random variation is thus corroborated.

Consider the term in the sub-title of the book (also given in the opening sentence of this chapter): "Empirical modeling". What do we mean by that, and how is "empirical modeling of a response variation" distinguished from "just" modeling?

Let us consider a few examples. Chemical engineers are interested in modeling liquid density in an industrial process in terms of temperature, and modern physicists model the energy of a particle that travels near the speed of light in terms of its mass and its velocity. In quality engineering, practitioners of statistical process control (SPC) wish to model the patterns of random variation of a process variable, which they wish to monitor, in terms of the underlying process distribution (or some approximation thereof). Likewise, software reliability engineers are interested in modeling the number of errors to be found in the testing phase of the software development process in terms of the effort (labor hours) invested in testing. Finally, statisticians who wish to test the hypothesis that the variances in two statistical populations are equal, may conceive of a certain statistic, calculated from available sample data, by which to test this hypothesis. They would be interested in capturing the random variation of this statistic (given that the hypothesis of equal variances is true) so that various probabilities can be calculated. Deriving the statistical distribution that governs the random variation of the statistic will provide the needed model.

For all of these "Responses", models for their observed variation have already been developed. For the first response (liquid density), Rackett's equation (1970) is a formula, well-known to chemists and chemical engineers, that expresses liquid density in terms of temperature. The second example obviously relates to Einstein's well-known formula, derived from the Special Theory of Relativity. Quality-engineering practitioners have at their disposal some families of distributions (like Person, Johnson or Burr), that are flexible enough to deliver good representation for the variation of a process variable, either for purposes of SPC or for process capability analysis (PCA). Software reliability-growth models have been developed (based on the Poisson process) that assist software testers in deciding when to stop testing since the predicted number of errors left in the product does not justify further testing effort. Finally, the F-ratio distribution is known to provide a model for the common F statistic, used to test the equality of the variances of two unrelated statistical populations.

While scientific, engineering and statistical research is often conducive to the development of theories, from which certain models for the observed response variation can be developed, in many cases such a desirable result cannot be achieved. This may be due either to the complexity of the environment in which the response is operating, due to lack of satisfactory theory to explain how the response is related to various affecting factors, or simply due to the complexity of the procedure by which a useful model can be developed.

Consider the previous examples. The first two examples (Rackett's equation and Einstein's formula) demonstrate how explanatory models have been obtained from theoretic principles. The same is true for the final example: The F-ratio distribution is theoretically derivable as the exact distribution (under the null hypothesis) of the F statistic, used to test equality of variances. However, for the third and the fourth examples, application of theoretic principles is harder to accomplish. To model the distribution of a process variable, namely, to find the "true" underlying process distribution, a practitioner of SPC or PCA may opt to either establish that certain conditions exist that necessarily generate a specified distribution (a good example is the Poisson distribution), or s/he may be implementing a "distribution identifying procedure", now available in any statistical analysis package, in order to find the "true" process distribution. Both approaches incur considerable difficulties, which will be addressed later in the book (Chapter 5).

Alternatively, a practitioner may resort to "empirical modeling". Instead of attempting to find the exact ("Correct") distribution, one may use a member of some known parameter-rich family of distributions that can be shown to fit well the available process data. Goodness-of-fit criteria may then "justify" the adoption of the selected distribution as a satisfactory representation of the underlying "unknown" distribution. No claim is made that this is the true process distribution. No theory is assumed that may corroborate this "adoption". The whole analysis is solely data-driven.

Similarly, in developing software reliability-growth models, we may assume from theory that in testing a software product the rate of occurrence, or detection, of software errors follows a Poisson distribution. Yet there is no theory to explain how the mean of this distribution (most commonly, the mean of the *cumulative* number of errors detected) is changing as a function of the number of testing hours put in the testing effort. Consequently, for software reliability-growth models, *empirical* modeling of the mean of the Poisson, in terms of labor testing hours, is required, which will be shown to fit well available observations.

Our inability to always model a response by available relevant theory, especially in the engineering disciplines, has led to the development of "general systems of empirical modeling". A well-known "general methodology for empirical modeling of *systematic* variation", no doubt familiar to the reader, is linear regression analysis. Subject to some generally valid set of assumptions, known to form "the normal scenario" (an exact definition of this will be given later in the book, in Chapter 4), linear regression allows one to model a response's systematic variation in terms of the affecting effects so that good compliance (fit) with available observations is achieved. Yet in modeling via linear regression, we do not delve into the details of how theoretic principles govern the relationship between the response and the associated factors, nor are we interested in how theory may be used to develop this relationship.

"General methodologies for empirical modeling of *random* variation" also exist, though based on different principles. If one wishes to model a response's random variation without qualifying the theory underlying the observed variation, a well-known statistics-theory result may serve as the basis for the derivation of the required methodology. A statistics-theory theorem asserts, loosely speaking, that two distributions, each of which have defined and finite moments of all degrees (the mean and the variance, for example, are moments of first and second degrees, respectively), are identical if they share the same moments. Since as more moments are shared by two distributions the similar their "patterns of random behavior", any distribution with enough parameters can be used to model the underlying unknown distribution, provided the "fitted" distribution shares enough moments with the unknown distribution. This is achieved by setting the values of the parameters of the fitted distribution so that moment-matching with the unknown "true" distribution is obtained. For example, if the fitted distribution has three parameters, then matching of the first three moments can be obtained, where the moments of the true unknown distribution are estimated from available sample data.

To demonstrate how this principle has been applied in the past, and quite successfully, consider the well-known Shewhart control charts. When Shewhart used the normal distribution to build his "control charts", back in the mid-twenties of the previous century, what he had actually done was to fit the normal distribution to the presumably "unknown" distribution. This was done by *matching* of the first two moments of the normal distribution with those of the true (occasionally unknown) process distribution. For example, the commonly applied p control chart, used to monitor the stability of a "Percentage" variable, assumes that the process distribution is normal, though we know that the true underlying distribution is binomial. Ensuring that the mean and the variance of the normal distribution, used to construct the control limits, coincide with those of the binomial distribution, one can rest assured that in this case the control limits reflect well the true distribution of the monitoring statistic. Equality with the true process distribution (the binomial) of the first two moments has allowed us to use the normal distribution as an empirical model for the binomial distribution.

Shewhart charts constitute one example of "A general methodology for empirical modeling", this time of random variation, where equality of moments is the underlying principle that confers legitimacy on the methodology used. As judged by the general acceptability and degree of implementation of Shewhart control charts, this general empirical-modeling methodology is reasonably effective in capturing the underlying process variation.

When the process distribution is so skewed (asymmetric) so that matching of only the first two moments is not sufficient, matching of more moments may be required, for example, matching of the first four moments. In this case, as in the former, estimates of the "true" moments (of, recall, the unknown process distribution) are derived from sample observations of the monitored process.

Clements (1989) provides an example for the "moment matching approach", when the underlying unknown process distribution is asymmetric. For this case, Clements has suggested a methodology to obtain process capability indices, using a member of the four-parameter

Pearson family of distributions. Clements' approach for PCA is again "empirical modeling". No claim is extended that the modeled response necessarily follows a Pearson distribution. However, it is assumed that the latter is flexible enough to deliver good representation to the true (and unknown) process distribution. This assumption is supported by how Clements' method for calculating process capability indices (PCI) fits a Pearson distribution to the available data. By determining the parameters of the fitted Pearson distribution so as to preserve the first four moments of the true (unknown) process distribution (the moments are estimated from the data), the fitted distribution will most probably deliver good representation for the unknown process distribution. It may then be used to calculate the PCI, even though the true process distribution is still unknown (on some serious pitfalls of Clements method, refer to Shore, 2004).

Empirical modeling, this time of *random variation,* was once again the tool of choice.

We have demonstrated above how general systems of response modeling are used to model both systematic variation (the example given was linear regression) and random variation (two-moment and four-moment matching procedures were used as examples). The underlying motivation for all these methodologies is the realization that quite often theory is not sufficiently developed, or at times non-existent, to provide the basis for modeling the response variation. What characterizes these methodologies is a pragmatic approach. The latter maintains that a model is acceptable as long as it fits the data well (as judged by appropriate statistical criteria), even though there is no subject-area theory behind it. "Empirical modeling" may be an appropriate term to define the category to which all these methodologies belong.

What particular features set empirical modeling apart from scientific and statistical modeling? Consider modeling of systematic variation first.

Scientific relational models depart in a significant way from corresponding models derived by empirical modeling. In scientific modeling, random errors are rarely specified. As even a superficial examination of textbooks in the exact sciences would reveal (Halliday *et al.,* 1993), most models addressed in these books are assumed to be deterministic, with no allowance for random errors. There might be a good reason for this and a historical one. The good reason is that in many cases systematic variation associated with scientific models spans several orders of magnitude. The sheer magnitude of the systematic variation renders the component of random variation negligible by comparison. Random variation is therefore ignored. The historical reason is that scientific models and statistical models of random variation (read, statistical distributions) unfortunately developed relatively independently of each other. But this is a story that needs to be told elsewhere.

By contrast, in *empirical* modeling of *systematic* variation, we assume that the latter reflects variation in the *parameters* of a *statistical distribution* that governs the response variation. Thus, it is assumed that the response has an associated *error distribution*, which depicts its pattern of random fluctuations.

Earlier we have alluded to how linear regression models systematic variation. Perhaps it is instructive to return to this example in the current context. With linear regression analysis, we assume that the response is normally-distributed. However, we also assume that the mean is not constant, but varies in a systematic fashion due to variation transmitted to the mean by various affecting factors (or effects). Since systematic variation in the response is attributed to variation in some influencing factors, a relational model that ties together these two sources of variation is the end-result of the modeling effort. Unlike with common scientific modeling, both systematic and random components of variation are part and parcel of the final model.

In linear regression it is assumed that only the mean varies in a systematic fashion, while the variance is constant. This assumption is discarded by another general approach for empirical modeling of systematic variation, Generalized Linear Modeling (GLM). Therein, we assume that there is a certain specified relationship between the variance and the mean, so that systematic variation of the mean naturally affects also the response variance. In this case, as in the former, an assumed error distribution conveys a component of random variation, though modeling systematic variation is the prime goal of the modeling effort.

When we turn now to modeling of *random variation,* we find that *theoretical* statistical models also differ appreciably from *empirical* models of random variation. Let us address the former first.

Theoretical modeling of *random* variation commonly results in the derivation of a new distribution. Typically, statisticians derive new distributions in the context of some test statistics, the distributions of which they wish to identify in order to determine limit values in the framework of hypothesis testing or parameter estimation. A good example is the F-ratio distribution, which was originally derived in order to test homogeneity of variances. At times, a distributional characterization is attempted, which lay down the conditions that necessarily generate a certain distribution. In still other cases, a steady-state distribution is sought of some response in order to obtain good representation of its random fluctuations. For instance, the Weibull distribution has been originally derived by Weibull, back in the 1950's, to depict random fluctuations in the strength of materials.

In all these cases, it is assumed that the derived model describes the *true* random dispersion of the response, and a single distribution is the true response distribution. Another feature typical to most theoretically derived statistical distributions, exploring how various parameters may be affected by response-related factors (namely, characterizing an optional systematic variation) is not an integral part of the general specification of the model.

Empirical modeling of *random* variation departs in some major ways from how random variation is modeled within the science of statistics, as just described.

First, the criteria to judge whether the model is adequate is not based on the mathematical properties of the model (as done within statistics), but rather on goodness-of-fit criteria. This implies that different models may be used indiscriminately and they still all acquire a measure of legitimacy. Thus, in the quality engineering literature we may find the use of the Pearson family for process capability analysis (the Clements method, alluded to earlier), the use of the Johnson family of distributions to analyze quality data originating in non-normal populations (Polansky *et al.,* 1999, and references therein), or no specification at all of a fitted distribution in certain non-parametric control charts for non-normal populations (for example, Grimshaw and Alt, 1997, and references therein).

Secondly, and not unrelated to the first point, a major endeavor in statistics is the characterization of distributions, which attempt to formulate conditions leading necessarily to a certain *true* distribution (a well-known example is the memory-less property of the exponential distribution). Modeling in statistics thus implies that the *true* distribution is being modeled, and only a single model may at a certain point in time be the correct one. In *empirical* modeling of *random* variation- we do not care at all about the "true" distribution. As long as results from appropriate goodness-of-fit criteria support the adoption of the "fitted" empirical model, we are happy with it. An often-cited aphorism, attributed to Professor Box, comes to mind: "All models are wrong, some are useful".

Given the central role that empirical modeling plays in all branches of engineering, but most notably in quality and reliability engineering, it may come as a surprise that indeed no serious efforts have thus far been devoted to the formulation of requirements that empirical modeling should fulfill. In light of the myriad of methodologies and approaches available to the practitioner, who wishes to engage in empirical modeling, it seems highly desirable that in order to be able to evaluate and compare the quality of alternative methodologies, a set of requirements should be agreed upon, and compliance with these requirements be used as a measuring stick to judge, assess and compare the quality of various approaches to empirical modeling of systematic variation and of random variation.

This is done in Chapter 6, where requirements are defined, and evaluation of compliance with these requirements of major methodologies for empirical modeling is carried out.

In the pursuing Part I of the book, we describe current main stream relational models and current major methodologies for modeling systematic variation and random variation. Chapter 2 presents a sample of models that one can find in various scientific and engineering disciplines, where a monotone convex relationship exists between the affecting factor and the response. Features common to all these models are explored in Chapter 3, and "The Ladder of fundamental monotone convex functions" is introduced. The ladder later becomes one of the cornerstones for the axiomatic development of the RMM model.

Chapters 4 and 5 are dedicated to an overview of major methodologies for empirical modeling of systematic variation and of random variation, respectively. Having become acquainted with the current available arsenal of methodologies, we proceed in Chapter 6 to formulate requirements for empirical modeling, and accordingly assess the quality of existing methodologies.

References

[1] Clements, J. A. (1989). Process capability calculations for non-normal distributions. *Quality Progress,* 22(2), 49-55.

[2] Grimshaw, S. D., Alt, F. B. (1997). Control charts for quantile function values. *Journal of Quality Technology,* 24(1), 1-7.

[3] Halliday, D., Resnick, R., Walker, J. (1993). *Fundamentals of Physics, Extended, with Modern Physics.* 4th Ed. John Wiley&Sons.

[4] Pearson, E. S., Johnson, N. L., Burr, I. W. (1979). Comparisons of the percentage points of distributions with the same first four moments, chosen from eight different systems of frequency curves. *Communications in Statistics (Simulation and Computation),* B8(3), 191-229.

[5] Polansky, A. M., Chou, Y-M, Mason, R. L. (1999). An algorithm for fitting Johnson transformations to non-normal data. *Journal of Quality Technology,* 31(3), 345-350.

[6] Rackett, H. G. (1970). Equation of state for saturated liquids. *Journal of Chemical and Engineering Data,* 15(4), 514-517.

[7] Shore, H. (2004). Non-normal populations in quality applications- A revisited perspective. *Quality and Reliability Engineering International,* 20(4), 375-382.

Part I

CURRENT MODELS AND MODELING METHODOLGIES

Chapter 2

Relational Models in Engineering and the Sciences (Monotone Convex/Concave Relationships)

2.1. Introduction

Modeling systematic variation results in a relational model. Most scientific and engineering models are relational in nature. They attempt to explain observed variation in the response as a result of variation transmitted to it by some affecting factors. The constructed model is expected to convey the *relationship* between these two sources of variation, hence the term "relational".

In this chapter, current mainstream scientific and engineering relational models are introduced, which have been developed over the years in various unrelated scientific and engineering disciplines, like chemistry, chemical engineering, physics, electric engineering or reliability engineering (hardware and software). The motivation for this survey is twofold.

First, the survey demonstrates that there are indeed certain regularities that can be observed as common traits in a large spectrum of current relational models. These common properties are alluded to in Chapter 3 as "empirical observations". Later, they provide the required basis for the development of the general RMM model in Part II of the book.

Secondly, recognizing (and perhaps appreciating) the similarity between the features of the new RMM model and those of existing

15

models may help persuade potential users that the RMM approach is an effective platform for empirical modeling. For prospective users, who may recognize in the introduced models some familiar ones which they may have encountered in their daily routine, the current survey may serve as a bridge towards better understanding of the RMM approach, to be developed and demonstrated in later chapters.

All the relational models included in the survey are special cases of the RMM model. For some, this will be demonstrated in Chapter 12.

Two guidelines were pursued in selecting the models.

First, all models are confined to a single independent variable (or factor). When multiple factors affect the response, extension to the multiple-factor case is implemented via the "Linear Predictor" (LP), a linear combination of response-affecting factors that replaces the single factor. This is analogous to a similar term used in generalized linear models (GLM). In the current survey we do not relate to multiple-factor models. However, the latter serve as the prototype for developing the RMM approach in Chapter 7.

Secondly, only models with a monotone convex/concave relationship between the response and the affecting factor are included. This restriction is shared by all major methodologies for empirical modeling (like GLM), and we pursue the same in the selection of the models.

Seven subject areas are covered in the survey: Chemistry and chemical engineering, physics, electrical engineering, hardware re-liability engineering, software reliability-growth modeling and non-linear growth models. The related models are introduced in a uniform format:

- Definition of the response, the affecting factor and the model's parameters;
- Specification of the model (including reference to the source).

No theoretical background for the derivation of the models is provided as these may easily be retrieved from the quoted source reference or from other relevant dedicated references.

Three main sources served for this survey:
- Daubert (1998), for chemistry and chemical engineering;

- Halliday *et al.* (1993), for physics and electric engineering (two further sources for particular models are also used);

- Myers *et al.* (2002), for growth models.

A main consideration in selecting these sources is that they provide mainstream models that have been widely recognized as such by the respective community of scientists or engineers. For example, Daubert (1998) surveyed models developed over the years for chemical responses that are affected by temperature. Based on currently available data-sets, he has recommended models that best fit the data. We relate exclusively to these models. Similarly, Halliday *et al.* (1993) is a basic textbook in physics (with related engineering applications), for undergraduate college students. As such, it is expected to present mainly mainstream models. Myers *et al.* (2002) likewise present mainstream growth-models used in biology, economics and engineering.

It is well appreciated that the models selected for this survey do not necessarily constitute a representative sample of mainstream models. Other models could have been selected with equal justification. However, we consider this of secondary importance as this survey is intended primarily as a *demonstration* for the "empirical observations", addressed in Chapter 3, and for the general nature of the RMM model, as the latter is developed in Part II of the book.

Once the reader is convinced of being exposed to a large enough sample of models, s/he may skip the rest and turn to Chapter 3, without risking losing essential information that might be needed later on in the development of the RMM methodology.

2.2. Chemistry and Chemical Engineering

Several empirical or theory-based relationships that have appeared in the literature are considered. We start with temperature dependence of vapor pressure. This relates to modeling of vapor pressure, P, in terms of temperature, T (on a Kelvin scale). Perhaps the most well-known and widely used is the over a century old Antoine equation (1888), derived from theoretical principles (Clapeyron, 1834) that remain valid today:

$$\log(P) = A + B/(T+C), \quad B<0. \tag{2.1}$$

Riedel (1954) modified (2.1) by adding two terms, to obtain

$$\log(P) = A + B/T + (C)\log(T) + DT^E. \qquad (2.2)$$

Equation (2.2) is judged by Daubert (1998) to be the most adequate today for modeling temperature dependence of vapor pressure. However, this is basically Antoine equation, improved empirically by adding two polynomial terms.

For surface tension, S, Daubert (1998) recommends the use of Guggenheim's empirical equation (1945)

$$S = A(1-T/T_c)^C, \qquad (2.3)$$

and T_c is the critical temperature (a constant).

Correlation of heat of vaporization with temperature must be consistent with the exact Clapeyron equation $dP/dT = \Delta H_v / (T\Delta V)$, where V is volume, and ΔH_v is the heat of vaporization. Theoretical considerations require that ΔH_v will be modeled in terms of $t = 1-T_r$, where $T_r = T/T_c$ is temperature relative to its value at the critical point, T_c. Dauber (1998) introduces three current models

$$\Delta H_v = A\,[B^d - 1],$$

$$\Delta H_v = A\,t^B, \qquad (2.4)$$

$$\Delta H_v = Bt^A + Ct^{(2A)} + Dt^{(3A)} + ...,$$

with $d = t^C$.

For solid density, Daubert (1998) notes the following: "As the variation with temperature is not strong, a linear equation is normally adequate with a data-range of the lowest data point to the triple point of the compound. In a few cases with a large amount of data, a simple quadratic polynomial was recommended. For most situations use of a single exponential value over the entire temperature range is sufficient".

For liquid density, ρ_L, Rackett (1970) suggested an equation, which Daubert (1998) concludes "is recommended for correlation of liquid density". Rackett's equation is

$$\log(\rho_L) = A + B[1-T/C]^D, \qquad (2.5)$$

where {A, B, C, D} are parameters that need to be determined.

For solid heat capacity, Daubert (1998) recommends a simple polynomial in temperature, where "most data can be fitted with a linear equation, with a quadratic necessary for a few systems".

For liquid heat capacity, C_{PL}, Daubert (1998) recommends as a default the use of the following equation, originally suggested by Zabransky *et al.* (1990):

$$C_{PL} = A^2/t + B - 2(A)(C)t - (A)(D)t^2 - (C^3/3)t^3 - (C/2)(D)t^4 - (D^2/5)t^5, \quad (2.6)$$

with t= 1- T_r. This is a polynomial in t. We will address polynomials separately in Chapter 12, when we re-address the models of this chapter subsequent to the derivation of the RMM model.

For ideal gas heat capacity, Daubert (1998) notes that most suggested relationships are of polynomial form. However, he recommends the use of the following exponential equation (equation 14-1 therein):

$$C_P^o = A + B \exp(-C/T^D), \quad (2.7)$$

where {A, B, C, D} are parameters, which need to be determined.

For temperature dependency of liquid viscosity, μ_L, Daubert (1998) recommends

$$\log(\mu_L) = A + B/T + C\log(T) + DT^E. \quad (2.8)$$

This is again in a polynomial form.

For thermal conductivity, Daubert (1998) examines both liquid and low-pressure vapor thermal conductivity. All of the examined relationships are either polynomials in T or in $t = 1 - T_r = 1 - T/T_c$.

Finally, the famous Arrhenius equation describes the effect that temperature has on the rate of a simple chemical reaction:

$$R = A \exp[-E_a / (\kappa_B T)], \quad (2.9)$$

where R is the reaction or diffusion rate, T is temperature in degrees Kelvin, E_a is the effective activation energy (the amount of energy required to ensure that the reaction happens, not temperature-dependent), κ_B is Boltzmann's constant and A is a process-related constant.

2.3. Physics

Some well-known relationships, derived from Newtonian principles, from current cosmological theories or from quantum mechanics, are addressed. Three source references are used: Halliday *et al.* (1993), Benedek and Villars (2000) and Schroeder (1991).

In attempting to illustrate the trait of self-similarity typical to power laws, Schroeder (1991) emphasizes that "homogenous power laws, like Newton's universal law of gravitational attraction, abound in nature, dead and alive alike" (therein, p. 33). We start with some examples of power laws in Newtonian physics.

A well-known basic relationship, which binds kinetic energy, E_k, with velocity, V, is

$$E_k = M(V)^2/2, \qquad (2.10)$$

where M is a proportionality coefficient (the body mass). Similar quadratic relationships hold for an object moving in a circle (between centripetal force and velocity), for a spring [between the elastic potential energy and the spring's elongation (compression) from its rest position], and for a body that falls through air (between the drag force and the object's velocity).

Quadratic expressions abound in various areas of physics. There are also non-quadratic power relationships. For example, the radius of the earth, R, as a function of its mass, M, given its mean density, ρ, is

$$R = \{[3/(4\pi)](M/\rho)\}^{(1/3)}. \qquad (2.11)$$

Suppose that the temperature differential between an object and its surrounding is ΔT. Then, according to Newton's law of cooling, if at some instant t=0, the temperature difference is ΔT_0, then at some time later, t (>0), the temperature difference will be (Halliday *et al.*, 1993, p. 548):

$$\Delta T = \Delta T_0 \exp(-at). \qquad (2.12)$$

Every solid (and liquid) at finite temperature is in equilibrium with its own vapor. The saturation vapor pressure, P_{vapor}, depends on temperature, T, mainly through the Boltzmann factor (Benedek and Villars, 2000, p. 436, 442):

$$P_{vapor} \sim \exp[-E_b/(\kappa_B T)], \qquad (2.13)$$

where κ_B is the Boltzmann factor, T is on a Kelvin scale, and E_b is a particle binding energy (find details therein).

The "Barometric formula" is a well-known expression for the exponential decrease, with increasing height, of the density of a gas, assuming a "constant" temperature, T. Let M be the molecular weight, R the gas constant, and g the gravitational constant. Then, at height z the spatial density, ρ, depends on the ratio K= z/T, via the relationship (Benedek and Villars, 2000, p. 429)

$$\rho(K) = \rho(0) \exp[-(Mg/R)K]. \tag{2.14}$$

The similarity of (2.13) and (2.14) to the Arrhenius formula (2.9), with regard to dependence on T, is worth noting.

Let us consider some relationships from relativity and quantum physics. Rest energy plus kinetic energy for an object with rest mass of M_0, moving at a speed, v, near the speed of light, C, is (from Einstein's Special Theory of Relativity)

$$E(L) = M_0 C^2 (1- L)^{-1/2}, \tag{2.15}$$

where L is the relative speed squared: $L= (v/C)^2$.

Planck's radiation law is (Halliday *et al.*, 1993, p. 1140)

$$S(T_0) = (2\pi c^2 h/\lambda_0^5) \{\exp[(hc/\lambda_0 k)/(T_0)]-1\}^{-1}, \tag{2.16}$$

where S is spectral radiancy, λ_0 is the wavelength at temperature T_0, and h is the Planck constant.

The energy levels of the stationary states of the hydrogen atom are given by (Halliday *et al.*, 1993, p. 1147)

$$E(n) = - 13.6 \text{ eV} / n^2, \; n=1,2,3,.., \tag{2.17}$$

where n is the quantum number, and the numerical coefficient is the electron ground-state energy, measured in electron-volts (eV).

Radioactive decay rate, R, is given by (Halliday *et al.*, 1993, p. 1239)

$$R(t) = R_0 \exp(-kt), \tag{2.18}$$

where at time t=0 the rate is R_0.

The neutron generation time, t_{gen}, in a nuclear reactor, is the average time needed for a fast neutron, emitted in one fission, to be slowed down to thermal energies by the moderator, and to initiate another fission. Let P_0 be the power output of the reactor at time t=0. Then, the output at time t later, P(t), is given by (Halliday *et al.*, 1993, p. 1279)

$$P(t) = P_0 \exp[(k/t_{gen})t]. \tag{2.19}$$

2.4. Electrical Engineering

Coulomb's law specifies the electrostatic force of attraction (or repulsion), F, between two point charges, q_1 and q_2, separated by distance r. It is given by (Halliday *et al.*, 1993)

$$F(r) = k(q_1 q_2) / r^2 , \qquad (2.20)$$

where k is a constant.

A capacitor with capacity C, which is initially uncharged (q=0 at t=0), is charged by an ideal battery with an emf (electromotive force), E. At time t, the voltage across the capacitor, as measured by a resistor connected in parallel, with resistance R, is (Halliday *et al.*, 1993, p. 805)

$$V_R(t) = E \exp[-t/(RC)]. \qquad (2.21)$$

Similarly, a discharging capacitor, with an initial (full) charge of q_0, will have at time t a charge of

$$q(t) = q_0 \exp[-t/(RC)]. \qquad (2.22)$$

Similar equations apply for the rise (fall) of the current produced by an emf E, which is introduced into (removed from) a single loop circuit containing a resistor, R, and an inductor, L (Halliday *et al.*,1993, p. 903).

Suppose that electricity current of intensity I runs through a loop of radius R (area S). The magnetic field, B(d), produced at a point located a distance d along the loop axis is parallel to the axis and is given by (Halliday *et al.*,1993, p. 863)

$$B(d) = (\mu_0 / 2\pi) (IS) / d^3 = (\mu_0 / 2) (IR^2) / d^3, \qquad (2.23)$$

where μ_0 is a permeability constant.

2.5. Hardware Reliability Engineering

The relationship between fatigue life of metal, ceramic, and composite materials and applied stress is an important input to design-for-reliability. Much effort has been directed towards modeling this relationship. A recent paper by Pascual and Meeker (1999) attempted to model this relationship via the *Random Fatigue-Limit Model* (RFLM).

Let Y be the fatigue life and s the stress level. The model for Y, in terms of s, is

$$\log(Y) = \beta_0 + (\beta_1)\log(s-\gamma) + \varepsilon, \qquad (2.24)$$

where β_0 and β_1 are fatigue curve coefficients, γ is the (random) fatigue limit of the specimen, and ε is an error term.

Pascual and Meeker supported their theoretical analysis by applying it to fatigue data given by Shimokawa and Hamaguchi (1987). They have assumed that the fatigue limit is random, and for reasons associated with estimation of the model's parameters, they characterize the random variable $V = \log(\gamma)$, after standardization, as distributed either like the standardized smallest-extreme-value distribution or the standardized normal distribution.

When s is constant, Y experiences only random fluctuations, and these may be captured by the error distribution associated with this model. Pascual and Meeker derive this distribution, and investigate it briefly. The authors do not specify the relationship between the random fatigue limit and the random error term, ε. Therefore it is assumed that these two random variables are not correlated.

Chapter 15 is dedicated to RFLM and its relationship to the RMM model.

2.6. Software Reliability-Growth Modeling

Software reliability-growth models express the number (or rate) of software failures as a function of operating time or testing effort. Only the latter case will henceforth be addressed. It is commonly assumed that as more errors are detected and corrected no new errors are injected into the software as a result of the repair activities, and the reliability of the software would tend to grow. This assumption forms the basis for a widely used category of models generally referred to as reliability-growth models.

The most widely applied in this category are variations of a basic model in which the number of failures that have occurred by time t is modeled by a Non-Homogenous Poisson Process (NHPP), with mean value function $\mu(t)$. In other words, the probability function for the number of failures, N_i, in the time interval $[t_{i-1}, t_i)$, is given by

$$P(N_i=n) = \{[\mu(t_i) - \mu(t_{i-1})]^n / n!\} \exp \{-[\mu(t_i) - \mu(t_{i-1})]\}, \qquad (2.25)$$

that is, a Poisson probability function with time-dependent parameter, $\mu(t_i) - \mu(t_{i-1})$.

NHPP models differ by how they model the mean function, and these are divided into two main categories:

- Finite failure models, where it is assumed that $\mu(\infty)$ is finite;
- Infinite failure models, where the mean value is not bounded.

Examples of the former category are:

I. Goel and Okumoto (1979), who have assumed an exponential mean function:

$$\mu(t) = a[1-\exp(-bt)], \ a > 0, b > 0. \tag{2.26}$$

II. The delayed S-shaped model (Yamada and Osaki, 1984):

$$\mu(t) = a[1-(1+bt) \exp(-bt)], \ b > 0. \tag{2.27}$$

Examples of the latter category are:

III. Musa *et al.* (1987), who have proposed the logarithmic Poisson model:

$$\mu(t) = (a) \log(1+bt), \ t > -1/b. \tag{2.28}$$

IV. Duane (1964), who has proposed the Power-Law Process (PLP):

$$\mu(t) = \alpha t^{\beta}, \ \alpha > 0, \beta > 0. \tag{2.29}$$

V. The log-power model introduced by Xie and Zhao (1993):

$$\mu(t) = (a)[\log(1+t)]^{b}, \ a > 0, b > 0. \tag{2.30}$$

VI. The generalized power family models (GPFM) by Knafl and Morgan (1996):

$$\mu(t) = \alpha[k(t)]^{\beta}, \tag{2.31}$$

where k is a positive and strictly increasing known function of t, with k(0)=0.

Arnoux *et al.* (2000) suggest criteria for choosing an appropriate GPFM model, using Musa's data (1979).

A good review of these models and others is given in Xie (2000), which also served as a major reference for the above short review.

2.7. Growth Models

Growth models are a special category of non-linear regression models. They are used to describe how the response grows with changes in a regressor variable. Typical applications are in biology, where plants and organisms grow with time, but there are also many applications in economics and engineering. Some such models are also used in software reliability-growth modeling.

We pursue here some mainstream models as introduced in Myers, Montgomery and Vining (2002, therein Section 3.6).

The logistic growth model is

$$y = \beta_1 / [1+\beta_2 \exp(-\beta_3 x)]. \tag{2.32}$$

The Gompertz model is

$$y = \beta_1 \exp[-\beta_2 \exp(-\beta_3 x)]. \tag{2.33}$$

The Weibull model is

$$y = \beta_1 - \beta_2 \exp(-\beta_3 x^{\beta_4}), \tag{2.34}$$

where $\{\beta_1, \beta_2, \beta_3, \beta_4\}$ are parameters, which need to be determined.

Apparently, the first model is a power-exponential model, the second an exponential-exponential model and the last is an exponential-power model.

References

Section 2.1-2.4

[1] Antoine, C. (1888). Thermodynamic vapor pressures: New relation between the pressures and the temperatures (Thermodynamique, Tensions des Vapeurs: Novelle Relation entre les Tensions et les Temperatures), *C.R. Hebd. Seances Acad. Sci.,107*, 681, 836, 1146.

[2] Benedek, G. B., Villars, F. M. H. (2000). *Physics with Illustrative Examples from Medicine and Biology- Statistical Physics*. 2nd ed., Springer-Verlag, NY.

[3] Clapeyron, E. J. (1834). Memoirs on the motive power of heat. *L'ecole Polytechnique, 14* (23), 153.

[4] Daubert. T. E. (1998). Evaluated equations forms for correlating thermodynamics and transport properties with temperature. *Industrial and Engineering Chemistry Research*, 37(8), 3260-3267.

[5] Guggenheim, E. A. (1945). The principle of corresponding states. *Journal of Chemical Physics*, 13, 253.

[6] Halliday, D., Resnick, R., Walker, J. (1993). *Fundamentals of Physics, Extended, with Modern Physics*. 4th ed., John Wiley & Sons.

[7] Rackett, H. G. (1970). Equation of state for saturated liquids. *Journal of Chemical and Engineering Data*, 15(4), 514-517.

[8] Riedel, L. (1954). A new universal formula for vapor pressure (Eine Neue Universelle Dampfdruck formel). *Chem. Ing. Tech.*, 26, 83.

[9] Schroeder, M. (1991). *Fractals, Chaos, Power Laws*. W. H. Freeman and Company.

[10] Zabransky, M., Ruzicka, V. Jr., Majer, V. (1990). Heat capacities of organic compounds in the liquid state. I. C_1 to C_{18} 1-Alkanols. *Journal of Physical and Chemical Reference Data*, 19(3), 719-762.

Section 2.5-2.6

[1] Arnoux, F., Gaudoin, O., Makni, C. (2000). The generalized power family in software reliability data analysis. *MMR'2000: Second International Conference on Mathematical Methods in Reliability- Abstracts' Book*, V. 1, 107-110.

[2] Duane, J. T. (1964). Learning curve approach to reliability monitoring. *IEEE Transactions on Aerospace*, AS-2, 2, 563-566.

[3] Goel, A. L., Okumoto, K. (1979). Time-dependent error-detection rate model for software reliability and other performance measures. *IEEE Transactions on Reliability*, 28, 206-211.

[4] Knafl, G. J., Morgan, J. (1996). Solving ML equations for 2-parameter Poisson-process models for ungrouped software-failure data. *IEEE Transactions on Reliability*, 48, 159-168.

[5] Musa, J. D. (1979). Software Reliability Data. *Technical Report*, Rome Air Development Center.

[6] Musa, J. D., Iannino, A., Okumoto, K. (1987). *Software Reliability- Measurement, Prediction, Application*. McGraw-Hill.

[7] Pascual, F. G., Meeker, W. Q. (1999). Estimating fatigue curves with the random fatigue-limit model. *Technometrics*, 41(4), 277-290.

[8] Shimokawa, T., Hamaguchi, Y. (1987). Statistical evaluation of fatigue strength in circular-holed notched specimens of a carbon eight-harness-satin/epoxy laminate. In *Statistical Research on Fatigue and Fracture (current Japanese Materials Research, Vol. 2)*, Tanaka, T., Nishijima, S., Ichikawa, M. (Eds.). Elsevier, London, 159-176.

[9] Yamada, S., Osaki, S. (1984). Nonhomogenous error detection rate models for software reliability growth. In *Stochastic Models in Reliability Theory*, Springer-Verlag, NY, 120-143.

[10] Xie, M., Zhao, M. (1993). On some reliability growth models with simple graphical interpretations. *Microelectronics and Reliability*, 33 (2), 149-167.

[11] Xie, M. (2000). Software reliability models- past, present and future. In *Recent Advances in Reliability Theory- Methodology, Practice and Inference*. Limnios, N., Nikulin, M. (Eds.). Birkhauser, Boston, MA, 325-340.

Section 2.7

[1] Myers, R. H., Montgomery, D. C., Vining, G. G. (2002). *Generalized Linear Models, with Applications in Engineering and the Sciences*. John Wiley & Sons.

Chapter 3

Shared Features and "The Ladder"

3.1. Introduction

The previous chapter presented mainstream models that have been developed over the years in a selected sample of scientific and engineering disciplines. The sole criterion for inclusion in this limited-scope survey is the existence of monotone convex/concave relationship between the affecting factor and the response. It may be reasonable to expect that a considerable portion of models that one would encounter within any randomly selected scientific or engineering discipline conforms to this rather general characterization.

From the models surveyed, two features seem to emerge that all models share in common. To understand these common traits, let us be reminded that the models were introduced as they appear in the relevant literature (or undergraduate textbooks). No allusion to possible random errors, associated with the models, has been made. As conveyed in the Introduction (Chapter 1), this total disregard, intended or otherwise, to the model's error structure is common to most models of systematic variation, which one may find in the relevant engineering and scientific literature.

It is therefore not superfluous to refer again to some of the models introduced earlier, to discuss their possible error structure, and to categorize, if possible, the particular structure associated with each model. This is done in Section 3.2. The explored "shared features" may then assist us (Section 3.3) in creating some general classification of models (of uniform convex/concave relationships), which would

29

eventually be useful in developing the RMM model in Part II of the book.

3.2. Shared Features

In this section we consider three examples, two of which have been included in the survey of the previous chapter. Two conclusions are desirable for each model: The implied error-structure and characterization of the model's structure.

Example 1 (Chemistry)

Let us re-address Antoine equation [Eq. (2.1)]. This expression does not provide indication as to the structure of the error. Assume first that temperature is constant (no systematic variation or random error is associated with temperature). Obviously, the measured pressure would still fluctuate randomly, and this random variation will be unrelated to temperature. Natural instability of pressure caused by factors other than temperature (for example, fluctuations in vapor density), as well as measurement imprecision, may contribute to this random variation.

Equation (2.1) may therefore be re-written, with the response-related random error, ε_2, as

$$\log(P) = A + B/(T+C) + \varepsilon_2. \qquad (3.1)$$

Other forms of random error may be conceived relative to P, assuming a constant T, for example, $\log(P+\varepsilon_2)$ (refer to Chapter 7 for details).

Now discard the assumption that T is constant, and assume that T itself fluctuates randomly. It is natural to assume that random deviations in temperature will be transmitted to the measured pressure, or, in other words, that a second source of random variation affects the response.

Denote the random deviation associated with temperature by ε_1. Temperature may then be re-expressed as $T = \mu_T + \varepsilon_1$, where μ_T is the expected value (mean) of the measured T. Antoine equation, with the errors added, becomes

$$\log(P) = A + B / (\mu_T + \varepsilon_1 + C) + \varepsilon_2. \qquad (3.2)$$

We relate to variation associated with T (either systematic, caused by changes in μ_T, or random, represented by ε_1) as "Externally-generated variation". Likewise, we relate to variation associated with ε_2 as "Self-generated variation". Both of these terms are self-explanatory, given the explanations expounded earlier.

Next, consider the special structure of this model. Observing (3.2), it is obvious that the relationship between the response (P) and the affecting factor (T) is monotone (always increasing), and that it may be characterized as Exponential-Power.

Example 2 (Chemical Engineering)

In Eq. (2.3) Guggenheim's empirical equation for surface tension, S, was given. Written here with the errors (as the latter were explained in Example 1), it becomes

$$S = A[1 - (\mu_T + \varepsilon_1)/T_c]^C (1 + \varepsilon_2), \qquad (3.3)$$

where T_c is the critical temperature (a constant). Note, that this time the error, ε_2, relates to the original response (not to the log transformed response). Furthermore, it is assumed to be multiplicative. As for the earlier example, other error structures may be assumed.

From (3.3), the relationship between the response, S, and T may be characterized as a Power relationship.

Example 3 (Electrical Engineering)

A thermistor is a semiconductor device with a temperature-dependent electrical resistance. It is used in medical thermometers and to sense over-heating in electronic equipment. Over a limited range of temperature ($T > T_a$, T_a given), the resistance, R, is (Halliday *et al.*, 1993, p. 547):

$$R(T) = R_a \exp[B(1/T - 1/T_a)], \quad T > T_a, \qquad (3.4)$$

where B is a constant, which depends on the particular semiconductor used, and R_a is the resistance at $T = T_a$. Re-written with the errors, (3.4) becomes

$$R(T) = R_a \exp\{B[1/(\mu_T + \varepsilon_1) - 1/T_a] + \varepsilon_2\}, \quad T = \mu_T + \varepsilon_1 > T_a. \qquad (3.5)$$

From (3.5), the relationship between the response, R, and T may be characterized as Exponential-Power relationship.

Analyzing the above three examples (similar analyses may in fact be applied to any subset of the models in Chapter 2), two observations emerge, which, once captured in a general model, may provide the basis for the desired empirical modeling methodology, developed in Chapter 7.

OBSERVATION A

Variation in the response may be attributed to two sources of variation:

***Self-generated variation.** This consists of *random* variation in the response, emanating from its natural instability or from measurement error, associated with measuring the response. This component of variation is represented in the above examples by the error term, ε_2.

***Externally-generated variation.** This comprises two components:
- **Systematic variation**, which reflects variation transmitted to the response via variation in the mean of the affecting factor (systematic changes of the temperature mean, μ_T, in the above examples). If several factors influence the response, it is assumed that systematic variation in the "Linear Predictor" (LP, a linear combination of the affecting effects) is generating systematic variation in the response.
- **Random variation**, which reflects variation transmitted to the response via random deviations in the affecting factor (random changes in temperature in the above examples), or random changes in the LP, for the multiple-factor case. This variation is represented in the models by an additive random error, ε_1

OBSERVATION B

Certain model structures, or patterns, which are supposed to capture uniformly convex relationships, repeatedly appear in models developed independently in a myriad of scientific and engineering disciplines. Two examples for such patterns were provided earlier: Power and

Exponential-Power relationships. Other relationships, like linear, exponential or exponential-exponential, abound in the engineering and scientific literature (as the survey of Chapter 2 clearly attests). In fact, these various basic functions may be arranged in a hierarchical order that reflects a gradually increasing "intensity" of convexity.

This hierarchy is captured in the "Ladder of Fundamental Uniformly Convex Functions". The "Ladder" will be expounded in detail in the next Section 3.3. Once a model is constructed that includes as special cases all the "steps" of the "Ladder", the universal nature of the new modeling methodology can be established to a large degree.

What are the implications of these two observations, which may weigh heavily in formulating the properties required from a general response modeling methodology?

To start with, it is obvious that relational models should reflect a distinction between a random error transmitted to the response via external factors, and a random error associated with the response's own natural inner instability. That distinction is rarely made in current modeling methodologies, be they reported applications of linear or non-linear regression, data transformational approaches or generalized linear models (GLM).

By contrast, the dual-error structure is an essential component of the RMM model. Furthermore, a large number of non-normal distributions may be adequately represented by the RMM error distribution, which is based on a *two-component* structure of normal errors. This will be demonstrated later on in two ways: By deriving, via the RMM error distribution, exact representation for existing distributions, transformations and approximations (Chapter 12), and by deriving from the RMM model highly accurate distributional approximations to existing distributions (Chapter 19).

Secondly, empirical modeling often relates to the log transformed response rather than to the original scale. A major justification for this is that the error associated with the response, ε_2, is often proportional to the order of magnitude of the response itself. Such is often the case in modeling a chemo-response, as evident from the models in Chapters 2 and 17. The general model, to be developed in Chapter 7, should allow the response error to be, in terms of the original scale, either additive or multiplicative, and not be confined to an additive error, as non-linear normal-theory regression models commonly assume.

A third implication relates to the "universal" character of the desired model. A major drawback of the GLM approach is the need to specify prior to data analysis certain properties of the model, like the link function (the transformation of the mean that generates additivity of effects), the error distribution or, equivalently, the mean-variance relationship. The required a-priori specification of the model structure may pose an undesirable restriction on the "universality" of the methodology, and ultimately result in unsatisfactory fit.

These restrictions constitute serious violations of the desirable properties required from a general methodology for *empirical* modeling. We will further elaborate on this in Chapter 6. By contrast, implementation of the new RMM approach requires no a-priori specification of either the model's structure or of its error distribution. Furthermore, RMM would accommodate as special cases all of the models surveyed above (and others, included in the "Ladder" of Section 3.3).

The above two general "observations" would form a large part of the foundations on which the RMM model is developed in Chapter 7.

3.3. "The Ladder of Fundamental Uniformly Convex/Concave Functions"

The models of Chapter 2 demonstrate that there are indeed some fundamental convex functions that repeatedly appear in models with monotone convex/concave relationships. These functions may be arranged in a hierarchical fashion, which expresses a growing "intensity of convexity", as one moves from one "step" in this hierarchy to the next. Thus, an exponential function is obviously more convex than a power function, and an exponential-power function is more convex than just an exponential function (a comparison of the relative magnitude of the second derivative, appropriately standardized to account for scale, may provide a good measure for this "intensity" of convexity).

We denote the hierarchy of monotone convex/concave relationships "The Ladder of Fundamental Uniformly Convex/Concave Functions".

The "Ladder" is presented below, where V stands for the linear predictor (η) plus the associated random error, ε_1. Also, for each function the second derivative is given to allow the reader assess how certain parameters determine whether the expression is uniformly convex or concave.

1. Linear increase. This is a convex relationship with the smallest rate-of-increase for the slope, representing a second derivative of zero:

$$V = \eta + \varepsilon_1.$$

2. Power increase. This represents a monotone increase of the form V^k, where k is a real-valued parameter ($k \neq 1$). The second derivative is

$$k(k-1)\, V^{(k-2)}.$$

3. Exponential increase. This represents a monotone increase of the form $\exp(V)$. The second derivative is

$$\exp(V).$$

4. Exponential-Power increase. This represents a monotone increase of the form $\exp(V^k)$. The second derivative is

$$kV^{(k-2)}\, \exp(V^k)\{(k-1) + kV^k\}.$$

5. Exponential-Exponential increase. This represents a monotone increase of the form $\exp[a\,\exp(V)]$, where "a" is a real-valued parameter. The second derivative is

$$a[1 + a\,\exp(V)]\,\exp[a\,\exp(V) + V].$$

6. Exponential-Exponential-Power increase. This represents a monotone increase of the form $\exp[a\,\exp(V^k)]$. The second derivative is

$$(ak)V^{(k-1)}\, \exp[a\,\exp(V^k) + V^k][(k-1)/V + kV^{(k-1)} + (ak)V^{(k-1)}\, \exp(V^k)].$$

This hierarchy is not uncommon in certain disciplines (like in the study of algorithmic complexity). Needless to reiterate, all the models of Chapter 2 and Section 3.2 belong to one of the basic patterns included in the "Ladder".

We now wish to construct a general relational model that reflects well the various steps of "the Ladder". More specifically, we wish any general methodology for modeling uniformly convex/concave relationships to have the desirable property of being able to accommodate at least the functions included in the "Ladder". A natural way to achieve this is by multiplying or adding the above functions, introducing enough parameters that would allow extraction of each of the above cases individually. This approach, however, violates the important principle of parsimony, which requires that the underlying model be as simple as possible (but not simpler, to quote Einstein). An alternative approach needs to be pursued. This approach is developed in detail in Part II of the book.

Prior to developing the RMM approach, it is appropriate that current methodologies for empirical modeling be surveyed. In the next two chapters we survey current approaches for empirical modeling of systematic variation (Chapter 4) and of random variation (Chapter 5). Chapter 6 (the last chapter in Part I of the book) lay out the requirements that a general methodology for empirical modeling of variation needs to fulfill. Current modeling methodologies are then assessed for compliance with these requirements.

Chapter 4

Approaches to Model Systematic Variation

4.1. Introduction

In Chapter 2 some main stream models in engineering and the sciences have been reviewed. There are several features that these models share in common. First, the modeled relationships have been monotone convex. Secondly, most of the models have been derived from some well established theory, perhaps with some additional correction-terms, derived empirically to enhance the goodness-of-fit. Thirdly, as elaborated in Chapter 3, all models are associated with a dual-error structure, and the various structures displayed by these models may be arranged in a hierarchical order, captured by "the Ladder".

Being acquainted with these models and their shared properties, it is perhaps appropriate that we now turn to review the "State-of-the-Art" of empirical modeling. In Chapter 1 the term "empirical modeling" was explained, and a distinction made between modeling of systematic variation and modeling of random variation.

In Chapter 5 current general approaches to empirically model *random* variation are surveyed. In this Chapter 4, we survey the same for empirical modeling of *systematic* variation.

Common to all approaches that belong to the latter category are two components of the modeling process:

(1) Modeling the *linear predictor* (LP)

(2) Modeling the relationship between the LP and the response.

37

To demonstrate these two components, consider two of the approaches surveyed in this chapter: Linear regression analysis and the Box-Cox (BC) power transformation (Box and Cox, 1964).

In linear regression, the relationship between the LP and the response is assumed *a-priori:* It is a linear relationship. Therefore, what is required of the modeling process is to determine the structure of the LP (which effects are included) and estimate its parameters.

Consider the BC transformation (with parameter $\lambda \neq 0$), applied to the response, Y:

$$(Y^{\lambda}-1)/\lambda = \eta + \varepsilon, \tag{4.1}$$

where ε is a random (normal) additive error, η is the LP,

$$\eta = \beta_0 + \beta_1 X_1 + \beta_2 X_2 +..+ \beta_k X_k, \tag{4.2}$$

and X_i (i=1,..,k) is the i-th predictor variable ("independent variable"), a factor ("main effect") or a parameter-free function of factors (like an interaction).

To model via (4.1), two phases of the modeling process are required:

(1) *Determination of the linear predictor*, namely, the structure of the agent that transmits systematic variation to the response. This amounts to determining what predictor variables to include in the LP (namely, performing statistical significance testing) and estimating the coefficients of the LP;

(2) *Determination of the non-linear relationship* between the LP and the response, Y. In our case, this requires estimating the parameter λ.

Although methods have been devised to conduct con-currently the two phases of fitting a linear model to a BC transformed response, in most statistical packages available on the market the user can perform the two phases only sequentially. In many cases, this results in less than desirable goodness-of-fit.

The distinction between these two phases carries over also to other general methodologies, like GLM and non-linear regression.

In this chapter we briefly review the three major methodologies for empirical modeling of systematic variation, currently available to the practitioner: Linear regression analysis (Section 4.2), the BC power-transformation (Section 4.3) and generalized linear models (GLM, Section 4.4). Special focus will be put on how the various approaches cope with the two phases of the modeling process just delineated.

The following presentation intends to be qualitative in character. It is not intended to provide a full-fledged description and development of the theory behind each of the approaches addressed. Numerous books have been written, which describe these in detail, and the reader is referred to some of these references in the appropriate sections.

In Section 4.5 some conclusions are given that are expanded in Chapter 6.

4.2. Linear Regression Analysis

This is perhaps the most widely implemented approach for empirical modeling of systematic variation. There are three major merits to linear regression that separates it from all other approaches.

First, there are closed-form expressions that, given sample data, allow calculation of point estimates and confidence-interval estimates, with no need for a numerical iterative solution procedure.

Secondly, suppose that the assumptions of the linear regression analysis hold (we will refer to these assumptions as *the normal scenario*, and detail them shortly). Then fitting a regression equation by the method of least-squares is equivalent to estimating via maximum likelihood (ML). Since in general ML estimators have better statistical properties (like minimum variance) than estimators obtained by least-squares, the equivalence between the least-squares procedure, used in linear regression analysis, and the ML procedure, is an important advantage.

Thirdly, linear regression requires linearity in the coefficients of the predictor variables (often called regressors, or independent variables), and not in the variables themselves. For example, the polynomial relationship

$$y = \beta_0 + \beta_1 X_1 + \beta_{11} X_1^2 + \beta_{12} X_1 X_2 \qquad (4.3)$$

can be analyzed via linear regression analysis, provided the normal scenario has been shown to hold. In fact, any function of factors (or main effects), which does not include parameters that need to be estimated within the linear regression analysis, qualifies as a regressor that can be included in the linear regression equation.

This advantage of linear regression becomes even more pronounced due to the Taylor series expansion. Under commonly encountered regularity conditions, the Taylor expansion can be applied to any non-linear relationship. Since this expansion results in a polynomial in the independent variables, a Taylor series expansion may be analyzed within the framework of linear regression analysis (provided the normal scenario holds). This is made use of in a commonly applied empirical methodology for modeling systematic variation in the quality-engineering discipline, called Response Surface Methodology (RSM). The latter uses a full quadratic equation to model the relationship between the predictor variables and the response:

$$Y = \sum_{j=1}^{k} \beta_j X_j + \sum_{j=1}^{k} \beta_{jj} X_j^2 + \sum_{i<j=2}^{k} \sum \beta_{ij} X_i X_j . \qquad (4.4)$$

As conveyed in Myers and Montgomery (2002, p.8), "there is considerable practical experience indicating that second-order models work well in solving real response surface problems".

In this short and qualitative description of the unique features of linear regression it has been emphasized that all analyses depend to a large extent on the validity of the normal scenario. Box and Cox (1964), define this scenario by the following three assumptions:

(1) Normally and independently distributed errors;
(2) Homogeneity of the error variance (sometimes referred to as homoscedasticity). This implies that the error variance is constant and does not depend either on values of the response or the predictor variables;
(3) Additivity of effects (namely, linearity of the regression equation).

A further assumption, which is often not explicitly stated [though included in (1)], is that response observations do not have any correlation structure, namely, response observations are statistically independent.

There was extended research about the sensitivity of linear regression analysis to departure from the assumptions that constitute the normal scenario. This topic is discussed at some length in any reference book which deals with linear regression (refer, for example, to Draper and Smith, 1998, or Cook and Weisberg, 1999). It seems that the most important assumption is that of the homogeneity of variance. Appreciable departure from this assumption would most probably invalidate conclusions derived from linear regression analysis. Having an error distribution that is not normal seems to be of less effect on the prediction capability of models of linear regression.

When the normal scenario is violated to such an extent that a linear regression model is inappropriate, two alternative approaches can be pursued: A data transformational approach, where transformations are applied to either or both the predictors and the response, or GLM. One of the most widely applied data transformation, which tends to achieve all three goals of the normal scenario via a single transformation of the response, is the family of power transformations, suggested by Box and Cox (1964).

4.3. Box-Cox Power Transformations

A family of data transformations was suggested by Box and Cox (BC) in their seminal paper of 1964. Numerous papers that have since been published attest to the effectiveness of the BC transformation in revoking the normal scenario and allowing thereby application of linear regression. There is currently no theory to explain why a power transformation is so effective in revoking the normal scenario, and indeed the original derivation of the power transformation was based solely on empirical evidence [this has been conveyed to us in separate personal communications by both Box (Madison, 1993) and Cox (Bordeaux, 2000)]. As will be shown in Chapter 12 (refer also to Shore, 2004), the inverse Box-Cox transformation is a special case of the RMM model. This perhaps provides an explanation as to why the BC transformation is so effective.

The single-parameter BC transformation is defined by

$$Y^{(\lambda)} = \begin{cases} (Y^{\lambda} - 1)/\lambda, \lambda \neq 0, \\ \log(Y), \lambda = 0. \end{cases} \tag{4.5}$$

Since the scale of $Y^{(\lambda)}$ changes with λ, no comparison of residual sum of squares, $RSS(\lambda)$, is possible when we wish to compare several values for the parameter λ (minimization of RSS is desired). Therefore, it is customary to "standardize" the transformation by dividing $Y^{(\lambda)}$ by $[gm(Y)]^{\lambda-1}$, where $gm(Y)$ is the geometric mean of the n observations, namely,

$$gm(Y) = (Y_1 Y_2 \ldots Y_n)^{1/n} . \tag{4.6}$$

Dividing the RHS of (4.5) by $[gm(Y)]^{\lambda-1}$ now allows comparison of RSS obtained with various values of λ (find justification for this, for example, in Draper and Smith, 1998, Section 13.2).

While the BC procedure aims at normalizing the errors via a ML procedure (which is equivalent to minimizing the RSS), one may wish to transform the response in order to stabilize the variance. Or else, one may wish to attain linearity. As Cook and Weisberg (1999) comment: "In a surprisingly large number of regressions, linearity and variance stabilization can be achieved with the same transformation" (therein, p. 317).

Comment. An interesting relationship between the BC parameter λ, and the response transformation needed for stabilizing the variance, can be established. Suppose that the relationship between the mean, μ_Y, and the variance, σ_Y^2, is of the type

$$\sigma_Y = \alpha \, (\mu_Y)^{\beta}. \tag{4.7}$$

Let $T(Y)$ be a response transformation which stabilizes the variance. Developing $T(Y)$ into a Taylor series around the response mean, μ_Y, we obtain

$$T(Y) = T(\mu_Y) + T'(\mu_Y)(Y-\mu_Y) + \ldots, \tag{4.8}$$

where differentiation of T is with respect to Y. Applying a variance operator to both sides we obtain, introducing for σ_Y from (4.7):

$$T'(\mu_Y) = \sigma_T / \sigma_Y = \sigma_T / [\alpha(\mu_Y)^{\beta}], \tag{4.9}$$

where σ_T^2 is the variance of the transformed Y, T(Y). Since by definition the variance of the desirable T(Y) does not depend on the mean (the transformation is supposed to stabilize the variance), we obtain by integrating of (4.9) with respect to the mean:

$$T(\mu_Y) = K\mu_Y^{-\beta+1}, \tag{4.10}$$

or

$$T(Y) = KY^{-\beta+1}, \tag{4.11}$$

where K is a parameter. It is easy to realize the relationship between the Box-Cox transformation parameter, λ, and β. If Y^λ is the required Box-Cox transformation, then to ensure that this transformation also stabilizes the variance, we need to have

$$\lambda = 1 - \beta. \tag{4.12}$$

A simple method to identify the required β may be developed from (4.7). Take log of both sides to obtain

$$\log(\sigma_Y) = \alpha_0 + \beta \log(\mu_Y), \tag{4.13}$$

where $\alpha_0 = \log(\alpha)$. By drawing m samples from the same population, and calculating the means and variances for these samples, we obtain m observations, $\{\mu_i, \sigma_i\}$ (i=1,2,..,m), where μ_i and σ_i are the i-th sample estimates of the mean and the standard deviation, respectively. With linear regression, an estimate of β may now be found from (4.13) that stabilizes the variance. If this estimate approximately maintains (4.12), where λ was estimated by the conventional Box-Cox maximum-likelihood procedure, then we can be assured to a large extent that the BC transformation also stabilizes the variance.

How well does the BC transformation fare as a general platform for empirical modeling of systematic variation? Assuming that (4.5) has attained the normal scenario, we may model the transformed response by

$$Y^{(\lambda)} = \eta + \varepsilon, \tag{4.14}$$

where η is the LP, and ε is an error term, normally and independently distributed with constant variance.

Modeling via (4.14) is relatively straightforward if the structure of the linear predictor is determined in advance, namely, the predictors to be included are known. In that case, a value of λ is determined, the sample response observations are transformed, linear regression is applied to estimate the LP, and the associated RSS calculated. The optimal parameter λ is then determined, which minimizes the RSS. Box and Cox (1964) have shown how the transformation parameter, λ, may be estimated concurrently with the other model parameters (overall mean and treatment effects). However, the structure of the LP is specified in advance.

It is altogether a different scenario if *stepwise* regression is desirable. In that case, a re-run of stepwise linear regression on transformed values may result in a different LP, that would render the selection of λ (in a previous iteration for a different LP) irrelevant. Regrettably, most statistical packages nowadays do not address this problem. Furthermore, most current reference books on linear regression also fail to address the relationship between response transformations and *stepwise* regression.

Several recent publications attempted to generalize the BC transformation, or suggest other approaches for normalization of data. In a recent paper, Smith (2003) considers four methods for computing the best value for the single parameter of seven families of transformations. Li and Moore (2002) investigate a general Box-Cox transformation for use in multiple linear regression. An algorithm is proposed to identify optimal general transformations based on kernel density estimation techniques. The reader is advised to relate to these publications and references therein for further details.

Some further evaluation of the BC transformation is given in Chapter 6, where current methodologies are assessed for compliance with requirements, expounded in that chapter.

4.4. Generalized Linear Models

When the standard assumptions of the linear regression model are not met, the method of least-squares, which is supposed to be a ML procedure while the normal scenario holds, ceases to be the most efficient method of estimation. In particular, if the errors are not normal, the residual variance will most probably also cease to be constant, and conclusions derived from the linear-regression model may be in doubt.

For example, confidence intervals, which rely heavily on the normality assumption, may be biased.

For such cases, a new methodology has been developed, Generalized Linear Modeling (GLM), which may be regarded as a generalization of linear regression since the latter is a special case of the former. In its original formulation, GLM was based on the assumption that the distribution of the response belongs to the exponential family of distributions. Most applications reported in the literature (refer to the afore-cited books and also to Myers *et al.*, 2002) pursue this assumption.

Derivatives of GLM, however, exist, which incorporate extensions of the core theory of GLM. These extensions depart from the original formulation in a number of ways. For example, in the methodology termed *generalized estimating equations* the assumption that response observations are uncorrelated is abandoned, and a correlation structure is assumed. Another extension is *quasi-likelihood,* originally suggested by Wedderburn (1974). This methodology allows fitting a model within the GLM framework without assuming that the response distribution is a member of the exponential family. The motivation for this approach is the fact that the GLM score-function, used in fitting a GLM model, relies on the response distribution only via the first two moments (the score function is solved, by iteratively re-weighted least squares, in order to derive maximum likelihood estimates). Wedderburn (1974) indicated that the use of quasi-likelihood produces asymptotic properties that are quite similar to those of maximum likelihood estimators. Thus, good efficiency is obtainable even when the exact likelihood is unknown (since the response distribution is unspecified).

In the following we introduce the basic elements of the original formulation of GLM, where the response distribution is assumed to be a member of the exponential family of distributions. The latter has a density function

$$f(y; \theta, \phi) = \exp\{[y\theta - b(\theta)]/a(\phi) + c(y,\phi)\}, \qquad (4.15)$$

where $a(.)$, $b(.)$ and $c(.)$ are specific functions.

The parameter θ is a natural location parameter, and ϕ is often termed a dispersion parameter. The function $a(\phi)$ is generally of the form $a(\phi)= \phi\omega$, where ω is a known constant. For some common members of the exponential family (like the binomial and Poisson) $\phi=1$. However, situations occur in which the actual response dispersion is larger than that expected by the assumed error distribution. This results in a situation of over-dispersion. For example, the binomial parameter, P, may vary, so that the actual variance is larger than that predicted by the binomial model. In these cases we have $\phi>1$.

The dispersion parameter ϕ is an additional parameter that the analyst will occasionally have to estimate from the data.

Some distributions most commonly encountered in practice belong to the exponential family. Examples are the binomial, the negative binomial, the Poisson, the normal, inverse Gaussian, gamma (including the exponential) and Pareto. However, some other distributions in wide use, like Weibull, are not members of the exponential family.

Let us demonstrate (4.15) for some of these distributions. For the normal distribution

$$f(y; \mu, \sigma) = (2\pi\sigma^2)^{-1/2} \exp[-(y-\mu)^2/(2\sigma^2)] =$$

$$\exp\{(y\mu - \mu^2/2)/\sigma^2 - (1/2)[y^2/\sigma^2 + \log(2\pi\sigma^2)]\}. \tag{4.16}$$

It is clear that this density is of the form (4.15), with $\theta=\mu$, $b(\theta)=\mu^2/2$, $a(\phi)=\phi$, $\phi=\sigma^2$, and

$$c(y,\phi) = -(1/2)[y^2/\sigma^2 + \log(2\pi\sigma^2)].$$

One may realize why θ was termed the natural location parameter and ϕ is a dispersion (or scale) parameter.

Similarly, for the Poisson distribution

$$f(y; \mu, \sigma) = e^{-\mu} \mu^y / y! =$$

$$\exp[y \log(\mu) - \mu - \log(y!)]. \tag{4.17}$$

For this distribution

$$\theta = \log(\mu), \; b(\theta) = e^{\theta} = \mu, \text{ and } c(y,\phi) = -\log(y!).$$

The natural location parameter is $\log(\mu)$ and the scale parameter is $\phi=1$.

We will not delve here into the details of how estimates of the parameters of the GLM model are derived. However, it is important for the reader to understand how a transition from a model of random variation (namely, a model for the response distribution) to a model of systematic variation is implemented by GLM. This transition is not unlike the one carried out in linear regression analysis. In the latter, it is assumed that the response has a normal distribution. Yet the assumption that the location parameter (namely, the mean) is constant is discarded. Instead, the mean is assumed to vary in a systematic fashion, and we model it by a linear combination of effects, denoted the linear predictor, η. According to this model, systematic variation in the effects included in the LP is transmitted to the response via changes in the mean.

Likewise, in modeling via GLM it is assumed that the natural location parameter, θ, is not constant but varies systematically. In fact, we commonly assume that $\theta = \eta$. With this assumption, we have indeed modeled the response mean. For example, for the Poisson we have

$$\theta = \log(\mu) = g(\mu) = \eta, \tag{4.18}$$

or

$$\mu = g^{-1}(\eta) = e^{\eta}. \tag{4.19}$$

The function $g(\mu)$ is denoted the *link function* since it provides the link between the response mean and the LP. As given in (4.18), namely, when $\theta = g(\mu) = \eta$, $g(\mu)$ is termed the *canonical link*. It has some good theoretical properties that will not be discussed here. Note that unlike transformational approaches, where the response is commonly assumed to be transformed, in GLM it is the response *mean* that is transformed in order to obtain the LP.

A practitioner need not necessarily use a canonical link function. For example, the Box-Cox transformation may be used to model the link between the mean and the LP:

$$\eta = \begin{cases} (\mu^{\lambda} - 1)/\lambda, \lambda \neq 0, \\ \log(\mu), \lambda = 0. \end{cases} \tag{4.20}$$

This form of the link function requires some preliminary explorative investigation to determine the best link (the best parameter, λ).

Once the link function is determined, the modeler is required to make another decision, frequently independently of the selected link function, and that is the distribution of the response. (Obviously if the *canonical* link is selected, these decisions are not carried out independently of each other). This decision determines the relationship between the response variance and the response mean. Remember that one of the major reasons for using GLM is that the response variance is not constant. A reasonable assumption to make is that the variance changes with the mean, which is typical to most non-normal distributions. For example, for the exponential we have $\sigma = \mu$. As a result, selecting the response distribution also determines the weights given to the various observations in the iteratively re-weighted least-squares procedure.

This procedure is used in GLM to estimate the model's parameters, and it minimizes the objective function (n is the sample size):

$$S = \sum_{i=1}^{n} (y_i - \mu_i)^2 / \sigma_i^2 . \tag{4.21}$$

Since the mean depends on the linear predictor through the relationship $\mu = g^{-1}(\eta)$, and since the variance, σ^2, depends on the mean, it is obvious that both the mean and the variance depend on the linear predictor, the coefficients of which we wish to estimate via minimization of (4.21). This explains why an iterative procedure is needed: The weights, $1/\sigma_i^2$, need to be updated with each successive iteration, as new estimates for the parameters of the LP are derived.

To sum up, using GLM implies three different decisions:
- The link function
- The response distribution (which is tantamount to determining the variance function, expressing the variance in terms of the mean)
- The predictor variables to include in the linear predictor.

The reader may realize that only a single decision is empirical in nature and determined solely by the data. If stepwise GLM is employed, then the data determine which predictors to include in the LP. The data also determine estimates for the LP coefficients. However, the other two decisions, the link and the response distribution, need to be taken prior to data analysis. It is true that these decisions may rely on some explorative analysis of the data, but the fact remains that selecting the link and the variance function is carried out *before* a GLM estimating procedure can be implemented.

Various application-oriented reference books have appeared in recent years that deal with GLM, either as the main subject or as an item addressed in the context of a wider subject. Some of these are Cook and Weisberg (1999), Wu and Hamada (2000), McCulloch and Searle (2001), Myers and Montgomery (2002) and Myers, Montgomery and Vining (2002).

4.5. Conclusions

General current approaches for empirical modeling of systematic variation have been outlined in the previous sections. They all share some features in common.

First, they all originated in relatively general models of random variation. Linear regression assumes that the response is normally distributed, and a systematic component of variation is introduced by modeling the mean as an LP, namely, a linear combination of effects that carry with them systematic variation. GLM likewise assumes (in its original and most widely applied formulation) that the response has a distribution, which is a member of the exponential family. Once this distribution is specified, this leads to a certain modeling of the variance in terms of the mean. Furthermore, if a *canonical* link is desired this also mandates a certain non-linear transformation of the mean in order to obtain the LP. For example, assuming a Poisson error distribution mandates a log transformation of the mean, and an assumed gamma distribution mandates a reciprocal link function.

All of these imply that an empirical model of systematic variation, employed to analyze relational data, has in fact originated in a model of random variation, where the latter has merely been modified in order to qualify it for the intended purpose.

A second shared feature relates to the modeling of the non-linear relationship between the LP and the response. We have indicated in the "Introduction" section that there are two components to the modeling effort: Modeling of the linear predictor and modeling its relationship to the response (or its mean). While the first component is carried out in all approaches under similar statistical principles (namely, using least-squares with its various variations), the second component differs appreciably between the approaches. In the response-transformation approach, the response is related to the LP via a non-linear relationship conveyed by a *data* transformation; In the case of the BC transformation, via the parameter λ. In GLM, the non-linear relationship between the response and the LP is represented by a link function, which expresses the non-linear transformation of the *mean*, needed to obtain additivity of effects (or, equivalently, the LP). In both approaches, the non-linear relationship is dictated by the selected or desirable response distribution. In the case of the BC transformation, the non-linear relationship is selected to revoke the normal distribution. In the GLM approach, the non-linear relationship is most often modeled in accordance with the assumed response distribution (if a canonical link is used).

Both features, the transition from a model of random variation to a model of systematic variation, and the ensuing dependence of the non-linear modeling on the selected response distribution, leave much to be desired.

First, one may ask, why a general methodology for modeling systematic variation needs to originate in models of random variation (the normal distribution, in the case of linear regression, or the exponential family, in the case of GLM). Expectedly, the resulting non-linear models would carry little resemblance to models of systematic variation, as we know them from various scientific and engineering disciplines (refer to the few examples given in Chapters 2 and 3).

Secondly, some good practices that one may expect to find in empirical modeling seem to be sorely lacking. For example, reliance on the available data in order to determine not only estimates of the models' parameters but also the very structure of the model is non-existent in GLM. Some of the most important decisions, like the response distribution and the link function, are specified *prior* to data analysis, and not as a result of this analysis, as could be expected in good empirical modeling.

The new RMM approach seems to provide the desirable solutions to these drawbacks. In Chapter 6 we delineate the requirements that good empirical modeling should fulfill, and in Chapter 13 RMM is assessed for compliance with these requirements.

References

[1] Box, G. E. P., Cox, D. R. (1964). An analysis of transformations. *Journal of the Royal Statistical Society,* Series B, 26, 211-243.

[2] Cook, R. D., Weisberg, S. (1999). *Applied Regression Including Computing and Graphics.* John Wiley & Sons.

[3] Draper, N. R., Smith, H. (1998). *Applied Regression Analysis.* 3rd Ed., John Wiley & Sons.

[4] Li, B., Moore, B. D. (2002). The general Box-Cox transformations in multiple linear regression analysis. *Communications in Statistics (Simulation and Computation),* 31(4), 673-687.

[5] McCulloch, C. E., Searle, S. R. (2001). *Generalized, Linear, and Mixed Models.* John Wiley & Sons.

[6] Myers, R. H., Montgomery, D. C. (2002). *Response Surface Methodology.* John Wiley & Sons.

[7] Myers, R. H., Montgomery, D. C., Vining, G. G. (2002). *Generalized Linear Models, with Applications in Engineering and the Sciences.* John Wiley & Sons.

[8] Shore, H. (2004). Response Modeling Methodology (RMM)- Current distributions, transformations and approximations as special cases of the RMM error distribution. *Communications in Statistics (Theory and Methods),* 33(7), 1491-1510.

[9] Smith, D. M. (2003). Computing single parameter transformations. *Communications in Statistics (Simulation and Computation),* 32(3), 605-618.

[10] Wedderburn, R. W. M. (1974). Quasil-Likelihood functions, generalized linear models and the Gauss Newton method. *Biometrika,* 61, 439-447.

[11] Wu, C. F. J., Hamada, M. (2000). *Experiments- Planning, Analysis and Parameter Design Optimization.* John Wiley & Sons.

Chapter 5

Approaches to Model Random Variation

5.1. Introduction

A "model of random variation" relates to the modeling of the response random fluctuations. The characteristic pattern of dispersion observed in these fluctuations is captured by a statistical distribution, assumed to have constant parameters. Thus, any model of random variation implies that the response's random fluctuations have been given a representation by a well-defined statistical distribution.

A response which experiences only random variation is commonly related to as a random variable (r.v.). A function that assigns a unique quantitative value to any result that may occur randomly within a specified random phenomenon is defined as a r.v.. Naturally, different r.v.s may be defined for the same random phenomenon. For example, an insurance company that deals with claims that arrive at random from its clients may be interested in the number of incoming claims for a certain period (a r.v.), but also in the periodic total sum of money needed to pay claimants (another r.v.). The same random phenomenon (a stream of incoming claims) has led to defining different r.v.s.

The science of Statistics tells us that for any r.v. there is a "true" underlying statistical distribution (with constant parameters). In fact, it is the existence of a state of "statistical control" for the random phenomenon under investigation that allows the observed dispersion to be captured by a statistical distribution with constant parameters. For the insurance company example, the number of incoming claims most probably follows a Poisson distribution with a constant parameter, λ.

Similarly, if the distribution of a single claim (expressed in monetary terms) may be described by, say, an exponential distribution, then the total sum per period is a random sum of r.v.s, the distribution of which may also be derived.

For both r.v.s it is implicitly assumed that within the population of clients the rate of road accidents and their severity is in a state of statistical control (statistically stable). Therefore statistical distributions can be defined for the above two r.v.s, which capture the observed random variation.

Relating to current general approaches for *empirical* modeling of *random* variation, we give up on the claim that the exact distribution of the r.v. is identified in the modeling process. The assertion attributed to Box, and quoted earlier: "All models are wrong, some are useful", is again valid. What this implies is that in attempting to develop a general approach for empirical modeling of random variation, one must recognize that there is no attempt to pinpoint the true underlying distribution. Not because methods to accomplish that objective are not available. However, these methods usually require a large sample size in order to lower to acceptable levels the probability of misspecification. Allied hypothesis-testing procedures, needed to validate the selected distribution (like Kolmogorov-Smirnov D statistic or Anderson-Darling A^2 statistic), also require a large sample size to achieve acceptable levels of statistical power in the statistical testing.

Since these testing procedures require that the "true" underlying distribution be first identified (otherwise no hypothesis testing can be carried out), the problem with the reliability of the process of identifying the true distribution is compounded.

With empirical modeling of random variation, an altogether different approach is pursued. We wish to have at our disposal a *single* methodology that, based on collected data, would provide a good model, or approximation, for the true underlying distribution. Once this approximation has been derived and demonstrated to deliver acceptable goodness-of-fit (relative to the available data), we may use it instead of the "true" distribution, having high enough degree of confidence that the empirical model "mimics" well the true random dispersion underlying the observed data.

There are four categories of general approaches to empirically model random variation. Each category emanates from a different set of assumptions, which provides the theoretical basis for its implementation.

(1) *"The true distribution is from a specified family of distributions"*

This assumption renders the approach empirical in the sense that we do not attempt to establish that the assumption is true, but rely on the general prevalence of the assumed family of distributions as justification to fit this family to the data on hand. No statistical testing is carried out to establish the "correctness" of our assumption.

An example is the Pearson family of distributions, which includes the beta, gamma, normal, Student's t and F distributions as special cases. If this family is selected for modeling, then all we have to do is estimate its parameters. The probability would be high that the true underlying distribution indeed belongs to this family of distributions. Clements' procedure for process capability analysis in a non-normal environment, related to in Chapter 1, in essence follows the same logic.

(2) *Modeling via a parameter-rich family of distributions*

The main consideration here is that the family of distributions selected for modeling has enough parameters to retain the flexibility necessary to model differently-shaped distributions. Unlike the previous approach, the question of which family is actually selected for modeling is secondary. By employing a parameter-rich family of distributions, it is guaranteed that some of its moments match those of the underlying (unknown) true distribution. This is accomplished by "Moment matching", where the parameters of the "adopted" distribution are determined so that the first few moments will be equal (match) those of the true (unknown) distribution. The "true" moments are estimated from available data. In Section 5.3 we discuss why equality of moments provides good assurance that the underlying true distribution is modeled well. We will also discuss why in practice this approach suffers serious drawbacks.

Note that this approach is not altogether independent of the previous one: Most general systems of distributions also tend to be parameter-rich.

(3) *Modeling via a general transformation that transforms data into a recognized and well-studied distribution (like the normal)*

This approach transforms data from skewed distributions into a recognized and easy-to-work-with distribution. A good example is the Box-Cox (BC) transformation, which transforms data from a continuous r.v. into normality, even though the origin distribution is not known (or specified). Similarly, the arc-sin transformation transforms binomial data to normality. Another form of functional transformation to normality is provided by the Johnson families of distributions. (These families are shown in Chapter 12 to be special cases of the RMM model.) Also in use are series expansions, which provide polynomial transformations to normality, like the Cornish-Fisher (CF) expansions. However, these are not as commonly used as the BC transformation (find details on the CF expansions, for example, in Stuart and Ord, 1987, Section 6.25-6.26).

(4) *Heuristic methods*

These are procedures that have empirically been demonstrated to deliver good representation to variously-shaped distributions, although no matching of moments is carried out to guarantee that (hence the term "heuristic").

In following sections, these general methodologies are surveyed and their merits and drawbacks discussed. In particular, we focus on what is missing in current approaches, which the RMM error distribution, developed in Chapter 9, can fill in as a general platform for modeling *random* variation.

In Section 5.2 we survey some general parameter-rich families of distributions, transformations and expansions commonly used to model random variation. Section 5.3 defines moments of a distribution and show why moments uniquely define a distribution (although there are exceptions). Equality of moments is then addressed as a general principle that serves to empirically model random variation. Heuristic methods are the subject of Section 5.4. In the last Section 5.5, an alternative to the four-moment matching approach is discussed. The new approach is pursued when moment-matching procedures for the RMM model are later developed.

5.2. Parameter-Rich Families of Distributions, Transformations and Expansions

We survey in this section some commonly applied general families of distributions. The term "general" refers to two aspects associated with these families. First, the families include as special cases some of the most widely used distributions. Secondly, these families have been shown to be flexible enough to occupy large portions of the $\{\sqrt{\beta_1}, \beta_2\}$ plane. While the latter is described in detail later on, when we delineate desirable properties of empirical modeling (refer to Requirement 7 in Chapter 6), it is perhaps appropriate that the $\{\sqrt{\beta_1}, \beta_2\}$ plane be briefly explained at this point.

The distributional parameters, $\sqrt{\beta_1}$ and β_2, refer to measures of the shape of the distribution. Two features of this shape is skewness (Sk), which relates to how asymmetric (skewed) the distribution is, and kurtosis (Ku), which relates to how long the tails of the distribution are, and consequently how non-concentrated it is around its mean. Skewness is measured by $\sqrt{\beta_1}$ and the kurtosis is measured by β_2 (or β_2-3). For the normal distribution, we have $\sqrt{\beta_1}=\beta_2-3=0$. For distributions skewed to the right (long right-tail), we have $\sqrt{\beta_1}>0$. For distributions with long tails (longer than the normal) we have $\beta_2-3>0$.

In terms of central moments (moments around the mean), $\{\mu_k\}$, and non-central moments (moments about zero), $\{\mu_k'\}$, skewness and kurtosis are calculated by the following formulae:

$$\text{Sk} = \sqrt{\beta_1} = \mu_3 / \sigma^3 = [\mu_3'-3\mu_2'\mu_1'+2(\mu_1')^3]/\sigma^3, \qquad (5.1)$$

$$\text{Ku} = \beta_2-3 = \mu_4/\sigma^4-3 =$$

$$[\mu_4'-4\mu_3'\,\mu_1'-3(\mu_2')^2+12\mu_2'(\mu_1')^2-6(\mu_1')^4]/\sigma^4, \qquad (5.2)$$

where $\mu_1'=\mu_1$ is the mean, $\mu_2'=E(Y^2)$ is the non-central second moment, and $\mu_2=E[(Y-\mu)^2]=\sigma^2$ is the central second moment (the variance). For a symmetrical distribution $\mu_3=E[(Y-\mu)^3]=0$.

When a given parameter-rich family of distributions occupies large portions of the $\{\sqrt{\beta_1}, \beta_2\}$ plane this implies that by a proper choice of its parameters, this family of distributions can attain most combinations of skewness and kurtosis values that one may encounter in practice. For example, for the exponential distribution we have $\sqrt{\beta_1}=2$, $\beta_2-3=6$. If by a proper choice of the values of its parameters the four-parameter Burr distribution can be shown to have the same skewness and kurtosis as the exponential, this implies that a member of the Burr family of distributions can deliver good representation to the exponential distribution. Of-course, the mean and the variance are also preserved.

To understand why matching of skewness and kurtosis is important, note that two distributions sharing the same first four moments (namely, have equality of means, variances, and the skewness and kurtosis measures) almost surely display similar patterns of dispersion. The two distributions may therefore, for all practical purposes, interchangeably replace one another. For empirical modeling of random variation, this implies that if two distributions, the modeling distribution ("the fitted distribution"), and the modeled distribution ("the true distribution"), share the same first four moments, the fitted distribution would represent well the "true" distribution. This point will be discussed in further detail in Section 5.3.

In the following, we introduce some of the most commonly used general families of distributions. In introducing these families we pursue a survey of general families of continuous distributions given in Johnson, Kotz and Balakrishnan (1994, henceforth JKB). A similar survey for discrete distributions is given in Johnson, Kotz and Kemp (1992, Ch. 2). The reader is referred to these sources for further details. Other sources will also be appropriately alluded to.

5.2.1. The Pearson family of distributions

A first attempt to develop a general system of distributions (or *systems of frequency curves*, as it is often referenced) was made by Pearson (1895). The system was so designed that for every member the probability density function satisfies a second order differential equation expressed in terms of the cumulative distribution function (CDF):

$$d^2F/dy^2 = [-(\alpha+y) /(\beta_0 + \beta_1 y + \beta_2 y^2)](dF/dy), \qquad (5.3)$$

where F is the CDF (with argument y), and "α" and "β_i" (i= 0, 1, 2) are real-valued parameters. The shape of the distribution that satisfies (5.3) varies considerably according to the values of the parameters. Pearson classified these distributions into seven types, each fulfilling certain conditions with respect to the parameters β_0, β_1 and β_2.

The following distributions are members of the Pearson family (a partial list): The normal distribution, the beta distribution, Student's t distribution, the gamma distribution and the Cauchy distribution. A general classification of Pearson's distributions into seven classes (including parameter characterization) is given in JKB (1994). Note that Pearson took as a starting point the skewed binomial and hyper-geometric distributions, which he smoothed in an attempt to construct skewed continuous density functions. Contrary to the impression occasionally conveyed by modern writers, Pearson himself did not begin his derivation of the Pearson system with the above differential equation (Rodriguez, in Kotz and Johnson, 1983, V.3, p.214).

5.2.2. Other families of distributions (Burr, Tukey's g- and h-systems, generalized Lambda, Shore, the exponential family)

While the Pearson family has played a dominant role in reported endeavors to deliver representation for data from an unknown statistical population, other parameter-rich families exist that have been used in the past for the same purpose with reportedly good results regarding goodness-of-fit. Families of distributions which have proved to deliver good representation for differently-shaped continuous distributions are, among others, the Burr distributions, the S-distribution (Voit, 1992), the generalized Lambda distribution (Karian and Dudewicz, 2000), Tukey's g- and h-systems (Tukey, 1962, 1977), and the extended-power family of distributions (Albert, Delampady and Polasek, 1991). Some of these will briefly be introduced, concentrating in particular on those that have proved over the years to be useful for empirical modeling.

Burr (1942) suggested a number of forms for his general system of frequency curves. The expressions for the CDF in terms of the variable, y, for some of these forms, are

$$F(y) = [\exp(-y)+1]^{-k},$$

$$F(y) = [y^{-c}+1]^{-k}, \; y > 0,$$

$$F(y) = 1-(1+y^{c})^{-k}, \; y > 0. \tag{5.4}$$

Here k and c are positive parameters. These forms deliver simple inverse distribution functions. For example, for the last form, we obtain

$$y= [(1-P)^{-(1/k)} - 1]^{1/c}, \; y > 0, \tag{5.5}$$

where P=F(y).

In Hoaglin, Mosteller and Tukey (1985), Hoaglin introduces an approach, originally conceived by Tukey (1962, 1977), to model r.v.s in terms of transformations of the unit normal variable. This approach results in Tukey's g- and h- systems, which were suggested to model "skewed variables" and "elongated" symmetric variables, respectively. Combining the two systems to represent both skewness and elongation results in the gh system. The three systems of distributions are, respectively (Refer to Hoaglin, Mosteller and Tukey (1985), Chapter 11, for details)

$$Y_g(Z) = [\exp(gZ) - 1]/g, \; g \neq 0,$$

$$Y_h(Z) = (Z) \exp(hZ^2/2),$$

$$Y_{g,h}(Z) = (1/g) [\exp(gZ) - 1] \exp(hZ^2/2), \; g \neq 0. \tag{5.6}$$

Here Y is the quantile function expressed in terms of Z, a unit normal variable, and g and h are parameters. All these systems are shown in Chapter 12 to be special cases of the RMM model.

Ramberg and Schmeiser (1974) and Ramberg (1975) generalized Tukey's two-parameter Lambda distribution into a four-parameter generalized Lambda distribution (GLD). The latter has been thoroughly investigated in Karian and Dudewicz (2000, KD), who have dispalyed the quantile function of GLD by

$$y(P)= \lambda_1 + (1/\lambda_2)[P^{\lambda_3} - (1-P)^{\lambda_4}], \; 0 \leq P \leq 1. \tag{5.7}$$

In their book, KD describe various procedures to fit GLD to given data sets. In Chapter 11 we compare a solution, obtained via GLD (and taken from KD) to a solution derived by fitting the RMM error distribution.

Shore (2000ab) suggested several new parameter-rich families of distributions for empirical modeling of random variation. The simplest is

$$
y = \begin{cases} \mu + \sigma\,[(A-C)z - hC],\ z < 0, \\[2ex] \mu + \sigma\,[(A+C)z - hC],\ z \geq 0, \end{cases}
\tag{5.8}
$$

where $\{z,y\}$ are the respective P-th quantiles of the standard normal, Z, and the approximated r.v., Y, $\{\mu, \sigma, Sk\}$ are the mean, the standard deviation and the skewness measure, respectively, and the parameters are

$$
h = 0.7978,\ A^2 = 1\text{-}0.3635C^2,\ Sk = 2.3940C - 0.6523C^3.
\tag{5.9}
$$

Equation (5.8) is an approximation for the quantile function of Y in terms of the standard normal quantile, and this transformation preserves the first three moments of Y. Good estimates for the first three moments needs to be available for modeling purposes. Find details in Chapter 21.

A simpler approximate expression, easily derived from (5.8), is (Shore, 2000b)

$$
y = \begin{cases} \mu + \sigma[(1\text{-}0.4179Sk)z - (1/3)Sk],\ z < 0, \\[2ex] \mu + \sigma[(1 + 0.4179Sk)z - (1/3)Sk],\ z \geq 0. \end{cases}
\tag{5.10}
$$

Finally, (5.8) may be re-written as

$$
y = \begin{cases} A_1 z + B_1,\ z < 0, \\[1ex] A_2 z + B_2,\ z \geq 0, \end{cases}
\tag{5.11}
$$

where the parameters are determined by

$$
A_1^2 = \{(\sigma^2 + \mu^2) - M_2(Y) - 2[\mu - M_1(Y)]^2\} / [(1/2) - 2M_1^2],
$$

$$
B_1 = 2[\mu - M_1(Y) + A_1 M_1],
$$

$$
A_2^2 = \{M_2(Y) - 2[M_1(Y)]^2\} / [(1/2) - 2M_1^2],
$$

$$
B_2 = 2[M_1(Y) - A_2 M_1],
\tag{5.12}
$$

$M_1 = 1/\sqrt{2\pi} = 0.3989$ is the first upper partial moment of the standard normal distribution, $M_1(Y)$ and $M_2(Y)$ are partial upper moments of Y, namely,

$$M_k(Y) = \int_{Med}^{\infty} y^k f(y)dy, \qquad (5.13)$$

and Med is the median of Y.

Note, that given a sample of n observations, an estimate for $M_1(Y)$ is the sum of all observations larger than the median, divided by n, and an estimate for $M_2(Y)$ is the sum of all squared observations (for observations larger than the median) divided again by n (not n/2!). For example, for the data: y={7,2,15,3}, an estimate of $M_2(Y)$ is $M_2(Y)=(7^2+15^2)/4= 68.5$.

Equation (5.11), with parameters given by (5.12), preserves the first two moments of Y (partial and complete). It has been shown to provide good representation to a wide variety of distributions, and it complies with the requirement that the MSEs of high degree moments, like skewness and kurtosis, are small when sample data are used to estimate its parameters (find details in Shore, 2000a, and also refer to Requirement 6 in Chapter 6).

The piece-wise linear transformations, given in (5.8)-(5.12), are addressed in detail, with some numerical examples, in Chapter 21.

A second parameter-rich family of distributions is (refer to Shore 2002a)

$$y = \begin{cases} A_1[P/(1\text{-}P)]^{B1}, P < 1/2, \\ \\ A_2\{[P/(1\text{-}P)]^{B2} - 1\} + A_1, P \geq 1/2. \end{cases} \qquad (5.14)$$

This family of distributions looks similar to the generalized Lambda distribution (5.7). Fitting procedures were developed for (5.14) that again require estimates of only first and second degree moments, partial and complete (find details in Shore, 1998a,b, and also relate to Chapter 23).

Another type of efforts to derive general families of distributions aimed at assembling under a single "umbrella distribution" several existing distributions. These endeavors have not introduced new theory but tried instead to re-formulate known distributions in more general terms. Notable examples for this approach are the Darmois-Koopman class of distributions (the exponential-type class), power series distributions, the Polya-type distributions or the class of stable distributions (refer to JKB, 1994, for details).

Perhaps the most notable of these is the exponential family of distributions, commonly introduced by the density function

$$f(y) = \exp\{[y\theta - b(\theta)]/a(\phi) + c(y,\phi)\}, \qquad (5.15)$$

where $a(.)$, $b(.)$ and $c(.)$ are specific functions, and θ and ϕ are parameters (often referred to as the natural location parameter and dispersion parameter, respectively). Special cases of the exponential family of distributions are the binomial, the Poisson, the geometric, the normal, the inverse Gaussian and the gamma distributions (including the exponential case). The exponential family served as the platform for the original derivation of generalized linear modeling (GLM), an empirical approach for modeling systematic variation that has been addressed in Chapter 4 (also find therein further details about the exponential family).

5.2.3. Transformations (Johnson, Box-Cox) and expansions

This is a third type of endeavors that resulted in the formation of a general platform for empirical modeling of random variation. With this approach the transformation modifies the (occasionally unknown) distribution into a well-known and well-studied distribution, like the normal. Conversely, we may use a transformation of a widely-used r.v., like the normal, to deliver representation for a skewed (non-symmetrically distributed) r.v. A typical example for the latter is the log-normal distribution or the above g- and h-systems (5.6). In this subsection we concentrate on the former case, namely, transforming a r.v. into a well-recognized distribution.

Johnson (1949, 1954) suggested transformations to normality. The most well-known of these appear in Johnson (1949), where the transformations result in a unit normal variable, Z. These transformations are (JKB, 1994, p. 34):

The S_L family (from the log-normal distribution)

$$Z = \gamma + \delta \log Y, \; 0 < Y. \tag{5.16}$$

The S_B family

$$Z = \gamma + \delta \log[Y/(1-Y)], \; 0 \leq Y \leq 1. \tag{5.17}$$

The S_U family

$$Z = \gamma + \delta \sinh^{-1} Y = \gamma + \delta \log [Y + (Y^2+1)^{1/2}], \; -\infty \leq Y \leq \infty, \tag{5.18}$$

where $Y = (X-\xi)/\lambda$, X is the original r.v., to be modeled via a member of the Johnson family, and ξ and λ (>0) are location and scale parameters, respectively. Conventionally, δ is kept positive too.

The Johnson transformations have been widely used for empirical modeling in quality engineering applications (for example, Farnum, 1996/7, Chou, Polansky and Mason, 1998, and references therein). Tadikamalla and Johnson (1982) used the above three transformations to develop a system of frequency curves based on the logistic distribution.

The BC transformation (Box and Cox, 1964), discussed in Chapter 4, also provides a powerful platform for transforming data into normality, even though the distribution of the original data remains unknown. As shown in Chapter 12, both the Johnson and the inverse BC transformations are special cases of the RMM error distribution.

Finally, series expansions have been developed that attempt to provide general representations for various functions of a given distribution in terms of various functions of a known distribution (like the normal or the Poisson). Typical examples are the Gram-Charlier expansions, which approximate the CDF of an r.v. in terms of its cumulants and the standard normal CDF and density function. Another example are the Cornish-Fisher expansions (Cornish and Fisher,1937, Fisher and Cornish, 1960), where the quantile of an r.v. is expressed in terms of a weighted sum of powers of the corresponding standard normal quantile, and the weights are functions of the cumulants of the approximated distribution (find details in Stuart and Ord, 1987).

While this description of parameter-rich-distributions, transformations and expansions is by no means exhaustive, it obviously attests to an existing need for a general representation of distributions that will allow standardization of various distribution-dependent practices. In particular, these endeavors point to the need for a general platform for empirical modeling of random variation, one of the main themes of this book.

5.3. Moments and Their Role in Empirical Modeling of a Distribution

Moment-matching plays a central role in empirical modeling of random variation. With moment-matching we estimate the moments of the underlying (most likely unknown) true distribution using sample data, and then *match* the moments of a given parameter-rich distribution with these estimates.

The number of moments that can be matched depends on the number of parameters in the fitted distribution. For example, if we wish to fit the normal distribution (two parameters) then only matching of the first two moments (the mean and the variance) is possible. As we alluded to in the introductory Chapter 1, all of Shewhart's SPC control charts are based on a two-moment matching.

When the underlying unknown distribution is expected to be skewed (non-symmetric), matching of the first two moments will probably not be enough. In such cases a three- or four-parameter family of distributions (like Pearson) is employed to model the unknown distribution. The parameters of the parameter-rich distribution are determined so that moments of the fitted distribution coincide with the actual (true) first three or four moments, as explained above.

Given this widely applied practice, two natural questions arise:
- What is the theoretical justification for this practice?
- What is the proper number of moments that need to be matched?
In this section we attempt to answer both questions.

5.3.1. Why moment matching?

A thorough discussion of how far a set of moments (assuming they all exist) uniquely determine a distribution is given in Stuart and Ord (Section 4.20). Here we pursue a less statistically rigorous explanation that would still deliver a feeling for why moments determine a distribution.

A well known statistical function that characterizes a distribution is the moment-generating function (MGF). It is defined by

$$M(t) = E(e^{tY}) = \int_{-\infty}^{\infty} e^{ty} dF(y), \tag{5.19}$$

where $F(y)$ is the CDF of the r.v., Y. It can be shown mathematically that if the MGFs of two r.v.s, Y_1 and Y_2, are identical for all values of t in an interval around the point $t=0$, then the probability density functions of the two r.v.s must be identical.

Using a well-known series expansion of e^v,

$$e^v = 1+v+v^2/2!+v^3/3!+.., \tag{5.20}$$

we obtain from (5.19)

$$M(t) = E[1+tY+(tY)^2/2!+(tY)^3/3!+..]. \tag{5.21}$$

Although the series in (5.20) is *infinite*, it can be shown under fairly general conditions that the expected value of the sum is equal to the sum of the expected values (a result which is always true for a *finite* sum of r.v.s). This implies that one can write for (5.21)

$$M(t) = 1+ tE(Y) + t^2E(Y^2)/2!+ t^3E(Y^3)/3!+... \tag{5.22}$$

Denoting the k-th non-central moment, $E(Y^k)$, by μ_k', we may re-write (5.22) by

$$M(t) = 1+ t\,\mu_1' + (t^2/2)\mu_2'+ (t^3/6)\mu_3'+... \tag{5.23}$$

From (5.23) it can easily be recognized that the n-th derivative of $M(t)$ with respect to t, at the point $t=0$, is equal to μ_n'. This is why $M(t)$ is generally related to as "a moment-generating-function".

The upshot of all these expressions is that since $M(t)$ uniquely determines the distribution of the r.v., and since $M(t)$ may be represented as a series expansion in terms of its moments, it is reasonable to assume that if the first few moments in this expansion are shared by two r.v.s their characteristic patterns of dispersion, as captured by the respective statistical distributions, will also tend to be similar. Staurt and Ord (1987, Section 3.34) provide a mathematical demonstration of this assertion, and numerous empirical studies repeatedly corroborated this fact.

This leads to the second question: How many moments should two r.v.s share in common so that their distributions can mutually replace one another and still retain acceptable representability for the replaced distribution?

We address this question in the next subsection. However, a word of caution is necessary here. Although the MGF in most cases does define the distribution, not all r.v.s have defined MGF. Some distributions fail to have finite moments at all (for example, the Cauchy distribution) and therefore their MGFs also do not exist. In other cases, M(t) fails to exist for some values of t. For such cases, another function has been defined, the characteristic function, which is of great theoretical importance in Statistics. Unlike the MGF, the characteristic function uniquely defines both the moments and the distribution function. It does not, however, follow that the moments (or the MGF) determine the distribution in all cases, even when moments of all orders exist. Only under certain conditions will a set of moments determine a distribution uniquely.

However, to quote Staurt and Ord (1987) as they relate to the practical significance of this assertion, "fortunately for statisticians, these conditions are satisfied by most of the distributions commonly arising in statistical practice. For most ordinary purposes, therefore, knowledge of the moments is equivalent to a knowledge of the distribution function in the sense that it should be possible theoretically to exhibit all the properties of the distribution in terms of its moments" (therein, Section 3.33).

5.3.2. *How many moments to match*

The question of how many moments should a parameter-rich distribution, used for empirical modeling of random variation, preserve *vis-à-vis* the true distribution- is essential. As alluded to earlier, Shewhart control charts use the normal distribution for modeling, and therefore only the first two moments (mean and variance) are matched when a Shewhart control chart is developed.

Conversely, Clements (1989) developed a procedure to calculate process capability indices for non-normal populations, based on the Pearson family of distributions. His procedure requires matching of all first four moments.(The medians are used for matching instead of the means, but this makes little difference.)

From a theoretical point of view, as explained in the previous sub-section, as more moments are matched in fitting a certain parameter-rich distribution to given data, the better the fitted distribution represents the true distribution. Therefore it is crucial to determine the number of moments to match so as to ensure that the properties of the variation of the modeled r.v. have been adequately preserved by the fitted distribution.

To understand how the shape characteristics of the true distribution affect the number of moments that need to be matched, let us recall what each of the first four moments, namely, the mean, the variance, the skewness and the kurtosis measures, stand for. The mean is a location parameter, and the variance is a scale parameter that reflects the variation in the data. The third moment serves to determine the skewness of the distribution, that is, its departure from symmetry. The fourth moment serves to measure kurtosis, namely, the tendency of the distribution to have long tails. What all these imply is that the number of moments that need to be matched depends to a large extent on our knowledge of the basic properties of the distribution. For example, if the underlying distribution is known to tend to symmetry, then perhaps matching of the first two moments suffice, as done with Shewhart control charts. On the other hand, if we wish to implement a procedure that will be generally valid and not require a-priori information about the shape of the distribution, then matching of four moments may be essential.

That matching four moments is generally adequate in empirical modeling of random variation seems to be corroborated by an empirical study conducted by Pearson, Johnson and Burr (1979). The latter have compared the percentage points of eight different "systems of frequency curves", all having the same first four moments. Inspecting the extensive tables given therein of various percentage points of different distribu-tions, one may appreciate the closeness of the values of respective percentage points (quantile values) derived from distributions sharing the same first four moments. This fact has led to the use of different families of distributions, like Johnson (for example, Farnum, 1996/7), Pearson (for example, Clements, 1989) and others, both in process capability analysis (PCA, refer to Kotz and Johnson, 2002) and in statistical process control (SPC). What distribution is virtually used for fitting seems to have a marginal effect.

To sum up, the best guarantee that the true distribution has been properly modeled is to replace the unknown distribution by a known four-parameter distribution, and then determine the parameters of the latter so that equality of the first four moments is achieved

While this assertion is theoretically valid, there is a problem here. Though matching of the first four moments is a desirable result, we can rarely achieve it in practice. This is due to the fact that sample estimates of third and fourth moments are notoriously known for their large sampling errors. Using sample estimates of skewness and kurtosis in a fitting procedure may lead to extremely biased estimates of various quaniles that are calculated from the fitted distribution, like those used for PCA. Kotz and Johnson (1993) and Kotz and Lovelace (1998) repeatedly warn against this eventuality in their reference to PCA for skewed populations. This is also a major impediment to the application of Clements method, which is today the most widely applied procedure for PCA in a non-normal environment.

Unfortunately, users of this method are never pre-warned about this drawback, inherent to any PCA which relies on sample estimates of skewness or kurtosis.

Likewise, Karian and Dudewicz (KD, 2002, 2003) advocate the use of percentile-matching (rather than moment-matching) because "the relatively large variability of sample moments of order 3 and 4 can make it difficult to obtain accurate...fits through the method of moments" (KD, 2002, p. 154). Shore (1998ab, 2004) also makes repeated reference to this point. In particular, the large mean-squared-errors (MSEs) that result from using estimates of high degree moments were demonstrated, via Monte-Carlo simulation, with regard to PCA (Shore, 1998b).

The main conclusion is that the most important criterion to judge the merit of a new approach to deal with non-normality in the framework of empirical modeling is the degree to which the first three or four moments of the underlying distribution are preserved by the distribution that approximates it. If sample estimates are used, this criterion reduces to the question of the size of the MSEs of moments associated with the fitted (estimated) distribution. In practice, this criterion implies that if a large number of samples have been taken from the same non-normal population, a four-parameter distribution fitted for each sample (by any specified method), and moments calculated (numerically) from the fitted distribution, then the MSEs of these moments should be minimal.

The MSE criterion should also be the leading indicator in comparisons between different approaches. Regrettably, this is rarely the case. As a brief review of some of the references mentioned earlier would attest, comparison between methods based on the MSEs of moments of the fitted distribution is rarely practiced, and considerations of MSEs of moments are often missing in the literature. Furthermore, the wide use of a four-moment fitting procedure in reported quality and reliability applications provide ample evidence that this consideration is rarely given the weight that it deserves.

Alternatively, matching procedures have been developed where both partial and complete moments take part in the fitting procedure. Yet only low-degree moments are used in the fitting procedure (second degree at most). This alternative approach will be discussed in Section 5.5 and also in Section 20.3.

5.4. Heuristic Methods in Empirical Modeling of Random Variation

Heuristic methods have been defined in Section 5.1 as procedures that have been demonstrated to deliver good representation to variously-shaped distributions. However no matching of moments is carried out to guarantee that. This section presents a brief survey of some of the heuristic procedures that have appeared in the literature related to statistical process control (SPC).

This area has been selected since it has recently posed the most urgent need for the development of such procedures. In earlier applications of SPC, practitioners relied on periodic sampling of the process, where averages were most commonly used in Shewhart control charts. Based on the Central Limit Theorem, which predicts averages to be approximately normally distributed, the use of Shewehart charts was justified, and there was no need to model the real process distribution.

With the gradual transition to process monitoring in real-time, where individual observations (and not averages) are plotted in the control charts, the requirement for procedures that would reflect in the control limits the true process distribution (often a non-symmetrical one) became crucial. Control charts with asymmetrical control limits were needed. Yet one cannot expect practitioners of SPC to identify the true process distribution via statistical analysis. In other words, procedures for empirical modeling of random variation were needed.

Another area where empirical modeling of the process distribution is required is process capability analysis (PCA). Most process capability indices (PCIs) require specification of the end-points of the process distribution. These end-points are traditionally defined by the 0.135 and the 99.865 percentile values (corresponding to $z=-3$ and $z=3$, respectively, for a normal distribution). When the process distribution is skewed, the need to estimate these quantile values requires estimating the true process distribution, and here empirical modeling is again called for. One solution for this problem was suggested by Clements (1989), a solution which we have related to earlier.

The following brief survey presents some of the most recent heuristic procedures, proposed in the literature. It is not meant to be a comprehensive and exhaustive review, but only to provide the reader with a sample of the ideas implemented in such endeavors.

Quesenberry, in a series of papers (for example, 1991ab, 1995ab), suggested to use Q charts for SPC applications. The basic idea of the Q chart is to use the Q statistic for the control chart. Suppose that the statistic used to monitor the process is Y, and that the cumulative distribution function (CDF) is $F=F(y)$. For example, if Y is Poisson with parameter λ, then the CDF of a given sample value, y_i, is

$$F_i = \Pr(Y \leq y_i) = \sum_{y=0}^{y_i} \lambda^y e^{-\lambda} / y!, \, y \geq 0.$$

This value of F_i may be transformed into standard normal quantile value via the Q statistic:

$$Q_i = \Phi^{-1}(F_i)$$

where $\Phi^{-1}(F)$ is the inverse standard normal distribution function (namely, the standard normal quantile that corresponds to the CDF value of F). Since the statistic Q is normally distributed with mean zero and standard deviation 1, it can easily be plotted in a standard Shewhart chart. Properties of the Q charts for attributes and for variables have been extensively investigated by Quesenberry (see references above).

Grimshaw and Alt (1997) suggest that instead of monitoring parameters of a distribution (as in Shewhart charts), the control chart should monitor values of the quantile-function of the process distribution. They justify this on the basis of the observation that "often it is ignored that different probability models can have nearly equal values of summary statistics... however have quite different distribution shapes". Based on this observation, the authors suggest to estimate the in-control quantile-function for some selected values. In the monitoring phase, sample quantile-function values are estimated from process data, and a test statistic is used to test departure from the in-control values.

In a series of papers, the method of the weighted variance (or weighted standard deviation) was used in applications for both SPC and PCA in a non-normal environment. This method was originally suggested by Choobineh and Branting (1986), and was then adopted for quality engineering applications. Examples for the latter are Choobineh and Ballard (1987), Bai and Choi (1995, and also in Kotz and Lovelace, 1998), Wu *et al.* (1999), Chang and Bai (2001) and Chang, Choi and Bai (2002).

The weighted variance method essentially attempt to approximate the process distribution by two separate normal distributions: One fitted for the part of the process distribution above the mean, and another for the part below the mean. Both fitted normal densities share the same mean, however they have different standard deviations, $2\sigma_L^W$ for the lower part of the distribution, and $2\sigma_U^W$ for the upper part, where

$$\sigma_L^W = (1-P_Y)\sigma_Y,$$

$$\sigma_U^W = P_Y\,\sigma_Y,$$

$$P_Y = \Pr(Y \leq \mu_Y), \qquad\qquad (5.24)$$

and $\{\mu_Y, \sigma_Y\}$ are the mean and the standard deviation of the process variable, Y. P_Y is estimated from the data.

It is interesting to note that in the same year when Choobineh and Branting (1986) have introduced their original method, based on approximating the two parts of a process distribution by two different normal densities, the same method was suggested in Shore (1986a). The latter, however, differ from the former in two respects. First, "splitting" the approximated process distribution is done at the median (rather than at the mean). Secondly, Shore's method takes account, in an explicit way, of the skewness of the underlying process distribution. This is done by preserving the skewness measure when the approximation's parameters are determined (refer to an introduction of these approximations in Chapter 21, and also to Shore, 1986ab).

Failure to demonstrate, in reported applications of the weighted variance method, preservation of high moments (notably third and fourth moments) in the data-fitted approximation has been the subject of a critical comment in Shore (2004).

In a review of the literature on process capability indices for 1992-2000, Kotz and Johnson (2002) dedicate a special section to dealing with non-normality in PCA. The reader is referred to this section for further, often innovative, solutions that attempt to empirically model the process distribution without identifying it.

5.5. An Alternative Approach to Four-Moment Matching

In Section 5.3 we discussed the drawbacks of using four-moment fitting based on sample estimates of the first four moments. As an alternative, we related to a new approach where fitting was confined to moments of low degree (second degree at most). To accomplish this, yet retain the flexibility granted by a parameter-rich distribution, both complete and partial moments participate in the moment-matching procedure (partial moments refer to moments calculated either for the upper half or the lower half of the distribution). No estimates of skewness and kurtosis are employed in the fitting procedure. As have been repeatedly demonstrated (for example, Shore 1995, 1996, 1998ab, 2000a), MSEs associated with the moments of distributions thus fitted are smaller than the respective MSEs, associated with distributions fitted by matching of direct sample estimates of the first four moments. In fact, it has been shown (Shore, 1996) that this fitting procedure provides a better estimate of skewness (derived by numerical integration from the fitted distribution) than that derived from a direct sample estimate of skewness. This implies that better representation of the unknown distribution is accomplished, relative to four-moment fitting, irrespective of which distribution is used for the fitting.

For large sample size this relative advantage naturally vanishes.

When fitting procedures are developed for the RMM error distribution (in Chapter 10), and for the inverse normalizing transformations (in Chapter 20), only moments of second degree at most, partial and complete, are employed. Percentile-based fitting procedures are also developed.

References

[1] Albert, J., Delampady, M., Polasek, W. (1991). A class of distributions for robustness studies. *Journal of Statistical Planning and Inference*, 28, 291-304.

[2] Bai, D. S., Choi, I. S. (1995). X and r control charts for skewed populations. *Journal of Quality Technology*, 27, 121-131.

[3] Box, G. E. P., Cox, D. R. (1964). An analysis of transformations. *Journal of the Royal Statistical Society, Series B*, 26, 211-243.

[4] Burr, I. W. (1942). Cumulative frequency functions. *Annals of Mathematical Statistics,* 13, 215-232.

[5] Chang, Y. S., Bai, D. S. (2001). Control charts for positively-skewed populations with weighted standard deviations. *Quality and Reliability Engineering International*, 17, 397-406.

[6] Chang, Y. S., Choi, I. S., Bai, D. S. (2002). Process capability indices for skewed populations. *Quality and Reliability Engineering International*, 18, 383-393.

[7] Choobineh, F., Ballard, J. L. (1987). Control limits of QC charts for skewed distributions using weighted variance. *IEEE Transactions on Reliability*, 36, 473-477.

[8] Choobineh, F., Branting, D. (1986). A simple approximation for semi-variance. *European Journal of Operational Research*, 27, 364-370.

[9] Chou, Y.M, Polansky, A. M., Mason, R. L. (1998). Transforming non-normal data to normality in statistical process control. *Journal of Quality Technology*, 30(2), 133-141.

[10] Clements, J. A. (1989). Process capability calculations for non-normal distributions. *Quality Progress*, 22(2), 49-55.

[11] Cornish, E. A., Fisher, R. A. (1937). Moments and cumulants in the specification of distributions. *Review of the International Statistical Institute*, 5, 307-320.

[12] Farnum, N. R. (1996/7). Using Johnson curves to describe non-normal process data. *Quality Engineering*, 9, 329-336.

[13] Fisher, R. A., Cornish, E. A. (1960). The percentile points of distributions having known cumulants. *Technometrics*, 2, 209-226.

[14] Grimshaw, S. D., Alt, F. B. (1997). Control charts for quantile-function values. *Journal of Quality Technology*, 24(1), 1-7.

[15] Hoaglin, D. C., Mosteller, F., Tukey, J. W. (1985). *Explorative Data, Tables, Trends, and Shapes.* John Wiley & Sons.

[16] Johnson, N. L. (1949). Systems of frequency curves generated by methods of translation. *Biometrica*, 36, 149-176.

[17] Johnson, N. L. (1954) Systems of frequency curves derived from the first law of Laplace. *Trabajos de Estadistica*, 5, 283-291.

[18] Johnson, N. L., Kotz, S., Balakrishnan, N. (1994). *Continuous Univariate Distributions.* V. 1. 2nd Edition. John Wiley & Sons.

[19] Johnson, N. L., Kotz, S., Balakrishnan, N. (1995). *Continuous Univariate Distributions.* V. 2. 2nd Edition. John Wiley & Sons.

[20] Johnson, N. L., Kotz, S., Kemp, A. W. (1992). *Univariate Discrete Distributions.* 2nd Edition. John Wiley & Sons.

[21] Karian, Z. A., Dudewicz, E. J. (2000). *Fitting Statistical Distributions: The Generalized Lambda Distribution and Generalized Bootstrap Methods.* CRC Press, Boca Raton, Florida, USA.

[22] Karian, Z. A., Dudewicz, E. J. (2003). Comparison of GLD fitting methods: Superiority of percentile fits to moments in L^2 norm. *Journal of the Iranian Statistical Society*, 2(2), 171-187.

[23] Kotz, S., Johnson, N. L. (1982-). *Encyclopedia of Statistical Sciences.* John Wiley & Sons.

[24] Kotz, S., Johnson, N. L. (1993). *Process Capability Indices.* Chapman and Hall. London.

[25] Kotz, S., Johnson, N. L. (2002). Process Capability Indices- A Review, 1992-2000. *Journal of Quality Technology*, 34(1), 2-19.

[26] Kotz, S., Lovelace, C. R. (1998). *Process Capability Indices in Theory and Practice.* Arnold: New-York.

[27] Pearson, E. S., Johnson, N. L., Burr, I. W. (1979). Comparisons of the percentage points of distributions with the same first four moments, chosen from eight different systems of frequency curves. *Communications in Statistics (Simulation and Computation)*, B8(3), 191-229.

[28] Pearson, K. (1895). Contributions to the mathematical theory of evolution. II. Skew variations in homogeneous material. *Philosophical Transactions of the Royal Society of London, Series A*, 186, 343-414.

[29] Quesenberry, C. P. (1991a). SPC Q charts for a binomial parameter: short or long runs. *Journal of Quality Technology*, 23, 239-246.

[30] Quesenberry, C. P. (1991b). SPC Q charts for a Poisson parameter λ: short or long runs. *Journal of Quality Technology*, 23, 296-303.

[31] Quesenberry, C. P. (1995a). On properties of Q charts for variables. *Journal of Quality Technology*, 27, 184-203.

[32] Quesenberry, C. P. (1995b). On properties of binomial Q charts for attributes. *Journal of Quality Technology*, 27, 204-213.

[33] Ramberg, J. S. (1975). A probability distribution with applications to Monte-Carlo simulation studies. *Statistical Distributions in Scientific Work*, 2, G. P. Patil, S. Kotz and J. K. Ord (Eds.), Reidel Publishing Company, Dordrecht-Holland, 51-64.

[34] Ramberg, J. S., Schmeiser, B. W. (1974). An approximate method for generating asymmetric random variables. *Comm. ACM*, 17, 78-82.

[35] Shore, H. (1986a). Simple general approximations for a random variable and its inverse distribution function based on linear transformations of a non skewed variate. *SIAM J. Sci. Stat. Comput.*, 7(1), 1-23.

[36] Shore, H. (1986b). An approximation for the inverse distribution function of a combination of random variables, with an application to operating theaters. *Journal of Statistical Computation and Simulation*, 23(3), 157-181.

[37] Shore, H. (1995). Fitting a distribution by the first two sample moments (partial and complete). *Computational Statistics and Data Analysis*, 19, 563-577.

[38] Shore, H. (1996). A new estimate of skewness with MSE smaller than that of the sample skewness. *Communications in Statistics (Simulation and Computation)*, 25(2), 403-414.

[39] Shore, H. (1998a). Approximating an unknown distribution when distribution information is extremely limited. *Communications in Statistics (Simulation and Computation)*, 27(2), 501-523.

[40] Shore, H. (1998b). A new approach to analysing non-normal quality data with application to process capability analysis. *International Journal of Production Research (IJPR)*, 36(7), 1917-1933.

[41] Shore, H. (2000a). Three approaches to analyze quality data originating in non-normal populations. *Quality Engineering*, 13 (2), 277-291.

[42] Shore, H. (2000b). General control charts for attributes. *IIE Transactions (Quality and Reliability Engineering)*, 32 (12), 634-637.

[43] Shore, H. (2004). Non-normal populations in quality applications- A revisited perspective. *Quality and Reliability Engineering International*, 20(4), 375-382.

[44] Stuart, A., Ord, J. K. (1987) *Kendall's Advanced Theory of Statistics. V. 1: Distribution Theory.* Charles Griffin & Company, Ltd., London.

[45] Tadikamalla, P. R., Johnson, N. L. (1982). Systems of frequency curves generated by transformations of logistic variables. *Biometrika*, 69, 461-465.

[46] Tukey, J. W. (1962). The future of data analysis. *Annals of Mathematical Statistics*, 33, 1-67.

[47] Tukey, J. W. (1977). *Modern Techniques in Data Analysis*. NSF-sponsored regional research conference at Southeastern Massachusetts University, North Darmouth, MA.

[48] Wu, H. S., Swain, J. J., Farrington, P. A., Messimer, S. L. (1999). A weighted variance capability index for general non-normal process. *Quality and Reliability Engineering International*, 15, 397-402.

[49] Voit, E. O. (1992). The S-distribution: A tool for approximation and classification of univariate, unimodal probability distributions. *Biometrical Journal*, 34, 855-878.

Chapter 6

The Requirements and Evaluation of Compliance

6.1. Introduction

Empirical modeling implies that general theory-free methodologies are employed in the modeling process. These methodologies should therefore have a universal character about them which can render a certain degree of confidence that implementing them would result in acceptably adequate models. In particular, these methodologies should reflect general properties and "truths" that are known to be universally valid either with respect to scientific modeling of systematic variation or with respect to statistical modeling of random variation.

An example for the former is the property that quite often scientific modeling results in a model that has a monotone convex (concave) relationship between the affecting factor and the response (refer, for example, to Antoine equation, given in Chapter 2). An example for the latter is the known "truth" that when the Central Limit Theorem mandates that a response error-distribution approximates normality, a de-coupling of the mean from the variance should take place (changes in the mean do not incur changes in the variance).

Expressed in a different way, empirical modeling requires that the methodology used has been shown to be flexible enough to accommodate a large spectrum of scenarios, which a practitioner (a quality-engineering practitioner or otherwise) may encounter in reality. The properties needed from such a methodology, once specified, may then form the basis to assess the quality of current methodologies for

79

empirical modeling, and perhaps lead the way to devise better ones that more properly comply with the requirements.

As may be expected, not all requirements were created equal. Some may be more essential to the general effectiveness of a methodology for empirical modeling than others. However, if one wishes to define some general features that the requirements should reflect, we believe that the following characterization provides a good guide for that.

(1) Flexibility. The requirements should reflect the need for flexibility in any methodology of empirical modeling. For example, if we wish a methodology for *systematic* variation to be able to accommodate a wide spectrum of differently-shaped error distributions, then enough parameters should be included in the model to allow the error distribution to preserve such flexibility. This is analogous to the flexibility required from general families of distributions (like Pearson), employed in modeling *random* variation: These families are all characterized by a high number of parameters (relate to Chapter 5, where this property is addressed).

(2) Conformance to prevalent (highly likely) scenarios. The requirements should reflect the need of empirical-modeling methodologies to be able to reduce to commonly encountered scenarios as special cases. For example, one cannot conceive of a methodology for empirical modeling of systematic variation that does not include the linear model as a special case. Likewise, a methodology for modeling of random variation can not exclude the normal distribution as a special case.

(3) Compatibility with rules, laws, or empirical regularities that are known to be of a universal nature. We have mentioned before such two laws (regularities): Monotone convexity, for relational modeling, and the Central Limit Theorem, for modeling random variation.

(4) Easy transition from modeling systematic variation to modeling random variation and *vice versa*. The requirements should reflect a demand that a single methodology be used for both modeling of systematic variation and modeling of random variation (even though the methodology may have been initially developed for either sorts of variation modeling). For example, linear regression would not qualify since it cannot be used for empirical modeling of random variation (the assumed underlying distribution is normal, and no other distribution can be modeled via the error distribution associated with the linear regression model).

(5) Reflecting what is generally considered best practices in empirical modeling. The requirements should be representative of features generally known to be associated with good empirical modeling. This characterization refers to all conceivable characteristics not covered above. In particular, it may reflect some consensus regarding good empirical modeling that is shared by most engineering disciplines. For example, use of high-degree polynomials is common in chemical engineering, but not encouraged in quality engineering (second-degree polynomials are of the highest degree one can encounter in quality-related applications). Therefore, no requirement should lead to a general acceptance of such practice.

The set of requirements in the next section reflects the above general characterization. The rest of this chapter is structured as follows: In Section 6.2 we outline desirable requirements of a general empirical (data-driven) methodology for modeling variation (either random or systematic). We then evaluate, in Section 6.3, how compliant with these requirements are current major methodologies for empirical modeling. We separate the analysis to approaches for modeling *systematic* variation (as surveyed in Chapter 4; Refer to Section 6.3.1) and those for modeling *random* variation (as surveyed in Chapter 5; Refer to Section 6.3.2).

6.2. Desirable Requirements of a General Methodology for Empirical Modeling

In this section, we detail the set of requirements that would serve as a basis for the evaluation of the quality of any general approach for empirical modeling. While there is no supporting theory to provide criteria for inclusion in the set of requirements (some other sets, perhaps partially overlapping, may be conceived), we believe that this set is representative enough of the realities that a practitioner would encounter in her/his daily routine. In particular, the general characterization of the desirable requirements, outlined in the previous section, may assist in judging the quality of the requirements included in the set.

In the following, the requirements are expounded, arranged from the more important (essential) to the less important ones. However, they are all perceived to be of a universal enough character to warrant inclusion in this set.

Each requirement will first be introduced ("Description"), followed by an empirical or theoretical justification for its inclusion ("Reasoning"). "An Example", which concludes presentation of each requirement, demonstrates how it is satisfied by some current methodology. This may help understand the practical significance of the requirement.

Requirement 1: Provide monotone convex/concave relationship in modeling systematic variation.

Description. The relationship between the affecting factor and the response should include monotone convexity (or concavity), increasing or decreasing. If several factors (or effects) exist that transmit variation to the response, monotone convexity (concavity) refers to the relationship of the response with a *linear combination* of these factors (or effects). The latter was earlier denoted the "Linear Predictor" (LP).

Comment. For brevity, we will henceforth relate only to *convex* relationships, understanding that similar assertions may be made with respect to *concave* relationships.

Reasoning. The requirement relates to the empirical observation that a large portion of relationships observed in nature, and reflected in corresponding scientific and engineering models, is comprised of monotone convex relationships (refer to the survey in Chapter 2 and also to Shore, 2004a). All link functions used in the framework of GLM are likewise monotone convex functions. Examples for convex functions are X^a, e^x and $\exp(x^a)$.

An Example. Consider the BC normalizing transformation (Box and Cox, 1964):

$$(Y^\lambda - 1)/\lambda = \begin{cases} (Y^\lambda - 1)/\lambda = \eta + \varepsilon, & \lambda \neq 0, \\ \log(Y) = \eta + \varepsilon, & \lambda = 0, \end{cases} \tag{6.1}$$

where $\eta = \beta_0 + \beta_1 X_1 + \beta_2 X_2 + .. + \beta_k X_k$ is the LP, ε is the error (probably normal), and $\{X_i\}$ are the effects (predictor variables) included in the LP. Note, that the latter may contain non-linear terms, and the "linear" refer, as in linear regression, to linearity in the parameters only. Likewise, "convexity" relates to the LP and not necessarily to individual factors.

From relationship (6.1), the non-linear model for Y (>0) is

$$Y = \begin{cases} [\lambda(\eta + \varepsilon) + 1]^{1/\lambda}, & \lambda \neq 0, \\ \exp(\eta + \varepsilon), & \lambda = 0, \end{cases} \tag{6.2}$$

and in terms of the standard normal variable, Z,

$$Y = \begin{cases} [\lambda(\eta + \sigma Z) + 1]^{1/\lambda}, & \lambda \neq 0, \\ \exp(\eta + \sigma Z), & \lambda = 0, \end{cases} \tag{6.3}$$

where σ is the error standard deviation. Equations (6.2) and (6.3) are power and exponential relationships (dependent on the value of λ). Both are monotone convex relationships.

Requirement 2: The effects to include in the LP and the structure of the model are part and parcel of the empirical modeling process.

Description. Empirical modeling that may result in a non-linear model should be data-dependent not only with respect to the effects, included in the LP, but also with respect to the final structure of the (possibly non-linear) model.

Reasoning. This property is desirable, but seems hard to accomplish. The ultimate goal of all empirical modeling approaches is that important features of the resulting model be determined solely by the data on hand (no a-priori assumptions specified prior to data analysis). In particular, the structure of the final model should be determined by the analyzed data. Achieving this would expectedly deliver better goodness-of-fit, provided that the price paid is not a non-parsimonious model (Requirement 8).

An Example. In GLM modeling, the composition of the LP is determined by data analysis. However, a link function needs to be specified in advance. The link function is the (possibly non-linear) transformation of the response mean required to obtain the LP, and it determines the structure of the final model. Since the link is not determined by data analysis, Requirement 2 is not satisfied by GLM. Conversely, for a BC transformation, the parameter λ, which determines the non-linear structure of the model, is decided solely by data analysis, in compliance with Requirement 2.

Requirement 3: Provide linear relationship as a special case.

Description. An empirical modeling methodology should accommodate linear relationship (the response is equal to the linear predictor plus an error term).

Reasoning. The common prevalence of linear models, as judged by the wide and effective use of multiple linear regression analysis in empirical modeling, renders this property essential to any general methodology for empirical modeling.

An Example. Consider the BC transformation (6.1). For $\lambda=1$, we obtain a linear relationship between η and the response. Requirement 3 is fulfilled by the BC transformation.

Requirement 4: A dual-error structure.

Description. This property implies that an empirical model should include two error terms, possibly correlated: An error term that is associated with the LP, and another error term directly associated with the response.

Reasoning. This property is a logical conclusion of the definition of a relational model. For any such model, one may plausibly conceive of a scenario where the systematic variation plus the associated random error are non-existent (no variation is *transmitted* to the response). The response may still be expected to display random variation. This scenario requires separation of error terms, and the assumption of a dual-error structure: One error is assumed to originate in the inherent instability of the response and in measurement imprecision; The other error is assumed to be transmitted via the effects included in the LP, most commonly in the form of an additive error term. Examples for this "Separation of the errors" are given in the survey of Chapter 2. The reader is encouraged to re-study these examples in order to get better insight as to the meaning of the "Dual-error structure".

If the LP includes non-linear effects, like in $\beta_0 + \beta_1 X + \beta_2 X^2$, then obviously the additive error term should be considered as an approximation to the true error. Thus, if we write $X = \mu + \varepsilon$ (the mean plus error), then the above equation becomes

$$\beta_0 + \beta_1(\mu+\varepsilon) + \beta_2(\mu+\varepsilon)^2 = \beta_0 + \beta_1\mu + \beta_2\mu^2 + (\beta_1 + 2\beta_2\mu)\varepsilon + \beta_2\varepsilon^2.$$

The last term is negligible as we may assume that $\varepsilon^2 << \varepsilon$. Therefore a single additive error term seems an adequate description even for cases where the linear predictor comprises non-linear terms (like interactions or polynomial terms).

An Example. Apart from RMM (refer to the basic model developed in Chapter 7), no current major methodology for relational modeling incorporates a dual-error structure in its models. For example, in linear regression analysis it is commonly assumed that the effects in the linear predictor (the regression equation) are all deterministic, and the only error term is additive (refer for a discussion of the implications of this assumption, when it is not valid, to Draper and Smith, 1998, Section 3.4).

Alternatively, we may interpret the (single) error term in linear regression as reflecting the sum of two independent errors, one associated with the linear regression equation (the linear predictor) and another (additive) associated with the response. Obviously, in linear regression such independent errors would be indistinguishable from one another.

If one wishes to interpret a generalized linear mixed model (GLMM), which incorporates random effects in the linear predictor, as having *an error term* added to the linear predictor, then this model comes closest to incorporating a dual-error structure.

Requirement 5: Modeling systematic variation that spans several orders of magnitude should allow the allied error-distribution to change in a major way. In particular, the modeling methodology should allow for asymptotic normality and the resulting decoupling of the variance from the mean.

Description. Systematic variation by definition *may* change the error distribution. An empirical model should reflect this reality. In particular, asymptotic normality should be possible, with the resulting decoupling of the variance from the mean.

Reasoning. It is well known from both theory and practice that distributions may change as a result of systematic changes in their parameters. For example, the binomial with parameters (n,p) goes to the Poisson when n is very large and p is very small. Similarly, the Poisson distribution tends to normality when the parameter of a Poisson variable increases. The gamma distribution also approaches normality under a certain asymptotic condition. If systematic variation of a relational model spans a number of orders of magnitude, the error distribution will probably change with changes in the LP, and the hypothesized constant relationship between the variance and the mean may cease to be valid. Good empirical modeling requires that systematic variation in the LP would afford gradually changing error distribution, with non-constant shape characteristics, and non-constant relationship between the mean and the variance. In particular, asymptotic normality, mandated by the Central Limit Theorem and otherwise, should be an optional consequence of the transmission of systematic variation to the response. A major consequence of normality, the decoupling of the variance from the mean, should be apparent from the general structure of the model offered by the modeling methodology.

An Example. Consider the RMM model (Chapter 7) for the log-transformed response

$$W = \log(Y) = (\alpha/\lambda)[(\eta+\varepsilon_1)^{\lambda}-1] + \mu_2 + \varepsilon_2, \qquad (6.4)$$

where ε_1 and ε_2 are independent normal errors with zero mean and standard deviations $\sigma_{\varepsilon 1}$ and $\sigma_{\varepsilon 2}$, respectively. The density function of W has been developed (Shore, 2002) and shown to be either non-normal or normal (dependent on its parameters). It may be easily realized from (6.4) that as η increases, ε_1 gradually becomes negligible ($\varepsilon_1 \ll \eta$). This implies that with appreciable systematic variation in the linear predictor (η), the error distribution of W will not be constant. As η becomes large, W will tend to normality. Correspondingly, the variance of W will tend to stabilize and become dissociated (decoupled) from the mean.

Requirement 6: The model's error distribution needs to maintain a degree of flexibility, which would allow it to preserve some of the actual moments (preferably the first three or four) of the modeled distribution. The allied estimation procedures should also ensure preservation of moments.

Description. The number of degrees of freedom in the estimated model needs to ensure that at least the first three or four moments of the true unknown distribution be preserved in the fitted model. Furthermore, with systematic variation non-existent, the estimation procedure should be shown to result in the model's first three or four moments being associated with small mean squared errors (MSEs).

Reasoning. The error distribution of the model is the distribution of the random fluctuations in the response, when no systematic variation is transmitted via the linear predictor. Both the statistical literature (for example, Pearson, Johnson and Burr, 1979), and pure theoretical arguments mandate that two random variables with the same first three or four moments display similar patterns of random variation (refer to a discussion of this issue in Chapter 5).

For empirical modeling to deliver faithful representation of a modeled distribution, two conditions need to be fulfilled. First, the empirical model needs to have at least three parameters, not associated with the LP, so that flexibility is preserved to represent differently-shaped error distributions (at least the first three moments of the modeled distribution are preserved in the empirical model). Secondly, the allied estimation procedure results in a fitted (estimated) error distribution having first three or four moments with acceptably small MSEs. In practice, this means that if we estimate a moment by numerically calculating it from the fitted distribution, this moment's estimate is associated with small MSE. Current practices often fail to satisfy this requirement. In particular, fitting a parameter-rich distribution by matching its first four moments with the sample moments (as done in Clements' procedure, 1989) almost surely guarantee that the fitted distribution has high-degree moments with large MSEs (refer to Shore, 2004b, and to Chapters 5, 20 and 21, where this issue is also addressed).

An Example. Consider the inverse BC transformation, given by (6.3). Assume that the LP is constant, so that (6.3) in fact defines the associated error distribution. Without loss of generality we may assume $\eta=1$, to obtain from (6.2)

$$Y = \begin{cases} [1 + \lambda + \lambda\sigma Z]^{1/\lambda}, \lambda \neq 0, \\ \exp(1 + \sigma Z), \lambda = 0. \end{cases} \tag{6.5}$$

This equation expresses the random variable, Y, in terms of a standard normal variable, Z. In fact, for any specified quantile of the normal variable, the corresponding quantile of Y may be calculated from (6.5). For instance, setting $Z=0$ the response median, $(1+\lambda)^{1/\lambda}$ $(\lambda\neq0)$, is obtained. The reader may easily deduce that the error distribution associated with the BC transformation cannot accommodate differently-shaped error distributions since a single parameter, λ, is not sufficient to accommodate different sets of values for the first three or four moments. In conclusion, the inverse BC transformation, as given by (6.5), is inadequate for modeling random variation.

Requirement 7: Provide good coverage of the ($\sqrt{\beta_1}$, β_2) plane.

Description. $\sqrt{\beta_1}$ and β_2-3 are the traditional measures of skewness and kurtosis (corresponding, respectively, to the third and fourth cumulants of the standardized variable; For the normal distribution

$\sqrt{\beta_1} = \beta_2 - 3 = 0$; For a discussion of cumulants refer to Stuart and Ord, 1987, Section 3.12). The $(\sqrt{\beta_1}, \beta_2)$ plane was introduced in Chapter 5. Requirement 7 states that a methodology for empirical modeling should be shown to have an error distribution that occupies areas of the $(\sqrt{\beta_1}, \beta_2)$ plane similar to those occupied by distributions widely encountered in practice.

Reasoning. In empirical modeling of random variation, it is standard practice to demonstrate that when modeling is implemented via some parameter-rich family of distributions (like Pearson or the Generalized Lambda), the family used for fitting has members with $(\sqrt{\beta_1}, \beta_2)$ values that span large portions of the $(\sqrt{\beta_1}, \beta_2)$ plane. This property reflects the capability of the family of distributions to provide good representation to a large spectrum of shape characteristics, associated with distributions that a practitioner may encounter.

Likewise, the error distribution embedded in a general empirical model should be able to accommodate large portions of the $(\sqrt{\beta_1}, \beta_2)$ plane.

In practical terms, realization of this property is conditional on achieving Requirement 6. One cannot expect an error distribution to provide good coverage of the $(\sqrt{\beta_1}, \beta_2)$ plane unless it has a sufficiently large number of parameters. The "representation" feature, which Requirement 7 demands, is conditional on having enough parameters (Requirement 6). The statistical literature shows that the latter is not a sufficient condition for the fulfilling of Requirement 7. The following example demonstrates this.

An Example. Relate to the three-parameter gamma distribution, with density function (Johnson, Kotz and Balakrishnan, 1994, Chapter 17)

$$f(y) = (y-\gamma)^{(\alpha-1)} \exp[-(y-\gamma)/\beta] / [\beta^\alpha \Gamma(\alpha)], \ \alpha > 0, \ \beta > y > \gamma. \quad (6.6)$$

For this distribution, the skewness measure $(\sqrt{\beta_1})$ is equal $2\alpha^{-1/2}$ and the kurtosis measure $(\beta_2 - 3)$ is equal $6\alpha^{-1}$. Using this distribution as a platform for empirical modeling of *random* variation, Requirement 7 is not fulfilled. First, since both skewness and kurtosis depend on a single parameter, no flexibility is preserved to model distributions with varying values of skewness and kurtosis. Secondly, since $\alpha > 0$, negative skewness (long left-tailed distribution) cannot be achieved. However Requirement 6 is fulfilled [since (6.6) has three parameters].

Requirement 8: Parsimony in modeling.

Description. Let the model be as simple as possible. In practical terms, this implies simplest structure and a small number of parameters that needs to be estimated.

Reasoning. This property is the ultimate goal of any modeling effort. (Box, 1988, defines parsimony as "the provision of the simplest additive models", and he considers this one of the two most important criteria in data analysis from quality improvement experiments.)

Requirement 9: Compatibility with current proven-effective methodologies for empirical modeling.

Description. This property requires a demonstration that a new methodology is either similar to current proven-effective methodologies, compatible with them in some defined sense, or that a certain existing methodology may be derived from the new one as a special case.

Reasoning. If cumulative experience has shown that a certain current empirical-modeling methodology is commonly effective (namely, effective over a wide spectrum of possible scenarios), then either this methodology has some theoretical basis for its wide applicability or it is based on an empirical observation of a universal nature. An example for the former is GLM, where the exponential family of distributions serves as the basic model. An example for the latter is the normalizing BC transformation. To justify its introduction, a new methodology for empirical modeling should be shown to be either a generalization of at least one current methodology, or compatible with current methodologies in some sense.

An Example. The new RMM is based on *data* transformation to achieve additivity of effects (unlike GLM which requires transformation of the *mean*). It can be easily shown that RMM contains the inverse BC transformation as a special case (refer to Chapter 12). Thus, RMM is compatible with a certain data transformation (the BC transformation) that has been shown to be widely effective in relational modeling.

The Johnson family of distributions is commonly used to model random variation. The RMM error distribution has been shown to include the Johnson normalizing transformations as special cases (refer to Shore, 2004c, and to Chapter 12).

GLM includes linear regression as a special case. It also fulfills Requirement 9, but only in the limited sense expressed by Requirement 3.

Requirement 10: Ease of application.

Description and Reasoning. Self evident.

6.3. An Evaluation of Compliance of Current Methodologies

6.3.1. Modeling systematic variation

The set of desirable requirements just delineated may serve to assess to what degree current major methodologies for empirical modeling are satisfactory. In other words, compliance with the set of requirements may assist in judging the quality of current methodologies.

In this sub-section, we evaluate the three major general methodologies for empirical modeling of systematic variation: Linear regression, data transformations (intended to revoke the normal scenario in a non-normal environment) and GLM. (By "normal scenario" we imply the three conditions that are conducive to the application of linear regression, namely, additivity of effects, constancy of variance and normal errors).

Non-linear regression analysis is excluded from this comparison. The reason is that the latter assumes a non-linear model, with an additive normal error, but with no particular theory regarding the nature of the suggested model. Therefore it cannot be considered to be a general methodology in the same sense that the other methodologies compared are.

The new RMM will be assessed for compliance in Chapter 13.

A summary of a comparative evaluation of all three major approaches addressed in this sub-section is given in Table 6.1. Non-compliance is marked by (-) (minus), *partial* and *full* compliance are marked by a single plus (+) and a double plus (++), respectively. For the sake of brevity, a detailed evaluation is not elaborated here, and the following remarks are intended merely to assist in understanding the reasons for the various assertions in the table. Some of the points in the table are obviously open to debate. This is to be expected given the largely qualitative nature of the evaluation displayed in Table 6.1.

Linear Regression Analysis

Although not commonly presented in such a fashion, linear regression is indeed the outcome of a transition from a relatively general model of random variation to a general platform for modeling systematic variation. The transition has been accomplished by assuming that one of the parameters of the normal distribution, namely, the location parameter, is not constant but may be expressed as a linear combination of systematic effects. Thus, by allowing one of its parameters to vary systematically, a model of random variation (the normal distribution) becomes a model for systematic variation. Since only the location parameter varies, the various error distributions resulting from the transmission of systematic variation to the response differ in their location but not in their scale or shape characteristics.

A summary of the assessment of linear regression for compliance with the requirements is given in Table 6.1. We add here a few remarks.

With regard to Requirement 1, linearity is included in the possible set of monotone convex functions. However, this is a poor representation of the set of conceivable convex functions, included in the "Ladder" (Chapter 3).

With regard to Requirement 4, a dual-error structure may indeed be part of a linear regression model, but only if it is assumed that the model's two error terms are not correlated. In that case, due to linearity, the two error terms combine to form one normal error term, as is the case

Table 6.1. Comparison of Compliance with Requirements of Major Approaches for
Modeling Systematic Variation

Requirement	Methodology			
	Linear regression	BC Transformation	GLM	RMM
1 (Convexity)	-	+	+	++
2 (All Data Driven)	-	++	-	++
3 (Incl. Linear)	++	++	++	++
4 (Dual-error)	+	-	-	++
5 (Changing error dist.)	-	-	-	++
6 (Moments preserved)	-	-	-	++
7 (Representability)	-	-	+	++
8 (Parsimony)	++	++	++	++
9 (Compat. with current method.)	++	++	++	++
10 (Easy to apply)	++	+	+	+

for linear regression. Stated differently, the single error term in linear regression is compatible with a dual-error structure only if the two error terms are assumed independent (or uncorrelated).

Linear regression does not accommodate skewness or kurtosis different from those of the normal distribution (Sk=Ku=0). Compliance (or rather lack thereof) is consequently affected with regard to Requirements 5, 6 and 7.

Data Transformation

The second major approach to empirical modeling is data transformation, which aims to revoke the normal scenario. In particular, where the original data suggest that the normal scenario does not exist, the BC transformation, applied to the response, has been shown to be effective in revoking the normal scenario, and thus allow application of linear regression. A few comments again complement the assertions displayed in Table 6.1, with better clarity in mind.

First, note from (6.2) that the BC transformation intends to normalize a random variable (a response), which has a monotone convex relationship with the linear predictor. However, the assumed relationship is confined to linear, power or exponential relationships only. Cases like exponential-power are not included. Requirement 1 is better pursued by the BC transformation than by linear regression, but in a limited sense only.

Requirement 2 is fulfilled since both the LP and the structure of the model are determined by data analysis.

Having a single error term, Requirement 4 is not fulfilled by the BC transformation.

Having a single parameter to determine the structure of the error distribution (namely, λ), Requirement 5 is also not satisfied since once λ is determined, changes in the order of magnitude of the linear predictor do not modify the shape characteristics of the associated error distribution. This affects also compliance with Requirements 6 and 7 because a single shape parameter (λ) does not allow flexible representation for diversely-shaped distributions.

Generalized Linear Models (GLM)

GLM has recently become the favorable choice for empirical modeling of systematic variation. Replacing the normal distribution of linear regression by the more inclusive exponential family of distributions, GLM may be presented, analogously with linear regression, as a transition from a relatively general model of random variation (the exponential family of distributions) to a general methodology for modeling systematic variation. This transition is accomplished in GLM by assuming that one of the parameters of the exponential family, namely, the location parameter, is not constant but is expressible as a linear combination of systematic effects (the linear predictor). Later developments (like "quasi-likelihood") have removed some of the restrictive features of the original formulation of GLM, and the response is no longer required to have an error distribution from the exponential family.

Assessment of GLM's compliance with the requirements of Section 6.2 is displayed in Table 6.1.

With regard to Requirement 1, there certainly is an improvement over the data-transformation approach in that specifying a link function is part of modeling, and it is allowed to take any shape. However, unlike the data-transformation approach, where the non-linear structure of the model is determined by the data and by data analysis (via estimating of λ), in GLM the link function is determined *in advance*, prior to data analysis. This is a setback relative to the data-transformation approach (Requirement 2 is not pursued).

Requirement 4 (a dual-error structure) is also ignored. In fact, no errors are defined in GLM in explicit form, and only the underlying response distribution is specified. Since the latter needs to be determined prior to data analysis, so is the relationship between the mean and the variance, and Requirement 5 is also not fulfilled. It is our contention that lack of compliance with Requirement 5 is a major setback to the general applicability of GLM.

The need to specify the error distribution (or, equivalently, the mean-variance relationship) in advance obviously also affects compliance with Requirements 6 and 7, at least with respect to the GLM original formulation where it is confined to the exponential family of distributions. Indeed, most current reported applications of GLM in engineering and the sciences are restricted to this family of distributions (Refer for example to Myers, Montgomery and Vining, 2002, and a survey of reported applications therein).

6.3.2. Modeling random variation

General methodologies to model random variation have been available for some time (see a review in Chapter 5). However, the need to empirically model random variation for quality-related scenarios in non-normal environments provided the needed impetus to develop new, and at times innovative, empirical methodologies. In particular, we refer to the need to develop control charts in real time for such scenarios where observations are uncorrelated however the process distribution deviates appreciably from normality. Similar scenarios exist when process capability needs to be assessed in a non-normal environment.

To appreciate the scale of the need for empirical modeling of random variation in quality engineering, consider a recent literature review of process capability analysis (PCA, Kotz and Johnson, 2002). In a separate section that deals with non-normality in PCA, the authors refer to over twenty (!) references which attempted to provide solutions to the non-normal nature of process distributions in PCA.

In this subsection, we evaluate compliance of current approaches to empirical modeling of random variation in view of Requirements 6 and 7, the only requirements of Section 6.2 that are relevant to such modeling.

The most commonly applied approach is moment matching. With this approach, the first few moments of the fitted parameter-rich distribution are made to coincide with the true moments, possibly estimated from sample data.

From a theoretical point of view, moment matching seems to be the most desirable. This has led to the use of different families of distributions in modeling random variation, as we have detailed in Chapter 5. Furthermore, in view of Requirements 6 and 7, moment-matching seems to be the appropriate approach to empirical modeling of random variation. As explained in Section 5.3, this is rarely the case. The most common approach to moment matching requires matching of the first four moments. However sample estimates of third- and fourth-degree moments are notorious for their large sampling errors. This can have serious adverse consequences, as related in Chapter 5.

In conclusion, the widely applied four-moment matching tends to deliver bad representation of the true moments, in violation of Requirement 6.

Alternative approaches to empirical modeling of random variation have relied on matching of percentile points (refer, for example, to Karian and Dudewicz, 2002, Ch. 4), a combination of moment and percentile matching (refer, for example, to the fitting of the four-parameter gamma distribution based on matching of the first three moments plus the first order-statistic, Cohen and Whitten, 1988), or various heuristic procedures, as reviewed in Chapter 5.

Regrettably, nearly all these approaches fail to demonstrate that the MSEs associated with the first few moments of the fitted distribution (or of the fitted quantile function) are small enough to guarantee that Requirement 6 is fulfilled (Shore, 2004b). Indeed, most of these approaches even fail to relate in any sense to the desirability of Properties 6 and 7.

In particular, for methods developed particularly for the quality-engineering discipline, there seems to be general ignoring of the need to show both the universality of the suggested methodology (in terms of its ability to accommodate a vast majority of distributional shapes that a practitioner may encounter in reality, Requirement 7) and its good "representability" *vis-à-vis* the actual error distribution (implying that the fitted distribution preserves well the actual moments of the modeled distribution, Requirement 6).

Finally, note that while current major methodologies for empirical modeling of *systematic* variation originated in relatively general models of *random* variation (like the exponential family), the same can not be extended in the opposite direction. In fact, as the brief survey above has shown, all current general approaches for empirical modeling of random variation have their roots in existing models of *random* variation. No general approach to model *systematic* variation has served as the platform to develop these methodologies. Consequently, no transition has been reported in the literature from empirical modeling of systematic variation to empirical modeling of random variation.

The unique advantages associated with such a transition, and why the new RMM embodies such a transition, will be discussed in Part II of the book, where the RMM approach is developed.

References

[1] Box, G. E. P. (1988). Signal-to-noise ratios, performance criteria and transformations (with discussion). *Technometrics*, 30, 1-40.
[2] Box, G. E. P., Cox, D. R. (1964). An analysis of transformations. *Journal of the Royal Statistical Society, Series B*, 26, 211-252.

[3] Cohen, A. C., Whitten, B. J. (1988). *Parameter Estimation in Reliability and Life Span Models*. Marcel Dekker, New-York.

[4] Clements, J. A. (1989). Process capability calculations for non-normal distributions. *Quality Progress*, 22(2), 49-55.

[5] Draper, N. R., Smith, H. (1998). *Applied Regression Analysis*. 3rd Ed., John Wiley & Sons.

[6] Johnson, N. L., Kotz, S., Balakrishnan, N. (1994). *Continuous Univariate Distributions. V. 1*. 2nd Edition. John Wiley & Sons.

[7] Karian, Z. A., Dudewicz, E. J. (2002). *Fitting Statistical Distributions- The Generalized Lambda Distribution and Generalized Bootstrap Methods*. Chapman & Hall/CRC Press. London.

[8] Kotz, S., Johnson, N. L. (2002). Process capability indices- A review, 1992-2000, *Journal of Quality Technology*, 34, 2-19.

[9] Myers, R. H. , Montgomery, D. C., Vining, G. G. (2002). *Generalized Linear Models, with Applications in Engineering and the Sciences*. John Wiley & Sons.

[10] Pearson, E. S., Johnson, N. L., Burr, I. W. (1979) Comparisons of the percentage points of distributions with the same first four moments, chosen from eight different systems of frequency curves. *Communications in Statistics (Simulation and Computation)*, B8(3), 191-229.

[11] Shore, H. (2002). Response Modeling Methodology (RMM)- Exploring the implied error distribution. *Communications in Statistics (Theory and Methods)*, 31(12), 2225-2249.

[12] Shore, H. (2004a). Response Modeling Methodology (RMM)- Validating evidence from engineering and the sciences. *Quality and Reliability Engineering International*, 20, 61-79.

[13] Shore, H. (2004b). Non-normal populations in quality applications- A revisited perspective. *Quality and Reliability Engineering International*, 20, 375-382.

[14] Shore, H. (2004c). Response Modeling Methodology (RMM)- Current distributions, transformations and approximations as special cases of RMM error distribution. *Communications in Statistics (Theory and Methods)*, 33(7), 1491-1510.

[15] Stuart, A., Ord, J. K. (1987). *Kendall's Advanced Theory of Statistics. V. 1: Distribution Theory*. Charles Griffin & Company, Ltd., London.

Part II

RMM - DEVELOPING AND EVALUATING THE GENERAL APPROACH

Chapter 7

The RMM Model

7.1. Introduction

Mainstream models developed over the years in various scientific and engineering disciplines were introduced in Chapter 2. All models had monotone convex/concave relationship with the response. In Chapter 3 we elaborated in further detail on three of the models, and derived important observations that are valid for all.

First, a particular classification has emerged of the components of variation shared by the models. These include two components of random variation, represented by two error terms, and a component of systematic variation, transmitted to the response via the linear predictor. With regard to the former components, earlier related to as the dual-error structure, it has become clear that one error term is associated with the linear predictor (and characterized as an additive error term), and another is directly associated with the response. The latter error can be assumed to exist irrespective of whether the linear predictor (with its associated additive random error) varies or is constant. Furthermore, current engineering and scientific models, reviewed in Chapter 2, seem to regard this error (either explicitly or implicitly) as additive in the original scale of the response or as additive in the log-transformed scale. As will be shown later in this chapter, most often these two seemingly different definitions of the error are approximately equivalent.

Secondly, a particular hierarchical classification has emerged of the functions that relate the linear predictor (with the associated additive random error) to the response. This classification has been captured by the "Ladder of fundamental uniformly convex/concave functions". Although monotone convexity had admittedly guided us in selecting the models included in the survey of Chapter 2, it was evident from the wide spectrum of models reviewed that uniformly convex/concave relationships permeate scientific and engineering modeling.

In Chapters 4 and 5, current general approaches to empirical modeling of a response variation were briefly outlined. These related both to systematic variation (Chapter 4) and to random variation (Chapter 5). Chapter 6 expounded desirable requirements that any general methodology for "empirical modeling" should satisfy. Assessing the degree of compliance of current methodologies with these requirements allowed an evaluation of their quality. Inherent drawbacks of existing modeling approaches were indicated. In particular, it was emphasized that the models derived in the framework of current approaches do not really capture some of the fundamental characteristics that are commonly shared by mainstream models developed independently in scientific and engineering disciplines.

RMM is an attempt to fill in the void created by the shortcomings of current methodologies for empirical modeling.

In Section 7.2, five assumptions are made which allow derivation of the basic RMM model in an axiomatic fashion. This model will later serve as the core platform for the development of the new Response Modeling Methodology (RMM). Note, that the axiomatic derivation detailed herewith is somewhat different from the original derivation (Shore, 2002) in that a more general model is assumed that allows the errors to be correlated.

Expressions for the response moments are derived in Section 7.3. The last Section 7.4 explores the relationship between the coefficient of variation of the RMM model and the linear predictor (LP). It is shown that as LP increases the variance tends to stabilize (be independent of the mean).

In Chapter 8 we will develop estimation procedures, and in Chapter 9 the RMM error distribution will be derived and its properties explored.

7.2. An Axiomatic Derivation of the RMM Model

Developing the general model, we attempt to capture the behavior of a random response, which experiences two independent sources of variation: Externally-generated variation and Self-generated variation. The former variation is assumed to comprise both systematic variation, transmitted to the response via the LP (denoted η), and random variation, transmitted to the response by an additive error, ε_1. Self-generated variation reflects the inner natural instability of the modeled response, as well as measurement imprecision. It is represented by a random error, ε_2. The two sources of variation are assumed to interact with one another.

All these features are summarized in a set of five assumptions, which form the axiomatic basis for the development of the general model.

In the following sub-sections, the RMM model is developed, starting with the assumptions (Sub-section 7.2.1) and concluding with the mathematical expression that expresses the response in terms of the LP, the random errors, and the model's parameters (Sub-section 7.2.5).

7.2.1. The model assumptions

Five assumptions produce the basis for the new RMM model.

Assumption A. The modeled response is a non-negative r.v., Y (≥ 0), which experiences self-generated random variation. This variation is an increasing function of the size of the response, where the latter is represented by the response median, M. As a first approximation, it is assumed that the dispersion of self-generated random deviations is *proportional* to M. This assumption is a natural outcome of our assertion that self-generated variation reflects the inner instability of the response, as well as response-measurement errors. Thus, a linear (proportional) relationship seems a plausible choice. (Note that M can be considered the response median only with regard to self-generated variation; If also systematic variation is transmitted to the response, M no longer represents the response median. This will be made clearer when the final RMM model is derived in Subsection 7.2.5.)

Self-generated deviations are produced by a random mechanism driven by a normal random variable.

Assumption B. The modeled response, Y (≥ 0), experiences externally-generated variation that includes transmitted systematic variation and transmitted random variation.

Systematic variation is transmitted to the response by the linear predictor (LP). The LP may represent a single affecting factor ("Main effect"), a weighted sum of factors or, more generally, a linear combination of predictors. The latter include factors ("Main effects") or functions of factors that do not contain unknown parameters.

Random variation is transmitted to the response by an additive normal error (additive to the LP).

Observed deviations in the response that may be attributed to externally-generated sources of variation thus result from both changes in the LP and from random fluctuations associated with the LP.

Assumption C. The two error terms (specified in Assumptions A and B) are assumed to derive from a bi-variate normal distribution with correlation ρ ($-1 \leq \rho \leq 1$).

Assumption D. The quantile-relationship between the LP, with the associated random error, and the response is monotone convex (second derivative is everywhere non-negative) or monotone concave.

As noted earlier, for the sake of brevity we refer throughout the book only to a convex relationship. However, any RMM-related claim with regard to a convex function equally applies to a concave function.

Assumption E. Self-generated deviations and externally generated deviations mutually interact in determining the value of the response. In other words, a multiplicative model is presumed.

7.2.2. The general model

The RMM model expresses the response, Y, in terms of a product of two separate functions, each modeling a different source of variation.

Externally-generated variation is modeled by $f_1(\eta, \varepsilon_1; \boldsymbol{\theta}_1)$. In this expression, η is the LP, and $\varepsilon_1 = (\sigma_{\varepsilon 1})Z_1$ is an additive error term, where Z_1 is a standard normal variable (namely, the error is normally distributed with zero mean and standard deviation $\sigma_{\varepsilon 1}$). $\boldsymbol{\theta}_1$ is a vector of parameters that need to be determined (estimated). It is assumed that $f_1(\eta, \varepsilon_1; \boldsymbol{\theta}_1)$ is monotone convex or concave.

Self-generated variation is modeled by $f_2(\varepsilon_2; \boldsymbol{\theta}_2)$. In this expression, $\varepsilon_2 = (\sigma_{\varepsilon 2})Z_2$ is a random error, Z_2 is a standard normal variable (namely, the error is normally distributed with zero mean and standard deviation $\sigma_{\varepsilon 2}$), and $\boldsymbol{\theta}_2$ is a vector of parameters that need to be determined.

As implied by assumption E, the response is modeled as

$$Y = f_1(\eta, \varepsilon_1; \boldsymbol{\theta}_1)\, f_2(\varepsilon_2; \boldsymbol{\theta}_2), \tag{7.1}$$

where f_1 and f_2 are as defined above, and the random variables $\{Z_1, Z_2\}$ are assumed to derive from a bi-variate standard normal distribution with correlation ρ. The functions f_2 and f_1 will be developed in the next sub-sections 7.2.3 and 7.2.4, respectively.

7.2.3. Deriving f_2

By Assumption A, the simplest model for self-generated random variation, which increases with M, is a proportional one:

$$f_2(\varepsilon_2; \boldsymbol{\theta}_2) = M(1 + \varepsilon_2) = M(1 + \sigma_{\varepsilon 2}Z_2) = M + (M\sigma_{\varepsilon 2})Z_2. \tag{7.2}$$

For the absolute value of the relative random error we may assume that

$$\left| f_2 - M \right| / M = |\varepsilon_2| \ll 1. \tag{7.3}$$

Therefore (7.2) may be approximately represented by

$$f_2(\varepsilon_2; \boldsymbol{\theta}_2) = M(1 + \varepsilon_2) = M(1+\varepsilon_2)^{\,(1/\varepsilon 2)\,\varepsilon 2}$$

$$\approx \exp[\log(M)+\varepsilon_2] = \exp[\mu_2 + \sigma_{\varepsilon 2}Z_2],. \tag{7.4}$$

where $\boldsymbol{\theta}_2 = \{\mu_2, \sigma_{\varepsilon 2}\}$, and μ_2 and $\sigma_{\varepsilon 2}$ are the mean and the standard deviation of the *normal* variable that generates self-variation. In other words, self-generated variation is modeled by a *log-normal* variable (a log-normal variable is any response which is normally distributed on a log-transformed scale).

As is evident by the survey of current mainstream models (refer to Chapters 2 and 3), both versions of the error, (7.2) and (7.4), are quite common. Thus, (7.4) is typically assumed when the log of the response is modeled, while (7.2) is adopted when the response is modeled on the original scale. If ε_2 fulfills (7.3) the two formulations are about equivalent.

7.2.4. Deriving f_1

By assumptions B and D, transmitted variation is modeled by the relationship between the response and the LP (with the associated random error), and this relationship is monotone convex. One can conceive of a myriad of functions with this characterization. Therefore we should ask ourselves whether the universe of all monotone-increasing convex functions that may potentially serve to model $f_1(\eta, \varepsilon_1; \theta_1)$ can be captured by a limited set of some fundamental convex functions, which may deliver the required representation for varying degrees of convexity?

In fact, we have already tackled this problem in formulating "The ladder of fundamental uniformly convex functions" (Chapter 3). Therein it was emphasized that given the number of engineering and scientific models that are compatible with "the Ladder", a general methodology for empirical modeling should include all the different "steps" of "the Ladder" as special cases.

A natural way to form a unifying relationship that includes all the "special cases" in the ladder is by multiplying or adding the fundamental functions specified in these cases. This would require introduction of enough parameters that may allow extraction of each of these "special cases" individually. Obviously, an approach like that violates the important principle of parsimony, which requires that the underlying model be as simple as possible. An alternative approach needs to be pursued.

Consider expressing f_1 by

$$f_1(\eta, \varepsilon_1; \theta_1) = \exp\{(\alpha/\lambda)[(\eta+\varepsilon_1)^\lambda - 1]\}$$

$$= \exp\{(\alpha/\lambda)[(\eta+\sigma_{\varepsilon 1}Z_1)^\lambda - 1]\}, \qquad (7.5)$$

with ε_1 as defined earlier, and $\theta_1 = \{\alpha, \lambda, \sigma_{\varepsilon 1}\}$ is a vector of real-valued parameters that need to be determined. We assume that the linear predictor (η) with its additive error is strictly positive so that λ can take both negative and positive values, without changing the monotone increasing character of the relationship. For any non-positive LP, a change of location can achieve the required property.

It is easy to realize that by a proper selection of parameters' values, all of the six cases included in the ladder may be derived from (7.5). This will be demonstrated shortly.

7.2.5. The RMM Model

Combining (7.1), (7.4) and (7.5), the general model for the relationship between the response and the various variation-generating agents (denote it by f) is

$$Y = f(\eta, \varepsilon_1, \varepsilon_2; \boldsymbol{\theta}) = f_1(\eta, \varepsilon_1; \boldsymbol{\theta_1}) \, f_2(\varepsilon_2; \boldsymbol{\theta_2})$$

$$= \exp\{(\alpha/\lambda)[(\eta+\varepsilon_1)^\lambda - 1] + \mu_2 + \varepsilon_2\}$$

$$= \exp\{(\alpha/\lambda)[(\eta+\sigma_{\varepsilon 1}Z_1)^\lambda - 1] + \mu_2 + \sigma_{\varepsilon 2}Z_2\}, \qquad (7.6)$$

or

$$W = \log(Y) = (\alpha/\lambda)[(\eta+\varepsilon_1)^\lambda - 1] + \mu_2 + \varepsilon_2$$

$$= (\alpha/\lambda)[(\eta+\sigma_{\varepsilon 1}Z_1)^\lambda - 1] + \mu_2 + \sigma_{\varepsilon 2}Z_2, \qquad (7.7)$$

where Z_1 and Z_2 derive from a bi-variate standard normal distribution with correlation ρ, and $\boldsymbol{\theta} = \{\alpha, \lambda, \mu_2, \rho, \sigma_{\varepsilon 1}, \sigma_{\varepsilon 2}\}$ is a vector of parameters that need to be determined.

To motivate the expression in (7.6), three examples of commonly encountered convex functions will be addressed. For simplicity, assume that there is no random variation ($\varepsilon_1=\varepsilon_2=0$). Then, as $\lambda \to 0$, the expression $(1/\lambda)[(\eta)^\lambda - 1] \to \log(\eta)$. This implies that for $\alpha=1$, $\lambda=0$, we obtain a linear relationship $Y= \eta$. For $\alpha \neq 1$ and $\lambda=0$, we obtain a power relationship $Y= \eta^\alpha$. For $\lambda \neq 0$, an exponential relationship is obtained.

The reader may likewise easily show that Antoine equation (an exponential-power relationship, refer to Chapter 3) is also a special case of (7.6). In fact, numerous scientific and engineering models, with monotone convex relationships, can be shown to be special cases of

(7.6). This is the subject of Chapter 12, where we demonstrate for some of the models in Chapter 2 that they are indeed special cases of the RMM model.

All the "cases", included in the "Ladder", are now shown to be special cases of (7.6):

- To obtain **Case 1 (linear relationship)**: Set $\lambda = 0$, $\alpha = 1$;
- To obtain **Case 2 (power relationship)**: Set: $\lambda = 0$, $\alpha \neq 1$;
- To obtain **Case 3 (exponential)**: Set: $\lambda = 1$;
- To obtain **Case 4 (exponential power)**: Let $\lambda \neq 1$, $\lambda \neq 0$.
- To obtain **Cases 5 (exponential-exponential)** and **6 (exponential-exponential-power)**:
    Re-introduce the linear predictor with the error in the form of f_1, namely, introduce for $\eta + \sigma_{\epsilon 1} Z_1$ in (7.5): $\exp\{(\beta/k)[(\eta + \sigma_{\epsilon 1} Z_1)^k - 1]\}$. Apparently, the values $\beta=1$, $k=0$, re-introduce the linear predictor as in (7.5). However, $k \neq 0$ introduces an exponential-exponential-power relationship.

Cases with convexity intensity like that of Cases 5 and 6 are rare in practice. We will therefore assume henceforth that the above values of $\beta=1$ and $k=0$ are the actual values in the general RMM model.

Note, however, that the introduction of two additional parameters, if re-iterated, may in fact allow us to achieve increasing degrees of convexity, to a desirable degree. Thus, the basic pattern of "linear, power, exponential" can be repeated at will with the introduction of additional parameters (as we have just demonstrated). With f_1 modeled by (7.5), the "Ladder" is rendered from a discrete hierarchy of fundamental convex functions into a "continuous conveyor" ("continuous convexity", if you will), that may carry the modeler in a continuous fashion from one step of the ladder to the next, while the ladder itself is unbounded from above. Jacob's ladder comes to mind.

Consider next the dual-error structure of the model. As indicated earlier, it is assumed that Z_1 and Z_2 originate in a bi-variate standard normal distribution with correlation ρ. The conditional distribution of Z_2, given $Z_1 = z_1$, is normal with mean ρz_1 and variance $1-\rho^2$. We may write for Z_2 in (7.6)

$$Z_2 \mid z_1 = \rho z_1 + (1-\rho^2)^{(1/2)} Z, \qquad (7.8)$$

where Z is a standard normal variable, independent of Z_1. Introducing for Z_2 from (7.8) into (7.6) and (7.7), we obtain, respectively,

$$Y = \exp\{(\alpha/\lambda)[(\eta+\sigma_{\varepsilon 1}Z_1)^\lambda - 1] + \mu_2 + \sigma_{\varepsilon 2}[\rho Z_1 + (1-\rho^2)^{(1/2)} Z_2]\}, \quad (7.9)$$

$$W = \log(Y) = (\alpha/\lambda)[(\eta+\sigma_{\varepsilon 1}Z_1)^\lambda - 1] + \mu_2 + \sigma_{\varepsilon 2}[\rho Z_1 + (1-\rho^2)^{(1/2)} Z_2], \quad (7.10)$$

where for uniformity in notation we replaced Z with Z_2, namely, Z_2 is now assumed to be a standard normal variable which is uncorrelated (independent) of Z_1.

Equivalently, by conditioning Z_1 on $Z_2 = z_2$, we obtain

$$W = \log (Y) = (\alpha/\lambda)\{[\eta + \sigma_{\varepsilon 1}[\rho Z_2 + (1-\rho^2)^{(1/2)} Z_1]]^\lambda - 1\} + \mu_2 + \sigma_{\varepsilon 2}Z_2. \quad (7.11)$$

Equations (7.10) and (7.11) display the RMM model in its most general form. Both models express the response in terms of two uncorrelated standard normal variables, and the two models are equivalent in the sense that one model can be derived from the other.

Some variations of the basic model, as represented by (7.10) or (7.11), may be developed by pursuing different definitions for the errors (this was briefly alluded to earlier and re-introduced here in a more consistent fashion). Consider the error-structure in (7.6) and (7.7):

For small error, $\varepsilon_2 \ll 1$, we may write

$$1 + \varepsilon_2 \approx \exp(\varepsilon_2).$$

For small error, $\varepsilon_1 \ll \eta$, we may write

$$(\eta+\varepsilon_1)^\lambda = \eta^\lambda(1+\varepsilon_1/\eta)^\lambda \approx \eta^\lambda \exp(\lambda\varepsilon_1/\eta).$$

These two possible versions for each error term give rise to four different error-structures for the basic model. Coupled with the two equivalent variations, where the correlated error-terms have been re-expressed in terms of two independent standard normal variables [(7.10) and (7.11)], the RMM model may be displayed in 8 different forms. These are introduced below. Only variations of the error-structures are shown. The reader may introduce for each variation the errors in terms of the non-correlated standard normal variates, similarly to the derivation of (7.10) and (7.11).

Model 1. The basic model (7.6):
$$Y = \exp\{(\alpha/\lambda)[(\eta+\varepsilon_1)^\lambda - 1] + \mu_2 + \varepsilon_2\}.$$

Model 2. Different presentation of ε_2:
$$Y = \exp\{(\alpha/\lambda)[(\eta+\varepsilon_1)^\lambda - 1] + \mu_2 \}(1+\varepsilon_2) . \tag{7.6a}$$

Model 3. Different presentation of ε_1:
$$Y = \exp\{(\alpha/\lambda)[\eta^\lambda \exp(\lambda\varepsilon_1 / \eta) - 1] + \mu_2 + \varepsilon_2\}. \tag{7.6b}$$

Model 4. Different presentation of both ε_1 and ε_2:
$$Y = \exp\{(\alpha/\lambda)[\eta^\lambda \exp(\lambda\varepsilon_1 / \eta) - 1] + \mu_2 \}(1+\varepsilon_2). \tag{7.6c}$$

For each model, we can introduce

Either for ε_2:
$$\varepsilon_2 = \sigma_{\varepsilon 2}[\rho Z_1 + (1-\rho^2)^{(1/2)} Z_2], \tag{7.12}$$

where $\varepsilon_1 = \sigma_{\varepsilon 1}Z_1$, and Z_1 and Z_2 are uncorrelated.

or for ε_1:
$$\varepsilon_1 = \sigma_{\varepsilon 1}[\rho Z_2 + (1-\rho^2)^{(1/2)} Z_1], \tag{7.13}$$

where $\varepsilon_2 = \sigma_{\varepsilon 2}Z_2$, and Z_1 and Z_2 are uncorrelated.

These different formulations of the errors introduce desirable flexibility into the structure of the quantile-function of the RMM error distribution, as elaborated on in Chapter 9.

7.3. The Response Moments

Exact moments of the response, as given by (7.9) or (7.10), may be obtained by numerical integration with respect to Z_1 and Z_2. Since the latter are independent, we may separate integration with respect to Z_1 from that with respect to Z_2. For the r-th non-central moment of Y (moment about zero) we obtain

$$\mu_r' = E(Y^r)$$

$$= \left\{ \int_{-\infty}^{\infty} \exp\{(r\alpha/\lambda)[(\eta+\sigma_{\varepsilon 1}z_1)^\lambda - 1] + r\sigma_{\varepsilon 2}\rho z_1 \} \; \phi(z_1)dz_1 \right\}$$

$$\times \left\{ \int_{-\infty}^{\infty} \exp[r\mu_2 + r\sigma_{\varepsilon 2}(1-\rho^2)^{(1/2)} z_2] \; \phi(z_2)dz_2 \right\}, \qquad (7.14)$$

where $\phi(.)$ is the standard normal density function (d.f.).

Alternatively, from an approximate highly accurate density function for W, developed in Chapter 9, an approximate d.f for Y is developed, whereof all moments of Y may be easily evaluated numerically.

Approximate *explicit* expressions for the moments of the response, Y, may be obtained with the help of a Taylor series expansion. To derive these moments, write (7.9) as a product of two functions, one in terms of Z_1 and the other in terms of Z_2:

$$Y = \exp\{(\alpha/\lambda)[(\eta+\sigma_{\varepsilon 1}Z_1)^\lambda - 1] + \sigma_{\varepsilon 2}\rho Z_1\}$$

$$\times \exp\{\mu_2 + \sigma_{\varepsilon 2}(1-\rho^2)^{(1/2)} Z_2\} = g_1(Z_1;\omega_1) \; g_2(Z_2; \omega_2), \qquad (7.15)$$

where $\omega_1 = \{\alpha, \lambda, \rho, \sigma_{\varepsilon 1}, \sigma_{\varepsilon 2}\}$, $\omega_2 = \{\mu_2, \sigma_{\varepsilon 2}, \rho\}$, and g_1 and g_2 are expressions that relate to Z_1 and to Z_2, respectively.

Since both expressions are exponential functions of the respective normal variables, and because of the statistical independence of Z_1 and Z_2, the r-th non-central moment of Y may be derived as the product of the means of g_1 and of g_2, where for the mean of g_1 we replace $\{\alpha, \sigma_{\varepsilon 2}\}$ by $\{r\alpha, r\sigma_{\varepsilon 2}\}$, and similarly for the mean of g_2 we replace $\{\mu_2, \sigma_{\varepsilon 2}\}$ by $\{r\mu_2, r\sigma_{\varepsilon 2}\}$. The r-th non-central moment of Y is

$$E(Y^r) = E\{[g_1(Z_1;\omega_1)]^r\} \; E\{[g_2(Z_2; \omega_2)]^r\}$$

$$= E\{\exp\{(r\alpha/\lambda)[(\eta+\sigma_{\varepsilon 1}Z_1)^\lambda - 1] + r\sigma_{\varepsilon 2}\rho Z_1\}\}$$

$$\times E\{\exp[r\mu_2 + r\sigma_{\varepsilon 2}(1-\rho^2)^{(1/2)} Z_2]\}. \qquad (7.16)$$

The function g_2 is a log-normal variable. For the r-th moment of g_2 we have (exact)

$$E\{[g_2(Z_2;\omega_2)]^r\} = \exp[r\mu_2 + (1-\rho^2)(r\sigma_{\varepsilon 2})^2/2].\qquad (7.17)$$

(Refer, for example, to Stuart and Ord, 1987, Section 6.30.)

Expanding g_1 into a Taylor series in terms of powers of Z_1, around zero, and taking expectation of the first four terms, we obtain for the r-th non-central moment of g_1:

$$E\{[g_1(Z_1;\theta_1)]^r\} \cong \exp[(r\alpha/\lambda)(\eta^\lambda - 1)]$$
$$\times \{1 + (1/2)[(r\alpha)\sigma_{\varepsilon 1}^2(\lambda-1)\eta^{\lambda-2} + r^2(\alpha\sigma_{\varepsilon 1}\eta^{\lambda-1} + \rho\sigma_{\varepsilon 2})^2]\}.\qquad (7.18)$$

From (7.17) and (7.18), an explicit approximate expression for the r-th non-central moment of Y is

$$E(Y^r) = \mu'_r(Y) \cong \exp[(r\alpha/\lambda)(\eta^\lambda - 1) + r\mu_2 + (1-\rho^2)(r\sigma_{\varepsilon 2})^2/2]$$
$$\times \{1 + (1/2)[(r\alpha)\sigma_{\varepsilon 1}^2(\lambda-1)\eta^{\lambda-2} + r^2(\alpha\sigma_{\varepsilon 1}\eta^{\lambda-1} + \rho\sigma_{\varepsilon 2})^2]\}.\qquad (7.19))$$

From the non-central moments, $\mu_r' = E(Y^r)$, the first four moments, namely, the mean, the variance, and the skewness (Sk) and kurtosis (Ku) measures are easily calculated by (Stuart and Ord, 1987, p. 87)

$$\mu = \mu(Y) = \mu'_1,$$

$$\sigma^2 = \mathrm{Var}(Y) = \mu'_2 - (\mu'_1)^2,$$

$$Sk = E[(Y-\mu)^3]/\sigma^3 = [\mu_3' - 3\mu_2'\mu_1' + 2(\mu_1')^3]/\sigma^3,$$

$$Ku = E[(Y-\mu)^4]/\sigma^4 - 3$$

$$= [\mu_4' - 4\mu_3'\mu_1' + 6\mu_2'(\mu_1')^2 - 3(\mu_1')^4]/\sigma^4 - 3,\qquad (7.20)$$

For the normal distribution $Sk = Ku = 0$.

For the mean and the variance we obtain, approximately:

$$\mu(Y) = \exp[(\alpha/\lambda)(\eta^\lambda - 1) + \mu_2 + (1-\rho^2)(\sigma_{\varepsilon 2})^2/2)]$$
$$\times \{1 + (1/2)[\alpha\sigma_{\varepsilon 1}^2(\lambda-1)\eta^{\lambda-2} + (\alpha\sigma_{\varepsilon 1}\eta^{\lambda-1} + \rho\sigma_{\varepsilon 2})^2]\},$$

$$\mathrm{Var}(Y) = \exp[(2\alpha/\lambda)(\eta^{\lambda} - 1) + 2\mu_2 + (1-\rho^2)(2\sigma_{\epsilon2})^2/2)]$$
$$\times \{1+(1/2)[(2\alpha)\sigma_{\epsilon1}^2(\lambda-1)\eta^{\lambda-2} + 4(\alpha\sigma_{\epsilon1}\eta^{\lambda-1}+\rho\sigma_{\epsilon2})^2]\} - [\mu(Y)]^2,$$

$$\mu(W) = E[\log(Y)] = (\alpha/\lambda)[(\eta)^{\lambda}-1] + (1/2)\alpha(\lambda-1)\sigma_{\epsilon1}^2\eta^{\lambda-2} + \mu_2, \qquad (7.21)$$

where $\mu(W)$ is again developed using a Taylor expansion approximation.

Assume momentarily that $\sigma_{\epsilon1}= \sigma_{\epsilon2}=0$. Examining the expressions for the means in (7.21), one may realize how versatile RMM is in modeling variously-shaped link functions (a link function is a transformation of the response mean needed to obtain the linear predictor). In particular, the RMM mean provides a large majority of the link functions used in the framework of GLM. Thus, for $\lambda=0$ and $\alpha=1$, we obtain a *linear* link function, while $\lambda=0$ and $\alpha=-1$ result in a reciprocal relationship between the linear predictor and the mean, typical to the gamma error distribution. A value of $\lambda=1$ results in a log link function, typical to the Poisson distribution.

One realizes that the values of the parameters, in particular that of λ, are critical in determining the final structure of the model. Since λ is estimated within the data-analysis procedure (refer to the estimation procedures developed in Chapter 8), the final form of the model would be determined solely by the data on hand. No specification of the model structure prior to data analysis is needed, in compliance with Requirement 2 (Chapter 6).

Chapter 8 introduces estimation procedures for the parameters of the RMM model.

7.4. Exploring the Relationship between the CV and η

For relational models it is a commonly encountered phenomenon that as the linear predictor increases the variance tends to stabilize and its relative value (relative to the mean) tends to decrease. In fact, the well-known property of the decoupling of the variance from the mean, as the mean becomes large, has already been alluded to in Requirement 5 (Chapter 6).

It is therefore of interest to examine whether the RMM model owns this commonly encountered feature. In Section 13.2 .we provide an intuitive explanation of why the RMM allows decoupling of the variance from the mean when the linear predictor becomes large. We base this intuitive explanation on the fact that the error term associated with the linear predictor, ε_1, becomes negligible and therefore the variance of $\log(Y)$ tends to stabilize. A more rigorous mathematical proof will now be developed.

Consider the version of the RMM model given by (7.6b). Rewriting it in terms of *independent* standard normal variables, Z_1 and Z_2 (assume $\rho=0$), we have

$$Y = \exp\{(\alpha/\lambda)[\eta^{\lambda}\ \exp\ (\lambda\sigma_{\varepsilon1}Z_1\ /\ \eta) - 1] + \mu_2 + \sigma_{\varepsilon2}Z_2\}. \qquad (7.22)$$

Relate to the log-normal variable in squared brackets. This is the only random variable in (7.22) having parameters that depend on the linear predictor. We will find out how its coefficient of variation (CV= σ/μ) changes with η. Denoting the log-normal variable by U, we have from (7.22)

$$U = \exp[\lambda\ \log(\eta) + (\lambda\sigma_{\varepsilon1}/\eta)Z_1]$$

$$= \exp[\mu_1 + \sigma_1 Z_1]. \qquad (7.23)$$

For a log-normal variable, U, with parameters μ_1 and σ_1, the mean and the variance are (refer to Johnson, Kotz and Balakrishnan, 1994, Chapter 14, or to Stuart and Ord, 1987, Section 6.30):

$$E(U) = \exp(\mu_1 + \sigma_1^{2}/2),$$

$$Var(U) = \exp(2\mu_1 + \sigma_1^{2})[\exp\ (\sigma_1^{2})-1], \qquad (7.24)$$

and for the squared coefficient of variation we obtain

$$(CV)^{2} = Var(U))/[E(U)]^{2}$$

$$= \exp(2\mu_1 + \sigma_1^{2})[\exp(\sigma_1^{2})-1]\ /\exp(2\mu_1 + \sigma_1^{2}) = \exp(\sigma_1^{2})-1. \qquad (7.25)$$

Introducing for σ_1^{2} in terms of the RMM parameters: $\sigma_1^{2} = (\lambda\sigma_{\varepsilon1}/\eta)^{2}$ [refer to (7.23)], we obtain, on differentiating (7.25) with respect to the linear predictor, η:

$$\partial(CV)^{2}\ /\ \partial\eta = -\ (2/\eta)(\lambda\sigma_{\varepsilon1}/\eta)^{2}\ \exp[(\lambda\sigma_{\varepsilon1}/\eta)^{2}]. \qquad (7.26)$$

Assuming that the linear predictor, η, is positive, it is obvious that CV is a decreasing function of the linear predictor. Furthermore, as the linear predictor tends to infinity, CV tends to zero, namely, towards a stable variance.

References

[1] Johnson, N. L., Kotz, S., Balakrishnan, N. (1994). *Continuous Univariate Distributions — Volume 1*. 2nd Edition. John Wiley & Sons.

[2] Shore, H. (2002). Modeling a response with self-generated and externally-generated sources of variation. *Quality Engineering*, 14(4), 563-578.

[3] Stuart, A., Ord, J. K. (1987). *Kendall's Advanced Theory of Statistics. V. 1: Distribution Theory*. Charles Griffin & Company, London.

Chapter 8

Estimating the Relational Model

8.1. Introduction

The RMM model has three sets of parameters:

- Parameters associated with the linear predictor (LP);
- "Structural" parameters;
- "Error" parameters.

Estimating procedures for all three sets are developed in this chapter.

An initial approximate ML estimating procedure for the RMM model was developed in Shore (2002). The procedure comprised two separate routines: Stepwise linear regression, to estimate the parameters of the linear predictor (LP), and weighted non-linear regression, to estimate the rest of the parameters of the RMM model. Alternating between the two routines, estimates for the RMM model were eventually derived. This estimating procedure was approximate in the sense that certain approximating assumptions were made in order to apply stepwise linear regression to appropriately transformed response values. A more efficient and rigorous estimation procedure is needed.

In developing the new estimation procedure, two variations of the RMM model are addressed. Recall that by pursuing various assumptions regarding the error structure and the definitions of the independent standard normal variables, associated with the errors, eight different variations of the RMM model are obtained (refer for details to Section 7.2.5). Here we address the basic model, as derived axiomatically in

117

Chapter 7, and a certain variation of the basic model. These are, respectively,

$$Y = \exp\{(\alpha/\lambda)[(\eta+\sigma_{\varepsilon 1}Z_1)^\lambda - 1] + \mu_2 + \sigma_{\varepsilon 2}[\rho Z_1 + (1-\rho^2)^{(1/2)} Z_2]\}, \quad (8.1)$$

$$Y = \exp\{(\alpha/\lambda)[\eta^\lambda \exp(\lambda\sigma_{\varepsilon 1}Z_1/\eta)-1] + \mu_2 + \sigma_{\varepsilon 2}[\rho Z_1+(1-\rho^2)^{(1/2)} Z_2]\}. \quad (8.2)$$

Equation (8.2) is a variation of (8.1), which may be more convenient to apply when search routines are employed to identify ML estimates for the parameters. More specifically, the term $(\eta+\sigma_{\varepsilon 1}Z_1)^\lambda$ in (8.1) may occasionally deliver imaginary values when an estimate of λ is searched. This problem is eliminated when model (8.2) is used.

The estimation procedures developed here may be easily extended to the other variations of the RMM model.

Let us specify the three sets of parameters:

I. Parameters associated with the LP:

$$\eta = \beta_0 + \beta_1 X_1 + ..+ \beta_k X_k \qquad (8.3)$$

II. Parameters associated with the non-linear structure of the model:

$$\{\alpha, \lambda, \mu_2\}$$

III. Parameters associated with the errors:

$$\{\rho, \sigma_{\varepsilon 1}, \sigma_{\varepsilon 2}\}$$

In developing the estimation procedures for these sets of parameters, a fundamental distinction between the approach taken by RMM and that of GLM should be noticed. For both approaches, estimates for the parameters of the LP (Set I) are determined from data analysis. However, with GLM the non-linear structure of the final model (as conveyed by the link function) is determined a-priori, before implementation of data analysis. Likewise, the relationship between the mean and the variance, which is equivalent to the determination of the error distribution, is established independently of data analysis.

This stands in stark contrast to the basic Requirement 2 of "empirical modeling" (Chapter 6). In RMM estimation, all decisions regarding the model are based on straightforward data analysis. The composition of the LP, the structure of the (possibly non-linear) relationship between the LP and the response and the "error parameters" are all estimated from the data.

The RMM estimation approach separates estimation of the three sets of parameters into three different procedures. The procedure to estimate the LP parameters (Set I) is conducted once, in a combined analysis based on Canonical Correlation Analysis (CCA) and stepwise linear regression. The two procedures to estimate Set II and Set III are conducted sequentially in an iterative fashion. The structural parameters that belong to Set II are estimated via weighted non-linear least-squares. The error parameters (Set III) are estimated via maximization of an approximate log-likelihood function.

We divide this estimation process into two phases, where estimation of the Set I parameters is referred to as Phase 1, and estimation of the parameters in Set II and Set III is denoted Phase 2. These phases will now be expounded in more detail.

In Phase 1, a non-linear transformation is applied to the response (not its mean!) so that the transformed response is equal to the LP plus a random error. However, the needed transformation is not specified (unlike in other transformational approaches). Instead, it is approximated by a Taylor series expansion, and the coefficients of the latter are estimated via Canonical Correlation Analysis (CCA; The reader is referred to Appendix A, where a brief introduction to CCA is provided, and to Appendix B, for a discussion of assumptions). As implemented here, CCA attempts to maximize the correlation between a linear combination of the terms in the above Taylor expansion and a linear combination of effects that may potentially transmit variation to the response (denoted the *initial* LP).

Several CCA analyses can be attempted, having Taylor expansions for the transformed response that differ in the number of terms included. In Section 8.3 we will address various issues associated with Phase 1, among them how to determine the number of terms in the Taylor expansion and how to determine the structure of the *initial* LP.

The canonical scores, obtained from the eventually-selected CCA, are used as response values in stepwise linear regression analysis, when the *final* LP is estimated.

In Phase 2 of the analysis, the RMM model is perceived as a non-linear model with one predictor (the LP), and ML procedures are implemented to determine estimates for the non-linear coefficients of the model (Set II) and for the error parameters (Set III).

In the following Section 8.2 Phase 1 of the analysis is depicted, and in Section 8.3 we relate to various issues associated with it. Section 8.4 describes Phase 2 of the estimation process. The estimation procedures are applied to two numerical examples in Section 8.5.

Some further numerical examples are given in Chapter 14, where we compare solutions obtained via RMM to reported solutions derived by some other major current approaches for empirical relational modeling.

8.2. Phase 1 - Estimating the Linear Predictor (LP)

8.2.1. Introduction and motivation

The new approach to estimate the LP, described in this section, is not limited to RMM. In fact, it may be implemented independently of any specified non-linear relational model. It is therefore appropriate that we first expound the "philosophy" behind the new estimation procedure.

Common to all current general approaches to empirical relational modeling is the existence of two components of modeling. First, it is assumed that the variation transmitted to the response from some affecting factors may be represented by a linear combination of effects, denoted the LP (as in GLM). Secondly, it is assumed that the observed variation in the response may be captured by some non-linear relationship between the LP and the response (or its mean).

To demonstrate these two stages, consider applying a normalizing transformation to the response (like in a Box-Cox transformation). Here we use stepwise linear regression to model the LP, and we use a maximum-likelihood procedure to determine the transformation parameter, λ. Although procedures have been devised to estimate concurrently both the coefficients of the LP and the parameter(s) of the response transformation, in most statistical packages currently on the market the two stages of modeling are carried out sequentially.

Similarly, in modeling via GLM, we determine, in advance, the structure of the final (possibly non-linear) model by assuming a certain link function (a transformation of the response *mean* needed to obtain additivity of systematic effects, as captured by the LP). Assuming also a certain relationship between the response mean and the response variance (or, equivalently, determining the error distribution), the *iteratively re-weighted least-squares* procedure is conducted to determine the structure of the LP (refer for details to Myers *et al.*, 2002).

As a third example, consider the RMM *approximate* estimation procedure, related to earlier. In estimating the "structure" parameters and the LP, the procedure iteratively alternates between weighted non-linear regression and stepwise linear regression, respectively.

Examining all these general methodologies for empirical relational modeling, it is apparent that the two components of modeling are not carried out independently of each other. To determine the structure of the LP using a normalizing transformation, one needs to know the nature of this transformation. Similarly, in GLM the a-priori assumed link function determines the structure of the relationship between the LP and the response mean. Specifying the link is therefore a pre-condition that enables estimation of the LP.

A troubling aspect of the inter-dependency between estimation of the LP and determining the non-linear model (either via estimation or otherwise) is that choice of the latter may affect the former and *vice versa*. GLM is a case in point. Different choices regarding the link function may lead to differently estimated LPs. The conclusions reached about the effects that transmit systematic variation to the response may therefore differ appreciably, and generate much confusion as to which effects deserve our attention as major "players" affecting the behavior of the investigated response.

This is demonstrated in an example, given in Myers *et al.* (2002). The latter refer to Poisson data, where two different links are assumed: A log link (p. 148) and a square-root link (p. 183). As related therein (p. 185), although the two links "give lack-of-fit performances that are approximately the same...it is interesting that the conclusions regarding the roles of the variables are somewhat different, suggesting that the analyst may wish to try more than one link".

In quality engineering, the implications of the latter observation may be particularly bothersome. While in scientific modeling, and at times also in engineering modeling, both the structure of the LP and the structure of the final (possibly non-linear) model may be of practical or theoretical interest, in quality engineering oftentimes the modeler is interested exclusively in identifying the effects that relate significantly to the response (namely, in the structure of the LP).

Such a scenario is encountered, for example, if the experiment and the pursuing analyses are intended only as diagnostic tools in determining significant effects, following which some corrective action can be taken. In this case, the exact structure of the relationship between the LP and the response may be of secondary importance or not important at all. Yet, if different link functions, derived in the framework of GLM, lead to inclusion of different effects in the LP, this would most probably imply different action-guidelines for improving quality. In other words, the mutual dependency of the two components of modeling, where one component is decided (not estimated), may lead to confusion regarding the best prescriptions for action in order to improve quality.

Box assertion, quoted earlier (Chapter 1), comes to mind: "All models are wrong, some are useful". But will a practitioner accept that two models are "equally wrong" (have equal goodness-of-fit), yet they tell different stories about which factors influence an investigated response?

From the foregoing discussion, it is clear that if a procedure is conceived that decouples estimation of the two components of modeling, this would spell an important development in empirical relational modeling. Separating estimation of the RMM parameters into Phase 1 and Phase 2 intends to achieve the required decoupling. These two phases correspond to the two components of modeling that we have alluded to earlier.

In Section 8.2 procedures for Phase 1 are developed. That phase is divided into two stages. In Stage I an *initial* LP is derived and the coefficients of the Taylor approximation for the transformed response are estimated. Canonical scores, used as response values in the next stage of the analysis, are the main output from Stage I. In the next Stage II, the final LP is estimated by stepwise linear regression.

The two stages of Phase 1 are described in Sub-sections 8.2.2 and 8.2.3, respectively. Section 8.3 discusses certain issues related to the implementation of Phase 1.

8.2.2. Stage I - Approximating a transformed response via a Taylor series expansion and estimating the parameters via CCA

Let Y be a response assumed to be expressed in standardized units so that it has zero mean and unit variance. Let $\{X_1, X_2,.., X_k\}$ be a set of K predictors (factors or parameter-free functions of factors) that may potentially transmit statistically significant variation to Y. Let η denote the LP (of a structure yet unknown), and assume that a preliminary examination of the data suggests that the relationship between Y and η does not justify linear regression analysis (the relationship is non-linear, the assumption of a constant error variance is violated by the data, or, more generally, no known normalizing transformation can revoke the normal scenario).

Let G(Y) denote an unspecified relationship between the response and the LP, namely, $G(Y) = \eta + \varepsilon$, where ε is the error. Note, that G(Y) is a response transformation required to achieve additivity of effects (it is not a transformation of the *mean*, like in GLM).

Developing G(Y) into a Taylor series expansion around zero (the response mean) and keeping the first p+1 terms, we obtain

$$G(Y) \cong G(0) + G_1(0)Y + G_2(0)Y^2/2 + .. + G_p(0)Y^p/p!, \qquad (8.4)$$

where $G_j(0)$ is the j-th derivative, calculated around Y=0, and p is the number of terms in the Taylor expansion, apart from the constant (considerations in determining p are discussed later on).

Define

$$YT_j = Y^j / j! \ (j=1,..,p), \qquad (8.5)$$

to re-express G(Y) as

$$G(Y) \cong B_0 + B_1 \, YT_1 + .. + B_p \, YT_p , \qquad (8.6)$$

with coefficients $\{B_j\}$ that need to be estimated from the data.

Let the LP be as in (8.3). We wish to determine the parameters $\{B_j\}$ (j=0,1,..,p) so that G(Y), represented by the RHS of (8.6), may serve as the response in a stepwise linear regression procedure that will determine the coefficients (significance and estimate) of the LP (8.3). In other words, our model is

{Transformed response, a Taylor series expansion} = LP + error, or

$$B_0 + B_1\, YT_1 + .. + B_p\, YT_p = \eta + \varepsilon = \beta_0 + \beta_1 X_1 + .. + \beta_k X_k + \varepsilon. \quad (8.7)$$

The estimating procedure comprises two stages.

In Stage I, CCA is implemented to determine coefficients, $\{\beta_i\}$ (i=0,1,..,K), of an *initial* LP, and the *final* coefficients of the "fitted" Taylor approximation, $\{B_j\}$ (j=0,1,..,p). This stage of the analysis is expected to deliver canonical variables that maximally correlate G(Y), as approximated by (8.6), with η (8.3). If more than one canonical correlation turns out to be significantly different from zero [namely, (8.7) is significant for more than one pair of canonical variables], this stage may be conducted iteratively in order to maximally extract in one pair of canonical variables the significant co-variation between G(Y) and the LP.

In the next Stage II (Section 8.2.3), we use canonical scores, obtained for the LHS of (8.7), as response values in a stepwise multiple linear regression analysis to determine the final structure of the LP.

This estimating approach is supposed to deliver the structure of the LP (up to a linear transformation) that maximally correlates with an unspecified non-linear transformation of the response. The latter, being unspecified, is represented by a Taylor expansion that has been "fitted" (estimated) to maximally correlate with a linear combination of the effects. We have referred to the latter as "the initial LP" because it is not necessarily identical to "the final LP", obtained in the linear regression analysis of Stage II (this will be elaborated on and demonstrated later in the chapter).

Two observations are worth noting in relation to Stage I of the analysis. First, in most statistical packages, the raw canonical scores (the canonical scores calculated from the un-standardized variables) are automatically standardized to have zero mean and unit variance. Correspondingly, G(Y) in (8.6), as approximated by the final canonical variate, will have zero mean and unit variance.

Secondly, the number of canonical variates (or canonical roots, eigenvalues) is restricted by the number of terms in the response Taylor expansion. More specifically, the number of canonical correlations computed by CCA will be Min{K,p}. For example, if expansion to degree four is employed (p=4), and the number of effects in the initial LP is seven (K=7), then at most four significant canonical variables can be derived. In most cases, we expect p<K, so that the number of terms in the Taylor expansion will most probably determine the number of canonical variates.

Several issues related to the above estimation procedure are addressed in Section 8.3.

8.2.3. Stage II- Stepwise linear regression analysis with canonical scores as response values

At this stage of the analysis, the final structure of the LP is determined. As response values, we use the standardized canonical scores, saved from the CCA of the earlier stage. Stepwise linear regression is applied, where all potentially significant predictors, including those that have not participated in Stage I, are candidates for inclusion in the LP. As an initial (preliminary) model, we may use a full factorial of second degree, or a full response-surface model (namely, a quadratic model that includes all second degree interactions). Once the LP has been derived in a stepwise fashion, we may perform diagnostics checking as commonly practiced in the application of stepwise linear regression.

Having being satisfied with the linear model, we may proceed to the second phase of the modeling effort, where a non-linear RMM model that relates the LP to the response is estimated (refer to Section 8.4).

8.3. Issues Related to Implementation of Phase 1

(A) "How many terms to keep in the Taylor approximation for the transformed response?"

The first decision to be made in implementing Stage I is how many terms to keep in the Taylor expansion of G(Y) (the approximate response transformation). Two opposing considerations play part in that decision. On the one hand, we wish to increase the number of terms in order to enhance the accuracy of the approximation. On the other hand, adding variables may increase noise, which would obscure the true relationship between the two sets of variables being correlated.

Examining a few numerical examples, it seems that no strict rules can be specified, and several CCAs should be run to determine the best composition of the Taylor approximation to G(Y). A major consideration is to have the first canonical correlation (that associated with the first eigenvalue) be highly significant while having the second canonical correlation (and those that follow) be highly insignificant (a chi-square sequential test is conducted to provide these significance levels; Refer to Appendix A for further detail). Successive CCAs may therefore be run with an increasing number of power terms, and the best CCA be selected.

The associated canonical scores may then be used as response values for the second stage of the analysis, when the final structure of the LP is decided. Note, that if a single first-degree term is selected for the Taylor approximation (p=1), this implies that no response transformation is needed.

Table 8.1 introduces the results of such analyses related to the two numerical examples introduced in Section 8.5 (Example 1, the "Wave Solder Experiment", and Example 2, the "Resistivity Data" example). The significance levels in the table refer to testing of H_0: "The canonical correlations in the current row and all that follow are zero". The heading "YT_j" implies that the CCA includes terms up to and including YT_j.

Table 8.1. Results from CCA with different number of terms in the Taylor approximation to G(Y) (the optimal choice is starred).

Can. Correlations	Example 1. Taylor expansion up to term:		
(Significance)	YT_2	YT_3*	YT_4
First Can. Cor.	.7901	.7921	.8127
(Significance)	(<.0001)	(<.0001)	(<.0001)
Second Can. Cor.	.5384	.5674	.6025
(Significance)	(.0296)	(.1354)	(.2670)
	Example 2. Taylor expansion up to term:		
	YT_2*	YT_3	YT_4
First Can. Cor.	.9046	.9047	.9049
(Significance)	(.0082)	(.0210)	(.0829)
Second Can. Cor.	.3301	.5757	.6010
(Significance)	(.7235)	(.4740)	(.6454)

Consider the first example ("Wave-Solder Experiment"). It is noticed that "YT_2" turns out to result in a significant canonical correlation. This implies that using only YT_1 in the CCA (no response transformation at all) is not adequate, and YT_1 is not examined. However, as a general rule, it should be. Also notice that a Taylor expansion up to YT_2 is not sufficient to capture the co-variation between the two sets of variables being correlated: Not only the first canonical correlation, and correlations extracted in subsequent iterations, turn out to be significant ($p<0.0001$), but removing the first canonical correlation still results in mild significance ($p=0.0296$). Examining a Taylor expansion up to YT_3, We realize that with the first canonical correlation the result is highly significant, while removing it deliver a highly non-significant result ($p=0.1354$). Therefore in this example, a Taylor expansion up to a third degree is selected.

Consider the second example ("Resistivity Data"). Here we realize that already a polynomial expansion up to YT_2 results in the first canonical correlation being highly significant, while leaving it out, the second canonical correlation and those that follow are highly insignificant ($p=0.7235$). We therefore select to use for this example a Taylor expansion of $G(Y)$ of second degree only.

Another consideration that should be addressed is the *increase* in the first canonical correlation when we add a term to the Taylor expansion of $G(Y)$. A large increase would justify inclusion of an additional term in $G(Y)$. This is demonstrated in the numerical examples of Chapter 14.

Alternatively, suppose that a Taylor expansion of a specified degree is desired and several canonical correlations turn out to be significant [namely, we have several canonical variables that can represent $G(Y)$]. In that case, one may iterate the CCA. This implies that for any successive implementation of CCA, the significant canonical variables of the previous iteration serve as the set of input variables for the current iteration. Usually, a single iteration will suffice to conclude the analysis with a single significant canonical correlation. The associated canonical variable may then be used as the response for Stage II of the analysis.

(B) *"Which effects to include in the initial LP (Stage I)?"*

This is indeed still an open question, which seems to require further exploration before a conclusive answer can be given. There are two options to pursue. One may wish to include all possibly affecting predictors (even those that may later turn out to be insignificant) or else include in the *initial* LP only main effects (again irrespective of which are significant). As demonstrated in the two examples of Section 8.5, inclusion in the *initial* LP (of Stage I) only main effects does not bar other effects, like interactions, from entering as significant predictors into the final LP (in Stage II). We may wish to consider the option of including only main effects in the *initial* LP if the number of potentially significant main effects is large. In that case, inclusion of second order terms (or higher) may turn the implementation of CCA too cumbersome (due to the large set of variables that comprises the initial LP).

Conversely, all potentially significant effects (predictors) may be included in the *initial* LP, knowing that if a certain effect is unrelated to the response its weight in the canonical variable that represents the *initial* LP will be small. Consequently, that effect will have negligible influence in determining the coefficients of the "fitted" Taylor approximation for $G(Y)$.

More rigorous guidelines regarding the composition of the *initial* LP have yet to be developed. Cumulative empirical experience would probably be helpful in developing these guidelines. In the numerical examples of Section 8.5 and Chapter 14, only main effects are included in the initial LP of Stage I.

(C)*"What are the assumptions and are they valid"?*

Incorporating a "fitted" Taylor approximation as the response in a linear regression model requires that the assumptions underlying this model be specified and shown to be valid. In particular, one has to show that the assumption relating to the homogeneity of the residual variance, an essential assumption to the implementation of un-weighted linear regression, is valid. Since the canonical variable used as the response in the linear-regression part of the analysis is a Taylor-series approximation for $G(Y)$, we have to corroborate the assumption that $G(Y)$ indeed has a homogenous variance. This claim can be supported by three arguments, one relating to current known transformations that revoke the normal scenario, another based on more general considerations, and a third based on diagnostic checking.

First, let us note that there are transformations that stabilize the variance. In Chapter 5, a power transformation that aims to stabilize the variance has been explained in detail.

Another transformation is derivable from the RMM model. Referring to (8.1), and given the normal error associated with the LP, ε_1, we have

$$E[\log(Y \mid \varepsilon_1)] = (\alpha/\lambda)[(\eta+\varepsilon_1)^\lambda - 1] + \mu_2, \tag{8.8}$$

or

$$G(Y) = \eta + \varepsilon_1 = \{[\log(Y) - \mu_2](\lambda/\alpha) + 1\}^{1/\lambda}, \tag{8.9}$$

where in (8.9) $\log(Y)$ estimates $E[\log(Y \mid \varepsilon_1)]$.

In accordance with the assumptions of the RMM model, $G(Y)$ in (8.9) is normal with mean η and constant variance, $\sigma_{\varepsilon 1}^2$. Given the wide prevalence of the RMM model in a wide array of scientific and engineering disciplines (Shore, 2004b), the normality and the constancy of variance of $G(Y)$ in (8.9) are plausible expectations. In other words, one can expect *some* transformation of Y to revoke the normal scenario quite frequently. The success of the BC power transformation, a special case of RMM, is probably the empirical proof for that.

Let us address considerations of a somewhat more theoretical nature. First, the use of the canonical variable as a response is conditioned on the corresponding eigenvalue (root), or the respective canonical correlation, to be significantly different from zero. Since the canonical variable is a representation of a Taylor expansion of an unspecified transformation of Y, maximization of the canonical correlation (and consequently determination of the coefficients of the canonical variable) is equivalent to the determination of the "best fitting" transformation, $G(Y)$, where the latter is approximated by the respective canonical variable. Stated differently, by maximizing the canonical correlation, CCA ensures that a $G(Y)$ is approximated that best conforms to the assumptions of the linear regression scenario.

A somewhat comparable approach was pursued by Lin and Vonesh (1989) in determining the parameter λ of the BC transformation. They have constructed a non-linear regression model, which is used to estimate the parameter λ in such a fashion that the normal probability plot of the data on the transformed scale is as close to linearity as possible. Analogously, the new approach identifies, via CCA, a Taylor approximation for a transformation of Y that best correlates with a linear combination of the candidate effects. This is tantamount to the choice of the best response transformation that can then be used to construct a valid linear regression model.

Finally, note that since linear regression analysis is used in Stage II, diagnostic checking for the validity of the normal scenario may always be conducted. Given the foregoing discussion, it is highly likely that the diagnostic checking would justify implementation of linear regression analysis.

(D) *"Is there a single correct LP"?*

The new approach implicitly assumes that there is only a single "correct" LP, and that this LP may be identified irrespective of the non-linear model employed to represent the relationship of the LP with the response (be it a BC transformation, a link function or a fitted RMM model).

Given the coupling between the two phases of the estimation process, which exists with most current approaches, one may reasonably argue that the LP is essentially dependent on the non-linear model selected, and that the two modeling components cannot be separated (decoupled). This dependency was earlier demonstrated by a quote from Myers *et al.* (2002), where a log link and a square-root link produced different LPs (different effects included), even though goodness-of-fit statistics were the same. Given the latter criterion, one may advocate that different LPs may be adopted as equally valid.

We believe otherwise. If certain effects, and not others, turn out to be significantly related to the response, then these effects reflect reality, and they should be addressed, and not others, in making decisions regarding the response (whether these decisions relate to prediction, control or quality improvements). Maximizing the correlation between the LP and the transformed response, while not being confined to a particular transformation, seems to be the right way to identify a single LP, independently of the non-linear model selected. The Taylor approximation is the right platform to achieve this objective.

The similarity of the LPs, derived by GLM and by the new approach (refer to comparative solutions in Chapter 14), seems to corroborate the validity of this proposition.

Whatever modeling approach one may wish to pursue, certain good practices should never be overlooked. An engineering and scientific exploration into the fundamentals of an investigated problem nearly always adds value. And no modeling methodology, however innovative, can ever serve as a substitute for this time-honored practice.

(E) *"Integration with existing methodologies"*

The new approach to estimate the LP is a stand-alone, with no specific prescription as to how one should proceed if a non-linear model is desired. Integration of the new approach with estimating of an RMM model is demonstrated in the numerical examples of Section 8.5.

How should one proceed if, for example, a GLM model is desired, or if we wish to fit a specified non-linear regression model, assuming homogeneity of variance?

Once Stage II is completed the LP assumes definite values for individual observations. The estimation problem of a non-linear model then becomes one of modeling a response with one regressor (predictor). All known methodologies may therefore be applied, assuming a single predictor. For example, GLM may be implemented, attempting different link functions. By decoupling decisions regarding composition and estimation of the LP from those regarding the non-linear relationship with the response, the new approach reduces a multi-variate problem into a single-predictor problem, adding simplicity to the implementation of existing methodologies.

(F) *"Handling observational data vs. data from designed experiments"*

The two examples of this chapter (Section 8.5) relate to data from experimental designs where predictors are orthogonal. Would the new approach be equally applicable to observational data where predictors may be correlated?

To answer this question and others, related to the underlying assumptions of CCA, one has to look into these assumptions and find out whether they are inherently violated by the new approach. Appendix B delivers a short discussion of the most important assumptions of CCA, and the major threats to the reliability and validity of results derived from CCA. This brief exposition is based on a similar discussion given in the STATISTICA™ web-based Electronic Manual. Reading Appendix B, it seems that there are some important implications for the practice of the new approach in estimating the LP.

First, robustness to the assumed multivariate normality is quite evident, especially if sample size is sufficiently large. This robustness notwithstanding, it seems that the need for normality suggests that the initial LP of Stage I should include only main effects. Estimates of main effects would most likely be normally distributed, in contrast to estimates of interactions or other parameter-free functions of main effects. Therefore, the likelihood of not deviating from the normality assumption would be higher if only main effects are included.

Secondly, certain minimal ratios between the number of cases and the number of variables in the analysis need to be observed (Assumption 2 in Appendix B). This implies that the number of effects included in the initial LP of Stage I should be kept to a minimum. Inclusion of only main effects fulfills this condition, consistently with the conclusion derived in the previous paragraph.

Thirdly, correlations between variables belonging to the same set do not pose difficulty as long as no variable is completely redundant (no variable is a linear combination of other variables). The latter case causes the correlation matrix to be ill-conditioned (cannot be inverted within the CCA). Thus, observational data are analyzable by the new approach and internal correlations within each set are properly accounted for.

Finally, outliers may affect greatly correlations between variables, and therefore similarly affect the derived canonical correlations. Increasing sample size is one possible safety-caution against the influence of outliers. A second possible solution is to examine scatter-plots in order to detect outliers. This option is available in any software procedure that conducts CCA.

8.4. Phase 2 - Estimating the RMM Model

8.4.1. Introduction

Sample-values of the LP, obtained in Phase 1, may serve as input to estimate a single-regressor (predictor) RMM model in Phase 2.

Consider the location of the LP. From (8.1), we realize that to calculate the mean the LP needs to be raised to the power of λ. When an ML estimate for the latter is searched, it is desirable that the LP retains only non-negative values. It is therefore suggested that a change in location be applied by adding the minimal (negative) sample value of the LP to all sample values. This simple translation would not undermine the validity of the analysis by which the LP has been derived in Phase 1. To realize that note that two linear transformations applied, respectively, to two correlated variables, do not alter their correlation (provided the linearly transformed variables co-vary in the same relative direction as the untransformed variables). Consequently, adding a constant to all values of the LP would not alter the significance of the results obtained in the stepwise linear regression analysis of Phase 1. In particular, the regression multiple-correlation would not be altered. Henceforth, it is assumed that all values of the LP are non-negative.

Suppose that a sample of n observations, $\{\eta_i, w_i\}$, is available, where w_i is the log-transformed response-value for observation i and similarly η_i is the value of the LP. Since the latter has already been estimated in Phase 1 of the analysis, we need to estimate two sets of parameters (as detailed in the introductory Section 8.1):

Set II. Parameters of the non-linear structure of the model $\{\alpha, \lambda, \mu_2\}$
Set III. Parameters associated with the error terms, $\{\rho, \sigma_{\varepsilon 1}, \sigma_{\varepsilon 2}\}$.

The estimation procedure of Phase 2 is accordingly divided into two stages:

- Estimating Set II via weighted non-linear least-squares; Weights used are reciprocal values of variance estimates, calculated via the *updated* quantile function of the RMM model (parameters' estimates are from the most recent iteration);

- Estimating Set III via maximizing the log-likelihood function.

Estimation is implemented by iteratively alternating between the two stages, where for each stage recently updated estimates from the preceding stage are employed. Detailed description of the two stages follows.

8.4.2. Stage I- Estimating the RMM parameters $\{ \alpha, \lambda, \mu_2 \}$

A weighted non-linear least-squares (W-NL-LS) would minimize

$$S = \sum_{i=1}^{n} [w_i - \mu_i(W)]^2 / [\sigma_i(W)]^2 , \qquad (8.10)$$

where $\mu_i(W)$ and $\sigma_i(W)$ are the mean and the standard deviation of observation w_i (i=1,2,..,n). Myers *et al.* (2002) quote the work of Wedderburn (1974), which suggests that use of weighted least-squares produces asymptotic properties that are quite similar to those of maximum likelihood (ML) estimators. Thus, if good estimates of the variances of individual observations are available, minimizing (8.10) would provide nearly ML estimates for the parameters of the RMM model.

Note that since the LP has already been independently estimated, only the RMM parameters, $\{\alpha, \lambda, \mu_2\}$, are estimated at this stage of the analysis. Various diagnostics checking, routinely available in any statistical package that include a module for NL-LS, may be used to judge the quality of the derived estimates. For example, correlations between estimates or estimates' confidence intervals may be derived. This will be demonstrated in the numerical examples of Section 8.5.

Let us develop expressions for the mean and the variance in (8.10). For the basic model and its variation [(8.1) and (8.2), respectively]

$$E(W) = E\{(\alpha/\lambda)[(\eta+\sigma_{\varepsilon 1}Z_1)^{\lambda} - 1]\} + \mu_2, \qquad (8.11)$$

$$E(W) = (\alpha/\lambda)\{\exp[\lambda\log(\eta) + (\lambda\sigma_{\varepsilon 1}/\eta)^2/2]-1\} + \mu_2, \qquad (8.12)$$

where in (8.12) the expression in squared brackets is the mean of a log-normally distributed random variable. Note that both (8.11) and (8.12) imply that the mean does not include the error parameters $\{\sigma_{\varepsilon 2}, \rho\}$.

Using a Taylor expansion for the expression in squared brackets in (8.11), and taking expectation with respect to Z_1, we may use the expression given in (7.21) to obtain

$$E(W) \cong (\alpha/\lambda)[(\eta)^{\lambda} - 1] + (1/2)\alpha(\lambda-1)\sigma_{\varepsilon 1}{}^2\eta^{\lambda-2} + \mu_2. \qquad (8.11a)$$

The LP (η) has already been estimated, and therefore only the four parameters $\{\alpha, \lambda, \mu_2, \sigma_{\varepsilon 1}\}$ need to be estimated by minimization of (8.10). We prefer to estimate *all* parameters associated with the errors, $\{\rho, \sigma_{\varepsilon 1}, \sigma_{\varepsilon 2}\}$, by a separate estimation routine that will be expounded shortly. This implies that in minimizing (8.10) only the RMM parameters, $\{\alpha, \lambda, \mu_2\}$, will be estimated. The parameter $\sigma_{\varepsilon 1}$ will be estimated in the pursuing stage, together with the other error parameters (refer to Sub-section 8.4.3). An initial estimate to use in (8.10) (for the first iteration only!) may be taken from the residual standard deviation of the linear regression analysis of Phase 1. Alternatively, one may opt to include $\sigma_{\varepsilon 1}$ as an additional parameter to be estimated via the NL-LS procedure.

The first iteration in the estimation procedure differs relative to successive iterations also in that no estimates of $\{\sigma_i(W)\}$ are yet available. This will be true unless repeat observations (replicates) exist so that "pure error" estimates of variance are available. In all other cases, un-weighted NL-LS is conducted for the first iteration.

In pursuing iterations, updated estimates of $\{\sigma_i(W)\}$ will be available so that weighted NL-LS can be used.

Appropriate determination of the weights in (8.10) is crucial for the effective estimating of the RMM "structure" parameters. The weights are reciprocal values of the variance of W for individual observations. To derive these weights, assume that all parameters of the RMM model are known (or that updated estimates for these parameters have been derived at the completion of Stage II of the analysis). Estimates of variances for individual observations may then be derived numerically either from the quantile function [(8.1) or (8.2)] or from the density function of W.

Both methods will now be developed for (8.1). Corresponding expressions may similarly be developed for (8.2).

Estimating the Variance of W (Based on Eq. 8.1)

Method I

Re-write (8.1) as

$$W = Q_1(Z_1; \omega_1) + Q_2(Z_2; \omega_2), \tag{8.13}$$

where

$$Q_1(Z_1; \omega_1) = (\alpha/\lambda)[(\eta+\sigma_{\varepsilon1}Z_1)^\lambda - 1] + \mu_2 + \sigma_{\varepsilon2}\rho Z_1,$$

$$Q_2(Z_2; \omega_2) = \sigma_{\varepsilon2}(1-\rho^2)^{(1/2)} Z_2, \tag{8.14}$$

with, respectively, parameters' vectors

$$\omega_1 = \{\alpha, \lambda, \mu_2, \sigma_{\varepsilon1}, \sigma_{\varepsilon2}\rho\},$$

$$\omega_2 = \sigma_{\varepsilon2}(1-\rho^2)^{(1/2)}. \tag{8.15}$$

Note, that ω_1 contains five parameters while ω_2 has only one parameter.

Since Z_1 and Z_2 are independent random variables so are Q_1 and Q_2, and for the r-th non-central moment of W we obtain

$$E(W^r) = \mu_r'(W) = \sum_{j=0}^{r} \binom{r}{j} M_j(Q_1) M_{r-j}(Q_2), \tag{8.16}$$

where $M_j(Q_1)$ and $M_{r-j}(Q_2)$ are the j-th and (r-j)-th non-central moments of Q_1 and Q_2, respectively, namely:

$$M_k(Q_1) = \int_{-\infty}^{\infty} \{(\alpha/\lambda)[(\eta+\sigma_{\varepsilon1}z)^\lambda - 1] + \mu_2 + \sigma_{\varepsilon2}\rho z\}^k \, \phi(z)dz \tag{8.17}$$

$$M_k(Q_2) = \int_{-\infty}^{\infty} [\sigma_{\varepsilon2}(1-\rho^2)^{(1/2)} z]^k \, \phi(z)dz$$

$$= [\sigma_{\varepsilon2}(1-\rho^2)^{(1/2)}]^k \, \mu_k(Z) \tag{8.18}$$

where $\mu_k(Z)$ is the k-th standard normal moment.

The variance of W for individual observations may be estimated from (8.16)-(8.18), and inserted into the minimization routine (8.10).

Method II

This method uses an approximate expression for the density function of W, f_W, given in (8.20) below. The variance of W may then be estimated from

$$\mu_r'(W) = \int_{-\infty}^{\infty} w^r f_W(w)\,dw,$$

$$\text{Var}(W) = \mu_2'(W) - [\mu_1'(W)]^2, \qquad (8.19)$$

where $\mu_r'(W)$ is the r-th non-central moment of W.

Note, that applying Method I may encounter difficulties if the expression $(\eta + \sigma_{\varepsilon 1} Z_1)$ becomes negative during numerical integration (imaginary values may be encountered, especially if η has values close to zero). Therefore, we may either select a translation constant for η that ensures non-negative values for $(\eta + \sigma_{\varepsilon 1} Z_1)$ (we have discussed earlier why such a translation is legitimate), or we may opt for using Method II in calculating weights for the W-NL-LS procedure.

8.4.3. Stage II - Estimating the RMM "Error Parameters" $\{\rho, \sigma_{\varepsilon 1}, \sigma_{\varepsilon 2}\}$

At this stage, the error parameters, $\{\rho, \sigma_{\varepsilon 1}, \sigma_{\varepsilon 2}\}$, are estimated via a ML routine, based on the approximate log-likelihood function of the RMM error distribution.

In Section 9.1, an expression for the density function (d.f.) of W is developed, which requires numerical integration. This may render application of an ML estimation procedure a numerically prohibitive undertaking. We therefore opt for a Taylor expansion which would eliminate the need for integration. The result is a relatively simple expression for the d.f. of W, which is based on the expected values of the first four terms in the Taylor expansion of the conditional density $f_{W|\varepsilon 1}(w|\varepsilon_1, \theta)$. The resulting expression appears in (9.10), where $\eta=1$. With η as an additional argument, we have:

$$f_W(w; \theta) = \int_{-\infty}^{\infty} f_{W|\varepsilon1}(w|\,\varepsilon_1; \theta)\, f_{\varepsilon1}(\varepsilon_1)d\varepsilon_1$$

$$\cong [2\pi(1-\rho^2)\sigma_{\varepsilon2}^2]^{-1/2} \exp[-(1/2)Z^2]$$

$$\times \{1+(1/2)\sigma_{\varepsilon1}^2[(1-\rho^2)^{1/2}\sigma_{\varepsilon2}]^{-2}[(\alpha\eta^{\lambda-1}+\rho\sigma_{\varepsilon2}/\sigma_{\varepsilon1})^2(Z^2-1)$$

$$+ \alpha\eta^{\lambda-2}(\lambda-1)(1-\rho^2)^{1/2}\sigma_{\varepsilon2}Z]\}, \tag{8.20}$$

where $\theta = \{\alpha, \lambda, \mu_2, \rho, \sigma_{\varepsilon1}, \sigma_{\varepsilon2}\}$, and

$$Z = \{w-(\alpha/\lambda)[(\eta)^{\lambda}-1]-\mu_2\} / [(1-\rho^2)\sigma_{\varepsilon2}^2]^{1/2}. \tag{8.21}$$

Note, that this is an approximate expression for the d.f. of W, when the RMM model is given by (8.1). When the RMM model is (8.2), the approximate expression for f_W is the same except for the last term in (8.20), where the term $(\lambda-1)$ is replaced with λ.

The results developed here may easily be adapted for other variations of the RMM model, as given in Chapter 7. Henceforth, we will relate only to model (8.1), with the d.f given by (8.20).

From (8.20) and (8.21), the log-likelihood function is

$$LL(\theta|\,w) = \sum_{i=1}^{n} \; \log \int_{-\infty}^{\infty} f_{W|\varepsilon1}(w_i|\,\varepsilon_1; \theta)\, f_{\varepsilon1}(\varepsilon_1)d\varepsilon_1$$

$$\cong -(n/2)\log[2\pi(1-\rho^2)\sigma_{\varepsilon2}^2] + \sum_{i=1}^{n} \; [-(1/2)Z_i^2]$$

$$+ \sum_{i=1}^{n} \; \log\{1+(1/2)\sigma_{\varepsilon1}^2[(1-\rho^2)^{1/2}\sigma_{\varepsilon2}]^{-2}$$

$$\times [(\alpha\eta_i^{\lambda-1}+\rho\sigma_{\varepsilon2}/\sigma_{\varepsilon1})^2(Z_i^2-1)+\alpha\eta_i^{\lambda-2}(\lambda-1)(1-\rho^2)^{1/2}\sigma_{\varepsilon2}Z_i]\}, \tag{8.22}$$

where for the i-th observation

$$Z_i = [w_i - (\alpha/\lambda)[(\eta_i)^{\lambda} - 1] - \mu_2] / [(1-\rho^2)\sigma_{\varepsilon2}^2]^{1/2}. \tag{8.23}$$

Maximization of (8.22) with respect to the error parameters provides ML estimates for $\{\rho, \sigma_{\varepsilon1}, \sigma_{\varepsilon2}\}$.

8.4.4. Summary of the estimation procedure (Phase 2)

Assume that the values of the LP for individual observations are known (estimated) from Phase 1 of the analysis:

I. Minimize S (8.10) to obtain initial estimates for $\{\alpha, \lambda, \mu_2\}$. Use un-weighted NL-LS for this step. As an estimate for $\sigma_{\varepsilon 1}^2$, use an estimate of the residual variance, obtained from the stepwise linear regression of Phase 1. Alternatively, find estimates for all parameters $\{\alpha, \lambda, \mu_2, \sigma_{\varepsilon 1}\}$.

II. For the derived estimates of $\{\alpha, \lambda, \mu_2\}$, maximize the log-likelihood function (8.22) to find estimates for the errors' parameters $\{\rho, \sigma_{\varepsilon 1}, \sigma_{\varepsilon 2}\}$.

III. Calculate variances for individual observations, using either Method I or Method II. Use "pure error" variances, if available. Use reciprocal values of the derived variances as weights for the next iteration (Step IV);

IV. Derive updated estimates for $\{\alpha, \lambda, \mu_2\}$, using W-NL-LS to minimize S (8.10). Use diagnostic checks embedded in the NL-LS procedure (most statistical packages offer this procedure with allied diagnostic checking).

V. Test the solution for convergence (various criteria may be used, for example, significant reduction in the residual variance). If convergence has not yet occurred- goto step II. Otherwise- end the estimation procedure.

Two detailed numerical examples follow. Examples that demonstrate in a more comprehensive fashion the estimation procedure, and also offer comparisons with solutions derived by other major approaches, are given in Chapter 14.

8.5. Two Numerical Examples

8.5.1. Example 1 - The Wave-Soldering Process

This example deals with attaching components to an electronic circuit card assembly by a wave-soldering process. The response is the number of defects in the solder joint. The process involves baking and preheating of the circuit card, and passing it through a solder wave by a conveyor. Condra (1993) introduced the results, and Hamada and Nelder (1997) re-analyzed, using GLM with a variance function equal to the mean (a Poisson model) and a log-link function. Later, Myers *et al.* (2002, p. 145) re-analyzed using GLM. These solutions will be compared to the RMM solution in Chapter 14.

The following presentation of the problem is taken from the latter source. The factors are (A) pre-bake condition, (B) flux density, (C) conveyor speed, (D) preheat condition, (E) cooling time, (F) ultrasonic solder agitator, and (G) solder temperature. Only the interactions ab, ac, ad, bc, bd, and cd, were considered to be important (refer to Hamada and Nelder, 1997). Therefore all analyses are restricted to these interactions.

Table 8.2 presents the experimental design and the observations. Each factor is at two levels and the experimental design is 2^{7-3} fractional factorial. Data point 11 has one missing value (originally 173), which was excluded by Hamada and Nelder (1997) due to its outlying behavior. Also note that the second data point in run 14 (2) erroneously appears as 1 in Myers *et al.* (2002) (the correct value is given here). Finally, the zero value (data point 9) will be replaced in the analysis by 0.295

Table 8.2. Example 1 ("Wave Solder Experiment")- Experimental Design and Observations

Run	Factors							Observations		
	A	B	C	D	E	F	G	y1	y2	y3
1	-1	-1	-1	-1	-1	-1	-1	13	30	26
2	-1	-1	-1	1	1	1	1	4	16	11
3	-1	-1	1	-1	-1	1	1	20	15	20
4	-1	-1	1	1	1	-1	-1	42	43	64
5	-1	1	-1	-1	1	-1	1	14	15	17
6	-1	1	-1	1	-1	1	-1	10	17	16
7	-1	1	1	-1	1	1	-1	36	29	53
8	-1	1	1	1	-1	-1	1	5	9	16
9	1	-1	-1	-1	1	1	-1	29	0	14
10	1	-1	-1	1	-1	-1	1	10	26	9
11	1	-1	1	-1	1	-1	1	28	-	19
12	1	-1	1	1	-1	1	-1	100	129	151
13	1	1	-1	-1	-1	1	1	11	15	11
14	1	1	-1	1	1	-1	-1	17	2	17
15	1	1	1	-1	-1	-1	-1	53	70	89
16	1	1	1	1	1	1	1	23	22	7

Table 8.3. Example 1- Main CCA Results

	Canonical Correlation	Adjusted Canonical Correlation	Approximate Standard Error	Squared Canonical Correlation
1	0.792078	0.750096	0.054939	0.627387
2	0.567448	0.501351	0.099966	0.321997
3	0.202167	0.010847	0.141416	0.040871

	Eigenvalue	Differ.	Propor.	Cumulat.	Ratio	F-Value	Num. DF	Den. DF	Pr > F
1	1.6838	1.2088	0.7649	0.7649	0.2423	3.25	21	106	<.0001
2	0.4749	0.4323	0.2157	0.9806	0.6503	1.52	12	76	0.1354
3	0.0426		0.0194	1.0000	0.9591	0.33	5	39	0.8903

(refer to Shore, 2001, p. 11, for justification).

These data are now re-analyzed, implementing the new approach.

First, the response is standardized (mean 29.64, STD 31.44) and three new variables are defined, $\{YT_j\}$ (j=1, 2, 3) (for choosing K=3, refer to an earlier discussion in Section 8.3). Applying CCA (using a SAS procedure), Table 8.3 is obtained. Note, that only main effects were included in the LP for the CCA.

Only the first canonical correlation, 0.7921, is significantly larger than zero (p<0.0001). Furthermore, the first eigenvalue accounts for 76.49% of the variation in the data. The corresponding equation for the first canonical variable (in terms of the raw coefficients) is

$$W_1 = 1.4562 \ YT_1 - 0.1523 \ YT_2 - 0.2740 \ YT_3. \tag{8.24}$$

Standardizing W_1 to have zero mean (adding 0.1786) and unit variance, it may now become the response for a multiple stepwise regression analysis, the results of which are displayed in Table 8.4. The following model is obtained, which includes all estimates significant at the 1% level (refer to Table 8.4):

$$(W_1 + 0.1786)/1 = 0.164A - 0.144B + 0.526C - 0.157E - 0.495G$$

$$+ 0.224(A+0.02)(C+0.02) - 0.431(B-0.02)(D-0.02), \tag{8.25}$$

with R^2-adj.=0.848 and model F-ratio of 37.7 (p<0.0001). The root mean squared error is 0.3896. Lack-of-fit test proves to be highly non-significant (p=.569).

Table 8.4. Example 1- Main Results from Linear Regression Analysis. Response is the Standardized Canonical Variate W_1

Summary of Fit

RSquare	0.871287
RSquare Adj	0.848184
Root Mean Square Error	0.389635
Mean of Response	5.2e-17
Observations (or Sum Wgts)	47

Analysis of Variance

Source	DF	Sum of Squares	Mean Square	F Ratio
Model	7	40.079191	5.72560	37.7142
Error	39	5.920809	0.15182	Prob > F
C. Total	46	46.000000		<.0001

Lack Of Fit

Source	DF	Sum of Squares	Mean Square	F Ratio
Lack Of Fit	8	1.0626289	0.132829	0.8476
Pure Error	31	4.8581801	0.156715	Prob > F
Total Error	39	5.9208090		0.5692
				Max RSq
				0.8944

Parameter Estimates

Term	Estimate	Std Error	t Ratio	Prob>|t|
Intercept	-0.000628	0.056937	-0.01	0.9913
A	0.1640337	0.056922	2.88	0.0064
B	-0.143827	0.056921	-2.53	0.0157
C	0.5263339	0.056922	9.25	<.0001
E	-0.156856	0.056938	-2.75	0.0089
G	-0.49514	0.056938	-8.70	<.0001
(A+0.02128)*(C+0.02128)	0.2242015	0.056938	3.94	0.0003
(B-0.02128)*(D-0.02128)	-0.431501	0.056955	-7.58	<.0001

Note the small values of the coefficients of A, B and E, relative to the coefficients of C and G. These effects also prove to be least significant. However, a "lack of fit" test for the reduced model turns out to be mildly significant (p=0.0187). Therefore the extended model (8.25) is adopted.

Now that an estimate of the LP, given by the RHS of (8.25), is available, we can proceed to Phase 2, to derive ML estimates for the RMM model. First, we add 0.84004 to each value of the LP to ensure that there are no negative values. Implementing un-weighted NL-LS, using W= log(Y) as the response, (8.11a) for the mean and an estimate for $\sigma_{\varepsilon 1}$ derived from the linear regression analysis of Phase 1, namely, $\sigma_{\varepsilon 1}$= 0.3896, the initial estimates obtained are

$$\alpha = 0.8708, \lambda = 1.000, \mu_2 = 3.034.$$

(Note that the expression for the mean, E(W), employed in the NL-LS procedure, does not include either ρ or $\sigma_{\varepsilon 2}$).

These estimates are introduced into (8.22), and the latter is maximized with respect to ρ, $\sigma_{\varepsilon 1}$ and $\sigma_{\varepsilon 2}$, to obtain

$$\rho = 0.1, \sigma_{\varepsilon 1} = 0.0955, \sigma_{\varepsilon 2} = 0.7096, LL^* = 51.2.$$

The updated value of $\sigma_{\varepsilon 1}$ does not alter the values of the RMM parameters given earlier, and therefore no further iterations were needed.

Calculating variances for individual observations (using Method II) shows the variance, $\sigma^2(W)$, to be stable around 0.5. Therefore no further iterations, using weighted NL-LS, are warranted, and we accept the above parameters' values as final.

From these results two observations are apparent. First, a log transformation applied to the response delivers the best-fit model, and, secondly, this transformation also stabilizes the variance.

For the RMM mean model [using E(W) in (8.11a)], we obtain for the response on the original scale a mean squared error (MSE) of 127.57. It is shown in Chapter 14 that relative to GLM solutions, attempted with different link functions, the MSE obtained from RMM is appreciably lower.

Diagnostics-related analyses, based on the NL-LS results, are obtainable as standard output from the software module used. For example, the following 95% confidence intervals for the RMM parameters' estimates were obtained:

$$\alpha = \{0.632, 1.11\}, \lambda = \{1,1\}; \mu_2 = \{2.81, 3.26\}.$$

8.5.2. Example 2 - The Resistivity Data

This example originally appeared in Myers and Montgomery (1997), and re-introduced in Myers *et al.* (2002, p. 176). The introduction herewith follows these references. The experiment has an un-replicated 2^4 factorial design, which was run at a certain step in a semiconductor manufacturing process. The response variable, resistivity of the test wafer, is a characteristic well-known to have a non-normal, long right-tailed distribution.

The design, with the observations collected, is given in Table 8.5. We analyze these data using the new approach.

A first run of CCA with one set of variables including main effects $\{X_1,..,X_4\}$ and the other set including $\{YT_1,YT_2\}$ (on the reasons for selecting a Taylor expansion of a second degree refer to Section 8.3 and Table 8.1), results in Table 8.6. There is a single significant eigenvalue, associated with a canonical correlation of 0.9046 (p=0.0082). The first canonical variable associated with the set of variables derived from the Taylor-expanded response transformation is

$$W_1 = 1.0842 \ YT_1 - 0.3096 \ YT_2. \tag{8.26}$$

Standardizing W_1 to have zero mean and unit variance, it may now become the response for a multiple stepwise regression, to yield the following model:

$$W_1 = -0.1451 + 0.293X_1 - 0.707X_2 + 0.409X_3$$

$$- 0.201X_1X_3 - 0.188X_2X_3 - 0.205X_3X_4 + \varepsilon, \tag{8.27}$$

with R^2-adj.=0.882, and model F-ratio of 19.7. With 6 and 9 degrees of freedom for the model and the error, respectively, the latter result is highly significant (p<0.0001). All main effects are highly significant too (p<0.008). The root mean square error is 0.3430.

Note that as with the first example, although only main effects participated in the *initial* LP of Stage I, significant interactions were

Table 8.5. Example 2- Experimental Design and Observations

x1	x2	x3	x4	y
-1	-1	-1	-1	193.4
1	-1	-1	-1	247.6
-1	1	-1	-1	168.2
1	1	-1	-1	205
-1	-1	1	-1	303.4
1	-1	1	-1	339.9
-1	1	1	-1	226.3
1	1	1	-1	208.3
-1	-1	-1	1	220
1	-1	-1	1	256.4
-1	1	-1	1	165.7
1	1	-1	1	203.5
-1	-1	1	1	285
1	-1	1	1	268
-1	1	1	1	169.1
1	1	1	1	208.5

Table 8.6. Example 2- Main CCA Results

```
                    Canonical Correlation Analysis

                   Adjusted      Approximate    Squared
       Canonical   Canonical     Standard       Canonical
       Correlation Correlation   Error          Correlation

    1  0.904635    0.882968      0.046898       0.818365
    2  0.330063    0.164909      0.230070       0.108942

Eigenvalue Differ. Propor. Cumulat. Ratio  F-     Num. Den.
                                           Value  DF   DF    Pr > F

1  4.5056  4.3833  0.9736  0.9736  0.1618  3.71   8    20   0.0082
2  0.1223          0.0264  1.0000  0.8911  0.45   3    11   0.7235
```

Table 8.7. Example 2- Comparison of observed values and RMM predicted values

Obs. #	Sample y	RMM	Obs. #	Sample y	RMM
1	193.4	195.7	9	220	214.4
2	247.6	243.9	10	256.4	267.3
3	168.2	155.2	11	165.7	170.1
4	205	193.5	12	203.5	212.1
5	303.4	306.1	13	285	279.4
6	339.9	319.0	14	268	291.1
7	226.3	205.4	15	169.1	187.4
8	208.3	214.0	16	208.5	195.3

included in the *final* LP, derived in Stage II of the analysis.

For Phase 2 of the analysis, a value of 1.627 is first added to the LP to ensure that all LP values, obtained in Phase 1, are non-negative. A first run of NL-LS, with W as the response, and an initial estimate of $\sigma_{\varepsilon 1}$= 0.3430, gives initial estimates of

$$\alpha = 0.2231, \lambda = 1, \mu_2 = 5.268.$$

The log transformation (λ=1) is again the right transformation to apply to the response.

Maximizing (8.22) to derive estimates for $\{\sigma_{\varepsilon 1}, \sigma_{\varepsilon 2}, \rho\}$, we realize that for $\rho>0$, non-admissible (negative) estimates for $\sigma_{\varepsilon 1}$ are obtained. For $-0.3 \leq \rho < 0$, ML estimates of both $\sigma_{\varepsilon 1}$ and $\sigma_{\varepsilon 2}$ are non-negative, however the value of the log-likelihood function hardly changes and stays at the constant value of about LL*= 23.66. Therefore, an arbitrary value is selected of ρ=-0.1, to obtain $\sigma_{\varepsilon 1}$=0.1913, $\sigma_{\varepsilon 2}$= 0.04168. Inserting the updated value of $\sigma_{\varepsilon 1}$ into the estimation procedure for the RMM parameters, $\{\alpha, \lambda, \mu_2\}$, does not alter their estimates, and therefore no further iterations were required.

Table 8.7 displays predicted response values next to actual (observed) values. In terms of goodness-of-fit, the results suggest acceptable goodness-of-fit. Further comparisons, related to other major approaches (BC transformation and GLM), will be given in Chapter 14.

Some further relevant references dealing with the estimation procedures, may be found in Shore (2004abcd).

References

[1] Condra, L. W. (1993). *Reliability Improvement with Design of Experiments*. Marcel Dekker, NY.

[2] Hamada, M., Nelder, J. A. (1997). Generalized linear models for quality-improvement experiments. *Journal of Quality Technology, 29*(3), 292-304.

[3] Lin, L. I., Vonesh, E. F. (1989). An empirical nonlinear data-fitting approach for transforming data to normality. *American Statistician, 43*, 237-243.

[4] Myers, R. H., Montgomery, D. C. (1997). A tutorial on generalized linear models. *Journal of Quality Technology, 29*(3), 274-291.

[5] Myers, R. H. , Montgomery, D. C., Vining, G. G. (2002). *Generalized Linear Models, with Applications in Engineering and the Sciences*. John Wiley & Sons.

[6] Shore, H. (2001). Modeling a non-normal response for quality improvement. *International Journal of Production Research, 39* (17), 4049-4063.

[7] Shore, H. (2002). Modeling a response with self-generated and externally-generated sources of variation. *Quality Engineering, 14*(4), 563-578.

[8] Shore, H. (2004a). Response Modeling Methodology (RMM)- Current statistical distributions, transformations and approximations as special cases of RMM. *Communications in Statistics (Theory and Methods), 33*(7), 1491-1510.

[9] Shore, H. (2004b). Response Modeling Methodology (RMM)- Validating evidence from engineering and the sciences. *Quality and Reliability Engineering International, 20,* 61-79.

[10] Shore, H. (2004c). The Random Fatigue-Limit Model as a special case of the RMM model- a comment on Pascual (2003a). *Communications in Statistics (Simulation and Computation), 33*(2), 537-539.

[11] Shore, H. (2004d). Response Modeling Methodology (RMM)- Maximum likelihood estimation procedures. *Computational Statistics and Data Analysis.* In press.

[12] Wedderburn, R. W. M. (1974). Quasi-Likelihood functions, generalized linear models and the Gauss Newton method. *Biometrics, 44*, 121-130.

Appendix A - Canonical Correlation Analysis- Background

In this appendix we deliver a brief background on canonical correlation analysis, based largely on the SAS user guide. Canonical correlation is a technique for analyzing the relationship between two sets of variables, where each set can contain a different number of variables. As such, canonical correlation is a variation on the concept of multiple regression and correlation analysis. In multiple regression and correlation, you examine the relationship between a linear combination of a set of X variables and a single Y variable. In canonical correlation analysis, you examine the relationship between a linear combination of the set of X variables with a linear combination of a set of Y variables. Simple and multiple correlations are special cases of canonical correlation in which one or both sets contain a single variable.

Consider the situation in which you have a set of p X variables and q Y variables. The canonical correlation procedure finds the linear combinations

$$v_1 = a_1 x_1 + a_2 x_2 + \dots + a_p x_p,$$

$$w_1 = b_1 y_1 + b_2 y_2 + \dots \dots + b_q y_q,$$

such that the correlation between the two canonical variables, v_1 and w_1, is maximized. This correlation between the two canonical variables is the first canonical correlation. The coefficients of the linear combinations are canonical coefficients or canonical weights. It is customary to normalize the canonical coefficients so that each canonical variable has a variance of 1. The procedure continues by finding a second set of canonical variables, uncorrelated with the first pair, which produces the second highest correlation coefficient. The process of constructing canonical variables continues until the number of pairs of canonical variables equals the number of variables in the smaller group.

Each canonical variable is uncorrelated with all the other canonical variables of either set except for the one corresponding canonical variable in the opposite set. The canonical coefficients are not generally orthogonal, however, so the canonical variables do not represent jointly perpendicular directions through the space of the original variables. The first canonical correlation is at least as large as the multiple correlation between any variable and the opposite set of variables.

Finally, statistical tests are applied to test a series of hypotheses that each consecutive canonical correlation and all smaller canonical correlations are zero in the population. This implies that testing is conducted sequentially, where at each stage the largest canonical correlation is removed in order to test the statistical significance of all the rest (smaller) canonical correlations.

Appendix B - The Assumptions of CCA and Major Threats to the Reliability and Validity of Results

(1) *Multivariate Normality*

Tests of significance of the canonical correlations are based on the assumption that the distributions of the variables in each set are multivariate normal. Little is known about the effects of violations of this assumption. For a sufficiently large sample size, results from a CCA are quite robust to violation of the normality assumption.

(2) *Sample Size Sufficient for Reliable Estimates of Factor Loadings*

While relatively small samples (e.g., n=50) may detect strong canonical correlations in the data (e.g., $\rho>0.7$) most of the time, reliable estimates of factor loadings (needed for interpretation) require that certain proportions between sample size and the number of variables in the analysis be observed. Various recommendations regarding sample sizes are given in the literature.

(3) *Sensitivity to Outliers*

Outliers can greatly affect the magnitudes of correlation coefficients. Since CCA is based on (computed from) correlation coefficients, they can seriously affect the canonical correlations. Evidently, the larger the sample size, the smaller the impact of outliers. Use of scatter-plots may be instrumental in detecting influential outliers.

(4) *Ill-Conditioned Correlation Matrix*

A basic assumption of CCA is that the variables in the two sets are not completely redundant. When there are perfect correlations in the correlation matrix, or if any of the multiple correlations between one variable and the others is perfect ($\rho=1.0$), then the correlation matrix cannot be inverted, and CCA cannot be performed. In extreme cases, when the assumption of lack of redundancy is "almost" violated (namely, many variables *approach* redundancy), the statistical CCA procedure may issue a warning even when no exact redundancy has been obtained. It is good practice in implementing CCA, as it is in multiple linear regression analysis, to avoid extreme case of multicollinearity.

Chapter 9

The RMM Error Distribution

9.1. Introduction

Variation in the linear predictor (LP) provides the component of systematic variation in the response total variation. When the LP is constant, fluctuations in the response are random and their dispersion is described by the error distribution.

In this chapter the RMM error distribution is derived and its properties explored.

Setting the LP, η, to a constant value, the RMM basic model (7.6) expresses the relationship between the response and the two (possibly correlated) errors, ε_1 and ε_2. From this relationship, various formulae can be obtained, like the density function (d.f.) and the cumulative distribution function (CDF). These are then used to derive moments and other distributional properties, like the median and the mode.

In Section 9.2 basic expressions associated with the RMM error distribution are developed. In Section 9.3 we derive expressions for the moments and explore some other properties of the error distribution. Both sections relate to the basic model, given by (7.6) or (7.9). In Section 9.4 we introduce other variations of the basic RMM error distribution, corresponding to variations of the RMM basic relational model, developed in Chapter 7 [refer to (7.6abc), and to (7.12) and (7.13)]. The properties of these variations are left for the reader to explore.

In Chapters 10 and 11, fitting and estimation procedures, respectively, are developed. In Chapter 12 we show that the RMM error distribution, with its different versions, comprises as exact special cases

151

some well-known families of distributions, response transformations and distributional approximations. Chapter 19 demonstrates that the RMM error distribution may indeed serve as a good platform to model random variation. This is done by deriving accurate approximations to some commonly applied statistical distributions.

9.2. Derivation of the RMM Error Distribution

Without loss of generality, introduce in (7.9) and (7.10) for the linear predictor $\eta=1$. The response is expressed in terms of the errors, on the original scale (Y) or in the log-transformed scale (W), by

$$Y = \exp\{(\alpha/\lambda)[(1+\varepsilon_1)^\lambda - 1] + \mu_2 + \varepsilon_2\}$$

$$= \exp\{(\alpha/\lambda)[(1+\sigma_{\varepsilon1}Z_1)^\lambda - 1] + \mu_2 + \sigma_{\varepsilon2}[\rho Z_1 + (1-\rho^2)^{(1/2)} Z_2]\}, \quad (9.1)$$

$$W = \log(Y) = (\alpha/\lambda)[(1+\sigma_{\varepsilon1}Z_1)^\lambda - 1] + \mu_2 + \sigma_{\varepsilon2}[\rho Z_1 + (1-\rho^2)^{(1/2)} Z_2], \quad (9.2)$$

where $\varepsilon_1 = \sigma_{\varepsilon1}Z_1$ and $\varepsilon_2 = \sigma_{\varepsilon2}[\rho Z_1 + (1-\rho^2)^{(1/2)} Z_2]$ are two error terms from a bi-variate normal distribution with correlation ρ, zero means and standard deviations $\sigma_{\varepsilon1}$ and $\sigma_{\varepsilon2}$, respectively, and Z_1 and Z_2 are two non-correlated (independent) standard normal variables.

For $\rho=\pm1$, (9.1) reduces to the "origin" Inverse Normalizing Transformation (INT), developed and numerically demonstrated in Shore (2000). In that case, $\mu_2 = \log(M)$, where M is the median of Y. We will develop this transformation and some off-spring transformations in Chapter 20. These are applied in Chapter 22 to construct SPC schemes for non-normal process variables.

Some expressions associated with the error distribution will now be developed. In particular, we wish to derive the d.f. and the CDF, from which various distributional properties can be explored.

To develop the d.f., note from (9.2) that the conditional distribution of $W \mid Z_1=z_1$ has a normal distribution with mean and standard deviation given by, respectively,

$$\mu(W \mid z_1; \theta) = (\alpha/\lambda)[(1+\sigma_{\varepsilon1}z_1)^\lambda - 1] + \mu_2 + \sigma_{\varepsilon2}\rho z_1,$$

$$\sigma(W \mid z_1; \theta) = \sigma_{\varepsilon2}(1-\rho^2)^{(1/2)}, \quad (9.3)$$

where $\theta=\{\alpha, \lambda, \mu_2, \rho, \sigma_{\varepsilon1}, \sigma_{\varepsilon2}\}$. From this, the marginal (unconditional) d.f. of W is

$$f_W(w; \theta) = \int_{-\infty}^{\infty} [\sigma(W|z_1; \theta)]^{-1} \phi\{[w - \mu(W|z_1; \theta)]/[\sigma(W|z_1; \theta)]\}\phi(z_1)dz_1, \quad (9.4)$$

and $\phi(.)$ is the standard normal d.f. The marginal CDF of W is

$$F_W(w; \theta) = \int_{-\infty}^{\infty} \Phi\{[w-\mu(W|z_1; \theta)]/[\sigma(W|z_1; \theta)]\}\phi(z_1)dz_1, \quad (9.5)$$

and $\Phi(.)$ is the standard normal CDF.

The density function of Y, denote it by $f_Y(y;\theta)$, is, in terms of (9.4),

$$f_Y(y;\theta) = (1/y)\, f_W[\log(y); \theta]. \quad (9.6)$$

Similarly the distribution function, $F_Y(y;\theta)$, is

$$F_Y(y;\theta) = F_W[\log(y); \theta]. \quad (9.7)$$

For $\lambda=0$, $\alpha=1$, $\sigma_{\varepsilon2}=0$, (9.7) is the normal distribution, and for $\alpha=0$ we obtain the log-normal distribution.

The distribution (9.7) is the error distribution of the basic RMM model. Given the wide spectrum of scientific and engineering models, having an error distribution depicted by this model (refer to Chapter 2 and elsewhere), this is probably one of the most common distributions that one may find in the scientific investigation of nature. Yet it is highly unlikely that this distribution can be found in any textbook for students of engineering and the sciences.

The exact d.f and CDF of W require at least a single numerical integration. To obtain explicit expressions that eliminate one numerical integration, a Taylor expansion can be helpful. Re-writing (9.4) in terms of the conditional d.f of $W|\varepsilon_1$, we obtain

$$f_W(w;\theta) = \int_{-\infty}^{\infty} f_{W|\varepsilon1}(w|\varepsilon_1;\theta)\, f_{\varepsilon1}(\varepsilon_1)d\varepsilon_1$$

$$= \int_{-\infty}^{\infty} [\sigma_{\varepsilon2}(1-\rho^2)^{1/2}]^{-1}\phi\{[w_i - \mu(W|\varepsilon_1;\theta)]/[\sigma(W|\varepsilon_1;\theta)]\}f_{\varepsilon1}(\varepsilon_1)\,d\varepsilon_1, \quad (9.8)$$

where $f_{W|\varepsilon1}(w|\varepsilon_1;\theta)$ is the conditional d.f. of W, $f_{\varepsilon1}(\varepsilon_1)$ is the d.f of the normal error, ε_1, and

$$\mu(W|\varepsilon_1; \theta) = (\alpha/\lambda)[(1+\varepsilon_1)^{\lambda}-1]+\mu_2+\rho(\sigma_{\varepsilon2}/\sigma_{\varepsilon1})\varepsilon_1 = \mu(W \mid z_1; \theta)$$

$$\sigma(W|\varepsilon_1; \theta) = \sigma_{\varepsilon2}(1-\rho^2)^{(1/2)} = \sigma(W \mid z_1; \theta) \quad (9.9)$$

Developing $f_{W|\varepsilon_1}(w|\varepsilon_1; \theta)$, in terms of ε_1, into a Taylor series expansion around zero, and then applying the expectation operator to the first four terms (namely, including in the approximation terms up to a third degree), we obtain the relatively simple explicit approximate expression for the d.f of W ($\eta=1$)

$$f_W(w;\theta) = \int_{-\infty}^{\infty} f_{W|\varepsilon_1}(w|\varepsilon_1; \theta)\, f_{\varepsilon_1}(\varepsilon_1)d\varepsilon_1 \cong [2\pi(1-\rho^2)\sigma_{\varepsilon2}^2]^{-1/2}\, \exp[-(1/2)Z^2]$$

$$\times \{1+(1/2)\sigma_{\varepsilon1}^2[(1-\rho^2)^{1/2}\sigma_{\varepsilon2}]^{-2}[(\alpha+\rho\sigma_{\varepsilon2}/\sigma_{\varepsilon1})^2 Z^2-1)$$

$$+ \alpha(\lambda-1)(1-\rho^2)^{1/2}\sigma_{\varepsilon2}Z]\}, \tag{9.10}$$

where

$$Z = (w-\mu_2) / [(1-\rho^2)\sigma_{\varepsilon2}^2]^{1/2}, \tag{9.11}$$

and in taking expectation for the Taylor expansion we introduced

$$E(\varepsilon_1) = 0,\; E(\varepsilon_1^2) = \sigma_{\varepsilon1}^2,\; E(\varepsilon_1^3) = 0. \tag{9.12}$$

From (9.10), all moments of W may be easily calculated by

$$\mu'_r(W) = E(W^r)= \int_{-\infty}^{\infty} w^r f_W(w;\theta)dw. \tag{9.13}$$

Also, from (9.10), an approximate expression for the d.f. of Y may be derived via (9.6).

9.3. Properties of the Error Distribution

The properties of the RMM error distribution are explored. In particular, we investigate the shape versatility by studying the behavior of the RMM error distribution in the $(\sqrt{\beta_1},\beta_2)$ plane, where $\sqrt{\beta_1}$ and β_2-3 are measures of skewness and kurtosis, respectively [the third and fourth standardized cumulants; Refer to (9.26) below for their definitions].

Recall that in Chapter 6 we have related to adequate coverage of the $(\sqrt{\beta_1},\beta_2)$ plane as one of the most important requirements for the universal applicability of a methodology for empirical relational modeling (therein, Requirement 7). It is therefore natural that studying the coverage associated with the RMM error distribution is of high priority in order to assess its versatility and that of the RMM relational model.

We start by deriving expressions for the moments of either Y or W (the log-transformed response). If the parameters in (9.1) are specified, then the non-central r-th moment of Y is calculated numerically from

$$\mu'_r(Y) = E(Y^r) = \int_{-\infty}^{\infty} (y)^r \, f_Y(y)dy, \qquad (9.14)$$

where $f_Y(y)$ is given by (9.6). However, this method requires double numerical integration, which can be a numerically prohibitive task.

Alternatively, use can be made of the fact that Z_1 and Z_2 in (9.2) are *independent* to derive either *exact* or *approximate* explicit expressions for the moments of W and Y.

An exact method was detailed in Chapter 8 for the relational model. It is repeated here with $\eta=1$. Re-write (9.2) as

$$W = Q_1(Z_1; \omega_1) + Q_2(Z_2; \omega_2), \qquad (9.15)$$

where

$$Q_1(Z_1; \omega_1) = (\alpha/\lambda)[(1+\sigma_{\varepsilon 1}Z_1)^{\lambda} - 1] + \mu_2 + \sigma_{\varepsilon 2}\rho Z_1,$$

$$Q_2(Z_2; \omega_2) = \sigma_{\varepsilon 2}(1-\rho^2)^{(1/2)} Z_2, \qquad (9.16)$$

with, respectively, parameters' vectors

$$\omega_1 = \{\alpha, \lambda, \mu_2, \sigma_{\varepsilon 1}, \sigma_{\varepsilon 2}\rho\},$$

$$\omega_2 = \sigma_{\varepsilon 2}(1-\rho^2)^{(1/2)}. \qquad (9.17)$$

Note, that ω_1 contains five parameters while ω_2 has only one parameter.

Since Z_1 and Z_2 are independent random variables so are Q_1 and Q_2, and for the r-th non-central moment of W we obtain

$$E(W^r) = \mu_r{}'(W) = \sum_{j=0}^{r}\binom{r}{j} M_j(Q_1)M_{r-j}(Q_2), \qquad (9.18)$$

where $M_j(Q_1)$ and $M_{r-j}(Q_2)$ are the non-central j-th and (r-j)-th non-central moments of Q_1 and Q_2, respectively:

$$M_k(Q_1) = \int_{-\infty}^{\infty} \{(\alpha/\lambda)[(1+\sigma_{\varepsilon 1}z)^{\lambda} - 1] + \mu_2 + \sigma_{\varepsilon 2}\rho z\}^k \, \phi(z)dz, \qquad (9.19)$$

$$M_k(Q_2) = \int_{-\infty}^{\infty} [\sigma_{\varepsilon 2}(1-\rho^2)^{(1/2)} z]^k \, \phi(z)dz. \qquad (9.20)$$

From (9.18)-(9.20) all non-central moments of W can be obtained. Note that (9.20) can be written by

$$M_k(Q_2) = [\sigma_{\varepsilon 2}(1-\rho^2)^{(1/2)}]^k \, \mu_k(Z) \tag{9.20a}$$

where $\mu_k(Z) = E(Z^k)$ is the standard normal k-th moment.

Approximate explicit expressions for the non-central moments of Y can be developed, similarly to the derivation of approximate expressions for the first two moments of the response in the relational model in Chapter 7 (7.21).

To derive these moments, write (9.1) as a product of two functions, one in terms of Z_1 and the other in terms of Z_2:

$$Y = \exp\{(\alpha/\lambda)[(1+\sigma_{\varepsilon 1}Z_1)^\lambda - 1] + \mu_2 + \sigma_{\varepsilon 2}\rho Z_1\}$$

$$\text{x} \exp\{\sigma_{\varepsilon 2}(1-\rho^2)^{(1/2)} Z_2\} = g_1(Z_1;\omega_1) \, g_2(Z_2; \omega_2) \tag{9.21}$$

where $\omega_1 = \{\alpha, \lambda, \mu_2, \sigma_{\varepsilon 1}, \sigma_{\varepsilon 2}\rho\}$, $\omega_2 = \{\sigma_{\varepsilon 2}(1-\rho^2)^{(1/2)}\}$, and g_1 and g_2 are two functions of Z_1 and Z_2, respectively.

Since both functions are exponential expressions of the respective normal variables, and because of the statistical independence of Z_1 and Z_2, the r-th non-central moment of Y may be derived from the means of g_1 and of g_2, where in the expression for the mean of g_1 we replace $\{\alpha, \mu_2, \sigma_{\varepsilon 2}\}$ by $\{r\alpha, r\mu_2, r\sigma_{\varepsilon 2}\}$, and similarly in the expression for the mean of g_2 we replace $\sigma_{\varepsilon 2}$ by $r\sigma_{\varepsilon 2}$. The r-th non-central moment of Y is

$$E(Y^r) = E\{[g_1(Z_1;\omega_1)]^r\}E\{[g_2(Z_2; \omega_2)]^r\}$$

$$= E\{\exp\{(r\alpha/\lambda)[(1+\sigma_{\varepsilon 1}Z_1)^\lambda - 1] + r\mu_2 + r\sigma_{\varepsilon 2}\rho Z_1\}\}$$

$$\text{x} \, E\{\exp[r\sigma_{\varepsilon 2}(1-\rho^2)^{(1/2)} Z_2]\}. \tag{9.22}$$

The function g_2 is a log-normal variable, and we have for the r-th moment of g_2 (exact):

$$E\{[g_2(Z_2; \omega_2)]^r\} = \exp[(1-\rho^2)(r\sigma_{\varepsilon 2})^2/2]. \tag{9.23}$$

(refer to Johnson, Kotz and Balakrishnan, 1994, Chapter 14, or to Stuart and Ord, 1987, Section 6.30):

Expanding g_1 into a Taylor series in terms of powers of Z_1, around zero, and taking expectation of the first four terms, we obtain for the r-th moment of g_1

$$E\{[g_1(Z_1; \omega_1)]^r\} \cong \exp(r\mu_2)$$

$$\text{x} \{1+(1/2)[(r\alpha)(\lambda-1)\sigma_{\varepsilon 1}^2 + (r)^2(\alpha\sigma_{\varepsilon 1}+\rho\sigma_{\varepsilon 2})^2]\}. \tag{9.24}$$

From (9.23) and (9.24), an explicit approximate expression for the r-th non-central moment of Y is

$$E(Y^r) \cong \exp[r\mu_2 + (1-\rho^2)(r\sigma_{\varepsilon 2})^2/2)]$$

$$\times \{1 + (1/2)[(r\alpha)(\lambda-1)\sigma_{\varepsilon 1}^2 + (r)^2(\alpha\sigma_{\varepsilon 1} + \rho\sigma_{\varepsilon 2})^2]\}. \qquad (9.25)$$

From the non-central moments, the skewness and kurtosis measures can be calculated by (Stuart and Ord, 1987, p. 87)

$$\sqrt{\beta_1} = E[(Y-\mu)^3]/\sigma^3 = [\,\mu_3' - 3\,\mu_2'\,\mu_1' + 2\,(\mu_1')^3\,]\,/\,\sigma^3,$$

$$\beta_2 - 3 = E[(Y-\mu)^4]\,/\,\sigma^4 - 3$$

$$= [\mu_4' - 4\,\mu_3'\,\mu_1' + 6\,\mu_2'\,(\mu_1')^2 - 3\,(\mu_1')^4]\,/\,\sigma^4 - 3. \qquad (9.26)$$

For the normal distribution, both measures are zero.

Equations (9.25) and (9.26) can now be used to explore areas in the $(\sqrt{\beta_1}, \beta_2)$ plane where the RMM error distribution has a realization, and therefore may represent distributions sharing the same skewness and kurtosis values. Since $r\mu_2$ in (9.25) is a scale parameter, it does not influence either the skewness $(\sqrt{\beta_1})$ or the kurtosis (β_2) measures (both are scale invariant). The shape of the RMM error distribution is therefore determined only by the five parameters $\{\alpha, \lambda, \rho, \sigma_{\varepsilon 1}, \sigma_{\varepsilon 2}\}$.

Let the parameter λ in (9.2) be re-expressed as $\lambda = \lambda_0/\sigma_{\varepsilon 1}$. This allows using values of λ_0 that are not influenced by the scale of $\sigma_{\varepsilon 1}$ (further justification can be found in Shore, 2000).

For any arbitrarily selected values of the error standard deviations, $\{\sigma_{\varepsilon 1}, \sigma_{\varepsilon 2}\}$, we now wish to explore the effect that $\{\alpha, \lambda_0, \rho\}$ have on the shape of the resulting error distribution. Attempting to obtain plots of response surfaces for $\{\sqrt{\beta_1}, \beta_2\}$ in terms of any two of the above parameters proved to be computationally prohibitive. Therefore, values of $\{\sqrt{\beta_1}, \beta_2\}$ for selected values of $\{\alpha, \lambda_0\}$ are given for $\rho=0$, in Table 9.1, and for $\rho= 0.9$, in Table 9.2. The values in these tables may help assess how the various parameters influence shape characteristics of the RMM error distribution.

The error standard deviations ($\sigma_{\varepsilon 1}$ and $\sigma_{\varepsilon 2}$) were arbitrarily selected. However, we considered in selecting these values results obtained when the "origin" INT was fitted to some known distributions (Shore, 2000, and tables therein). The values selected are $\sigma_{\varepsilon 1}= 0.1$, $\sigma_{\varepsilon 2}= 0.05$. It is

Table 9.1. Values of $\sqrt{\beta_1}$ (upper values) and β_2 for selected $\{\alpha, \lambda_0\}$, $\rho=0$

λ_0	α				
	−0.6	−0.2	0.2	0.6	1.0
−1.0	276.9	56.55	−2.135	−2.422	−2.031
	90337	6054	15.56	9.829	5.639
−0.6	7.425	0.3948	−0.8331	−1.372	−1.362
	170.9	13.74	10.89	5.306	3.531
−0.2	0.6296	−0.5023	−0.6385	−0.4376	−0.3918
	4.135	12.93	13.30	3.024	0.9473
0.2	−0.1588	−0.6206	−0.5896	0.1166	0.4742
	3.187	13.86	13.82	3.304	1.257
0.6	−0.6731	−0.6567	−0.4197	1.083	1.975
	3.121	12.58	12.45	5.620	9.089
1.0	−1.440	−0.8360	0.3566	4.974	11.17
	4.936	10.31	11.77	64.43	300.3

Table 9.2. Values of $\sqrt{\beta_1}$ (upper values) and β_2 for selected $\{\alpha, \lambda_0\}$, $\rho=0.9$

λ_0	α				
	−0.6	−0.2	0.2	0.6	1.0
−1.0	297.2	65.97	−1.326	−1.564	−1.389
	104354	7763	5.475	4.208	2.616
−0.6	15.25	−0.2718	−0.5111	−0.7926	−0.7892
	529.6	28.93	2.222	1.6316	1.160
−0.2	−10.93	−0.5842	−0.1232	−0.1087	−0.06455
	616.1	23.20	1.863	0.4309	0.1301
0.2	−29.57	−1.024	0.1560	0.4710	0.6733
	1921	22.00	2.002	0.8398	0.9625
0.6	−7.771	−1.612	0.5202	1.322	1.944
	233.7	25.68	2.531	4.053	8.062
1.0	−4.605	−2.192	1.306	4.280	10.12
	39.46	33.02	6.120	46.76	258.1

currently not completely clear how these parameters affect the shape parameters (skewness and kurtosis). It is, however, to be expected that as the error $\varepsilon_1 = \sigma_{\varepsilon 1} Z_1$ becomes small, namely, $\varepsilon_1 << 1$, $W = \log(Y)$ approaches normality. This property of the RMM error distribution may help explain why the log transformation is often used successfully to normalize observational data.

Examination of Tables 9.1 and 9.2 reveals that the new distribution is flexible in its capability to model differently shaped distributions. This property is further demonstrated in Chapter 19, where the RMM error distribution is fitted to various commonly applied distributions.

Further details regarding properties of the RMM error distribution can be found in Shore (2002).

9.4. Variations of the RMM Error Distribution

In Section 7.2.5 we related to the fact that there are altogether 8 variations of the RMM model (including the basic model that was axiomatically developed in Chapter 7). Introducing $\eta=1$ for these models, the corresponding 8 variations of the RMM error quantile function can be derived.

Model 1. The basic model (9.1):

$$Y = \exp\{(\alpha/\lambda)[(1+\varepsilon_1)^\lambda - 1] + \mu_2 + \varepsilon_2\}.$$

Model 2. Different presentation of ε_2:

$$Y = \exp\{(\alpha/\lambda)[(1+\varepsilon_1)^\lambda - 1] + \mu_2\}(1+\varepsilon_2). \tag{9.1a}$$

Model 3. Different presentation of ε_1:

$$Y = \exp\{(\alpha/\lambda)[\exp(\lambda\varepsilon_1) - 1] + \mu_2 + \varepsilon_2\}. \tag{9.1b}$$

Model 4. Different presentation of both ε_1 and ε_2:

$$Y = \exp\{(\alpha/\lambda)[\exp(\lambda\varepsilon_1) - 1] + \mu_2\}(1 + \varepsilon_2). \tag{9.1c}$$

For each model, introduce

Either for ε_2:

$$\varepsilon_2 = \sigma_{\varepsilon 2}[\rho Z_1 + (1-\rho^2)^{(1/2)} Z_2], \qquad\qquad (9.27)$$

where $\varepsilon_1 = \sigma_{\varepsilon 1}Z_1$, and Z_1 and Z_2 are uncorrelated.

or for ε_1:

$$\varepsilon_1 = \sigma_{\varepsilon 1}[\rho Z_2 + (1-\rho^2)^{(1/2)} Z_1], \qquad\qquad (9.28)$$

where $\varepsilon_2 = \sigma_{\varepsilon 2}Z_2$, and Z_1 and Z_2 are uncorrelated.

These variations extend the spectrum of real-life scenarios that RMM can model. In particular, various current models of random variation are shown in Section 12.2 to be special cases of some of these variations.

References

[1] Johnson, N. L., Kotz, S., Balakrishnan, N. (1994). *Continuous Univariate Distributions — Volume 1.* 2nd edition. John Wiley & Sons.

[2] Shore, H. (2000). General control charts for variables. *International Journal of Production Research*, 38(8), 1875-1897.

[3] Shore, H. (2002). Response Modeling Methodology (RMM)- Exploring the implied error distribution. *Communications in Statistics (Theory and Methods)*, 31(12), 2225-2249.

[4] Stuart, A., Ord, J. K. (1987). *Kendall's Advanced Theory of Statistics. V. 1: Distribution Theory.* Charles Griffin & Company, Ltd., London.

Chapter 10

Fitting Procedures (for the Error Distribution)

10.1. Introduction

A fitting procedure assigns values to the parameters of a *fitted distribution* so that it delivers the best possible representation for the *approximated distribution*. For example, in statistical process control, the p control chart (most often used to monitor the percentage non-conformance of a process) is the result of fitting a normal distribution to the binomial distribution. The former is the *fitted* distribution, which is the platform on which all of Shewhart control charts are based. The latter (the binomial) is the true distribution of the monitoring statistic, that is, the percentage observed in a sample of items.

The use of the normal distribution in all of Shewhart control charts provides a uniformity of practice that enables practitioners to use the same routine for constructing control charts even though they use monitoring statistics with differently shaped distributions, like the Poisson, the binomial or some continuous distributions that are not extremely asymmetrical (skewed).

In other scenarios, one may wish to use distribution fitting because the true underlying distribution is unknown and there is a need to use one that would represent well the unknown distribution. It is common practice in such cases to approximate the latter by fitting a parameter-rich distribution, which is flexible enough to represent a wide spectrum of diversely-shaped distributions. This is the case, for example, with Clements' method (1989) for process capability analysis, which attempts to represent the true (often unknown) process distribution by a member

161

of the Pearson family of distributions.

Occasionally, a stochastic optimization model is developed that its developer wishes it to be valid for a variety of distributional scenarios, and not be restricted by certain distributional assumptions. A good example for an application area in this case is stochastic inventory-analysis models. Here, one may wish to formulate an optimal policy for inventory management that remains valid irrespective of the distribution related to this policy, for example, the true lead-time demand distribution. Employing a flexibly shaped distribution to represent the latter, plus a routine for distribution fitting, may provide the necessary platform for formulating distribution-independent inventory policy guidelines (refer, for example, to Shore, 1986, for a demonstration of this approach).

Empirical modeling of random variation is closely linked to distribution fitting. In the former, we wish to identify a model of random variation (that is, some distribution) that will deliver good representation to the true *unknown* distribution (if the distribution was known no empirical modeling would have been needed). In the latter, we wish to replace a certain distribution by another (the fitted distribution) of our own choice, irrespective of whether the distribution that we wish to replace is known or otherwise. Thus, one may say that *empirical modeling of random variation* is a subset of the general approach of *distribution fitting*. A major difference between the two is that when the distribution that we wish to approximate is known, no sample data is needed. Otherwise, that is, in empirical modeling of random variation, available sample data are used, plus a selected parameter-rich distribution, to achieve good fit to an unknown statistical distribution. Some well-defined statistical inference criteria may assist us in determining the quality of the fit.

In this chapter distribution-fitting procedures for the RMM error distribution are developed, assuming that the distribution that we wish to approximate is completely specified (both the distribution and its parameters are known). In other words, we wish to find parameters for the RMM error-distribution that deliver good representation to a completely specified distribution. Application of the developed fitting routines to some known distributions is demonstrated, and we examine the resultant goodness-of-fit.

A more comprehensive study will be conducted in Chapter 19, where we use the RMM error distribution to approximate a large sample of commonly applied distributions, and assess the accuracy of the approximations obtained.

A major implication of the results of these two chapters is that if the RMM error distribution delivers good representation to a wide spectrum of differently shaped distributions in common use, then most probably it may provide a good platform for an empirical modeling of random variation, namely, for distribution fitting when the underlying distribution is unspecified.

In Chapter 11 estimation procedures for the RMM error distribution are developed.

10.2. Brief Review of Current Methodologies

In Chapter 5 we discussed current general methodologies for empirical modeling of *random* variation (or, more generally, for distribution fitting). We introduced a distinction between two approaches:

- The "Moment Matching" approach, where parameters of the fitted distribution are assigned values so that matching of the first few moments is achieved.
- The "Quantile Matching" approach (referred to also as "Percentile-Matching"). With this approach, the parameters' values are determined so that an approximate one-to-one transformation is obtained that allows calculating the quantile of one distribution, given the corresponding quantile of the other distribution. This is achieved by selecting a sample of quantiles from both distributions, and fitting a specified quantile function based on this sample.

Comment. The 0.90 quantile of a distribution is its 90-th percentile. For example, the median is the 50-th percentile.

As an example, consider the Johnson S_L family of distributions

$$z = \gamma + \eta \log(y-\varepsilon),\ \eta > 0,\ -\infty < \gamma,\varepsilon < \infty,\ y > \varepsilon, \qquad (10.1)$$

where $\{\gamma, \eta, \varepsilon\}$ are real-valued parameters that need to be determined. This quantile-function transforms the quantile of a random variable, Y, into the corresponding quantile of the standard normal variable, Z, where $F(y)=\Phi(z)$, and F and Φ are the respective cumulative distribution functions (CDFs) of Y and Z.

Suppose that we wish to fit a Johnson S_L member to the standard normal distribution. Equation (10.1) may be fitted via a moment-matching procedure (the exact way to do so is not detailed here). Since (10.1) has only three parameters, this implies that matching with the standard normal distribution of only the first three moments can be obtained.

Alternatively, one may wish to identify the "best" parameters for (10.1) via percentile matching. Using a sample of pairs of percentiles from both distributions, values of the parameters of (10.1) can be identified, which achieve the closest matching between the true percentiles of Z to those modeled by the fitted RHS of (10.1). A fitting routine most often employed is non-linear least-squares (NL-LS), however other fitting procedures also exist. (For the Johnson family, find details, for example, in Chou, Polansky and Mason, 1998.)

Suppose that (10.1) was fitted to a given distribution. As a result of the fitting procedure, the equality $F(y)=\Phi(z)$ is desired for all possible pairs of quantiles, $\{y,z\}$, that satisfy the fitted (10.1). If that equality was exact for all pairs, (10.1) would be an exact representation for the quantile relationship between Y and Z. Since the true distribution of Y is not governed by (10.1) (otherwise no fitting would be needed), the above equality holds only approximately.

The more parameters a fitted distribution has, the better its flexibility in representing a wide spectrum of differently shaped distributions. This assertion is true irrespective of which approach we choose for the fitting procedure. Also, coverage in the $(\sqrt{\beta_1},\beta_2)$ plane, an issue that we have related to in Section 9.3, is closely linked to the choice of the fitted distribution: A distribution with good coverage properties will also be versatile enough to deliver good representation to differently shaped distributions.

An Example

To provide a simple demonstration for both approaches, "Moment Matching" and "Percentile Matching", suppose that we wish to fit a normal distribution to a given two-parameter exponential distribution. In practice, this is not a particularly good idea since the normal is a symmetrical distribution while the exponential is extremely skewed. Nevertheless, we proceed with this example due to its simplicity and also because both distributions are well recognized.

Let Y be a random variable having a two-parameter exponential distribution with CDF

$$F(y) = 1 - \exp[-\lambda(y\text{-}a)], \; y > a, \tag{10.2}$$

and known parameters $\{a,\lambda\}$. It is easy to show that the mean is $(1/\lambda\text{-}a)$ and the variance is $1/\lambda^2$. We approximate this distribution by fitting a normal distribution with parameters $\{\mu, \sigma\}$.

Pursuing the *moment matching* approach, the fitting procedure results in the following parameters for the normal distribution (matching of the mean and the variance):

$$\sigma^2 = 1/\lambda^2 , \; \mu = (1/\lambda\text{-}a). \tag{10.3}$$

By the *percentile matching* approach, we wish to identify values for the parameters of the normal distribution (the fitted distribution) that will allow us to obtain good approximation to quantiles of the exponential distribution. The quantile relationship between the two distributions is given by

$$\Phi(z_p) = F(y_p) = 1 - \exp[-\lambda(y_p\text{-}a)] = P, \tag{10.4}$$

where $z_p = (x_p\text{-}\mu)/\sigma$, and x_p is the P-th quantile of the r.v. X, which has a normal distribution with parameters $\{\mu,\sigma\}$ (the distribution that we wish to fit).

To implement quantile matching, we derive a sample of pairs of corresponding quantiles from the two distributions, namely, quantiles that have the same value for their respective CDF. In our case, we derive pairs of quantiles, $\{z_p, y_p\}$, where z_p and y_p are the 100P-th percentiles of the standard normal and the exponential distributions, respectively. Obtaining these quantile values is easy: First determine a value of y. Then calculate, from the *known* exponential distribution, its CDF value,

P [use (10.2)]. Finally, find the corresponding standard normal quantile. Our model for the quantile of the exponential distribution (the approximated distribution), expressed in terms of the fitted normal distribution (the approximating distribution), is

$$y_p \cong x_p = \mu + \sigma z_p. \tag{10.5}$$

Linear regression may be used to find values of μ and σ that provide the best fit.

As alluded to earlier, in this example the fit is not expected to be good since the two distributions have widely different shape characteristics that would make prediction of y_p from z_p very poor. However, if the expression on the right is non-linear, then perhaps good fit may be attained. In Chapter 20, inverse normalizing transformations, a spin-off from the error distribution of the RMM model, are introduced. We demonstrate therein that if a certain *non-linear* expression of z_p is introduced on the RHS of (10.5) (instead of the *linear* function), then a quantile-matching procedure may be applied, similar to the simplified version introduced earlier, that would provide good fit even for the exponential distribution. Evidently, non-linear regression will replace the linear regression used in the simplified example above.

Having clarified the two main approaches to fit a distribution, we proceed to develop fitting procedures for the RMM error distribution.

10.3. Fitting via "Moment Matching"

Moment matching comprises solving a set of m equations, where on the left-hand-side of equation j (j=1,2,..,m) there is an expression, in terms of the distribution's parameters, for the j-th moment of the fitted distribution, and on the right-hand-side there is the value of the j-th moment of the specified distribution, which we wish to approximate. To obtain a unique solution, the fitted distribution needs to have at least m parameters. Solving for the m parameters assures us that the fitted distribution has the first m moments identical to those of the distribution that we wish to replace.

When the moments of the fitted distribution can be expressed in an explicit form (say, with no need for numerical integration), the fitting procedure involves solution of m (probably non-linear) equations via some root-finding routine. In other cases, no explicit closed-form equations for the moments, in terms of the distribution's parameters, are available. For example, some or all of the m equations may require numerical integration. In these cases, resorting to some more complicated numerical routines will be needed, possibly involving some approximations.

In Chapter 7 exact expressions to calculate RMM moments were developed. Since no closed-form formulae were obtainable, approximate expressions were derived, based on a Taylor expansion of the g_1 function (7.18). In Chapter 9 we have similarly derived an explicit approximate expression for the d.f. of the log-transformed response, W, based on a Taylor series expansion of the conditional d.f. of W (9.10). Both equations may be used in a moment-matching procedure. However, employing the approximate d.f still requires numerical integration (to calculate the moments), coupled with a root-identifying routine. These methods are optional and will not be pursued in this chapter (other methods that are numerically easier to implement are developed).

Another problem encountered in attempting to fit the RMM error distribution is the relatively large number of parameters in the model. In fact, to fit the RMM error distribution we need to determine the values of six parameters, $\{\alpha, \lambda, \mu_2, \rho, \sigma_{\varepsilon 1}, \sigma_{\varepsilon 2}\}$. This may look like a blessing since it implies that m=6, and we may fit the RMM error distribution to any distribution, achieving matching of the first *six* moments (at most).

However, common practice, which indeed has been corroborated in a comprehensive study by Pearson, Johnson and Burr (1979), mandates that matching of the first four moments in most cases suffice to achieve good representation for one distribution by means of another. This implies that if two of the RMM parameters are given definite values outside of the fitting procedure, no appreciable loss of accuracy should be expected.

Consider the RMM quantile function of the basic model, as given, for example, in (7.9) and (7.10). Re-introduced here for convenience (with $\eta=1$), we obtain

$$Y = \exp\{(\alpha/\lambda)[(1+\sigma_{\varepsilon1}Z_1)^\lambda - 1] + \mu_2 + \sigma_{\varepsilon2}[\rho Z_1+(1-\rho^2)^{(1/2)} Z_2]\}, \qquad (10.6)$$

$$W = \log(Y) = (\alpha/\lambda)[(1+\sigma_{\varepsilon1}Z_1)^\lambda-1] + \mu_2 + \sigma_{\varepsilon2}[\rho Z_1+(1-\rho^2)^{(1/2)} Z_2]. \qquad (10.7)$$

It is easy to realize that introducing $\rho=\pm1$ achieves three desirable goals. First, we eliminate one parameter (ρ) from the set of parameters that need to be determined. Secondly, Z_2 disappears from the quantile function. This turns the RMM quantile function into an Inverse Normalizing Transformation (INT), namely, an expression that delivers the quantile of the RMM error distribution in terms of the quantile of a single r.v., the standard normal Z_1. As a result, the parameter μ_2 now represents the log of the response median, M, that is, $\mu_2= \log(M)$ (for $Z_1=0$, $Y=M$). The parameter μ_2 is determined also outside the fitting procedure, and we are left with only four parameters that need to be determined by the fitting procedure.

The INT and various off-spring INTs will be addressed in detail in Chapter 20. Furthermore, fitting procedures for the INTs will be developed.

In this section, we develop a moment-matching procedure that delivers parameters' values for the non-reduced RMM error distribution [namely, for all the parameters in (10.6) and (10.7)]. The following procedure for moment-matching is an adaptation of the procedure introduced in Section 9.3 to calculate moments for the RMM error distribution.

Assume that the response, Y, has a known statistical distribution with density function $g(y)$ and distribution function $G(y)$. It is easy to show that $W= \log(Y)$, has density function

$$f_W(w) = \exp(w)\, g[\exp(w)]. \qquad (10.8)$$

Suppose that one wishes to represent W via the RMM model. For the standardized W (denote it by WS), we obtain from (10.7)

$$WS = (W-\mu) / \sigma =\{(\alpha/\lambda)[(1+\sigma_{\varepsilon1}Z_1)^\lambda - 1] + \mu_2 + \sigma_{\varepsilon2}\rho Z_1 - \mu\} / \sigma$$

$$+ [\sigma_{\varepsilon2}(1-\rho^2)^{(1/2)} /\sigma]Z_2$$

$$= QS_1(Z_1; \omega_1) + QS_2(Z_2; \omega_2), \qquad (10.9)$$

where μ and σ are the given (exact) mean and standard deviation, respectively, of W, QS_1 and QS_2 are separate expressions in terms of Z_1

and Z_2, respectively, and $\omega_1 = \{\alpha, \lambda, \mu_2, \sigma_{\varepsilon1}, \sigma_{\varepsilon2}\rho\}$, $\omega_2 = \{\sigma_{\varepsilon2}(1-\rho^2)^{(1/2)}\}$, are the parameters that need to be determined (six parameters altogether). Note that QS_1 and QS_2 differ from Q_1 and Q_2, respectively, given in (9.16).

Since Z_1 and Z_2 are independent standard normal variables, the r-th moment of WS is

$$M_r(WS) = E[(WS)^r] = \sum_{j=0}^{r} \binom{r}{j} M_j(QS_1) M_{r-j}(QS_2), \quad (10.10)$$

where $M_j(QS_1)$ and $M_{r-j}(QS_2)$ are the non-central j-th and (r-j)-th moments of QS_1 and of QS_2, respectively. For the latter we have

$$M_k(QS_1) = E[(QS_1)^k]$$

$$= \int_{-\infty}^{\infty} \{\{(\alpha/\lambda)[(\eta+\sigma_{\varepsilon1}z)^\lambda - 1] + \mu_2 + \sigma_{\varepsilon2}\rho z - \mu\}/\sigma\}^k \, \phi(z)dz, \quad (10.11)$$

$$M_k(QS_2) = \int_{-\infty}^{\infty} [(\sigma_{\varepsilon2}/\sigma)(1-\rho^2)^{(1/2)} z]^k \, \phi(z)dz$$

$$= [(\sigma_{\varepsilon2}/\sigma)(1-\rho^2)^{(1/2)}]^k \, \mu_k(Z), \quad (10.12)$$

where $\mu_k(Z)$ is the k-th moment of the standard normal distribution. For k odd, we have identically $\mu_k(Z)=0$. For k even, we have for the first few moments

$$\mu_2(Z) = E(Z^2) = 1, \ \mu_4(Z) = 3, \ \mu_6(Z) = 15, \ \mu_8(Z) = 105.$$

This implies that to calculate a moment of any desirable degree via (10.10) only numerical integration of (10.11) is needed.

To identify the parameter set, $\{\omega_1, \omega_2\}$, the following routine needs to be applied:

Find $\{\omega_1, \omega_2\}$ that minimize the objective function

$$OF = [M_1(WS)]^2 + \sum_{2}^{6} [M_r(WS)/\mu_r(WS) - 1]^2, \quad (10.13)$$

where $\mu_r(WS) = E[(WS)^r]$ is the given *exact* r-th moment of WS (the standardized W), namely, the r-th moment associated with the distribution that we wish to represent by the RMM model (these moments may be derived from available explicit expressions for the moments, or numerically calculated with the exact density function).

Note that since WS is the standardized variable, $M_1(WS)$ and $M_2(WS)$ are expected to be zero and 1, respectively. Also note that since there are six parameters, the sum extends over the first six moments in order to obtain a unique solution. Matching a smaller number of moments, the obtained solution may not be unique.

Alternatively, to match only the first four moments, introduce $\rho=\pm1$, $\mu_2= \log(M)$ (where M is the median of the response in the original scale). Forcing these parameters' values reduce the number of parameters which the minimization procedure needs to determine to only four.

Once the parameters of the fitted RMM quantile function have been identified, the resulting goodness-of-fit can be assessed by various measures. Three types of measures can be defined:

- Measures based on a comparison (exact versus fitted) of the first few moments (comparison of the first four moments would suffice for most applications)

- Measures related to errors in predicting quantile values from the fitted quantile function. A desirable P-th quantile, y_P, can be derived from the fitted RMM quantile function by solving for y_P:

$$P = F_Y(y_P), \qquad (10.14)$$

where F_Y is the RMM's CDF, given in (9.5) and (9.7).

If it is decided to forego one degree of freedom by setting $\rho=\pm1$, then a quantile may be easily calculated from (10.6) by following the path:

$$\text{Determine P --> Find } z_P \text{ --> Find } y_P \text{ from (10.6),} \qquad (10.15)$$

where z_P is the 100P-th percentile of the standard normal distribution.

- Traditional criteria for goodness-of-fit, like correlation, standard deviation of the errors, or "Minimum distance measures" to assess distributional fit (refer, for example, to Warick, 2004).

In addition to these goodness-of-fit criteria, various error plots may be employed to demonstrate the resultant goodness-of-fit. For example, a plot of the error of the respective density functions (fitted vs. exact), or a plot of the error of the quantile values. All comparisons may be made either with regard to the original scale (Y) or to the log-transformed scale (W).

For a good source relating to comparisons of fitted models refer to Burnham and Anderson (2002).

10.4. Fitting via "Quantile Matching"

Consider the response, Y, with a specified CDF, G(y). From this distribution, n quantiles are drawn with their corresponding CDF values, namely, the sample consists of $\{y_i, G(y_i)\}$ (i=1,2,..,n).

Values for the parameters of the RMM quantile function (10.6) are desired so that the objective function

$$OF = \sum_{i=1}^{n} (y_i - y_{fi})^2 , \qquad (10.16)$$

be minimized, where $\{y_{fi}\}$ is the set of the corresponding quantile values, calculated from the RMM quantile function (10.6).

Observing (10.6), it is easy to realize that there is no one-to-one correspondence between quantiles of the standard normal variables, Z_1 and Z_2, and the quantile value of Y. Although distribution fitting procedures for a bi-variate quantile function may be conceived, we are unaware of an existing quantile-matching method, where the fitted quantile function comprises two random variables [as in (10.6)].

In fact, a quantile function with two random variables is best suited for fitting to a bi-variate distribution, as implemented, for example, with the bi-variate generalized Lambda distribution in Karian and Dudewicz (2000). It is therefore preferable to obtain an RMM quantile function with a *single* r.v, which can then be fitted to a distribution with a *single* random variable.

Taking the expected value, with respect to Z_2, of the log-transformed response W= log(Y), we obtain from (10.7):

$$W = (\alpha/\lambda)[(1+\sigma_{\varepsilon 1}Z)^{\lambda} - 1] + \mu_2 + \sigma_{\varepsilon 2}\rho Z. \qquad (10.17)$$

This is a quantile function with a single standard normal r.v., Z, which can now be fitted via a "Quantile-Matching" procedure. The same result would have been obtained by setting, in (10.7), $\rho= \pm 1$. Note, that

(10.17) implies that the coefficients of Z, as obtained from the fitting procedure, may have different signs (that is, if $\rho=-1$).

An equivalent form for (10.7) is [refer to (9.28)]

$$W = \log(Y) = (\alpha/\lambda)\{[1+\sigma_{\varepsilon 1}[\rho Z_2+(1-\rho^2)^{(1/2)} Z_1]]^{\lambda}-1\}+\mu_2+\sigma_{\varepsilon 2}Z_2. \quad (10.18)$$

Introducing $\rho= \pm 1$, we obtain, corresponding to (10.17),

$$W= \log(Y)= (\alpha/\lambda)[(1+\sigma_{\varepsilon 1}\rho Z)^{\lambda}-1] + \mu_2 + \sigma_{\varepsilon 2}Z. \quad (10.19)$$

Like with (10.17), this again implies that as a result of distribution fitting the coefficients of Z may have different signs (that is, if $\rho=-1$). Since in practice one cannot tell whether (10.17) or (10.19) is fitted (they are equivalent), it can be expected that either of the coefficients of Z (but not both) may turn out to be negative. Furthermore, if required to provide initial parameter values for the fitting procedure, negative values for the coefficients of Z should not be ruled out.

We are now ready to implement a "Quantile Matching" procedure. The quantile function [(10.17) or (10.19)] has only five parameters to be determined. To preserve the median of the approximated distribution, we may wish to set $\mu_2= \log(M)$, where M is the (exact) median. This leaves us with four parameters that need to be determined $\{\alpha, \lambda, \pm\sigma_{\varepsilon 1}, \pm\sigma_{\varepsilon 2}\}$. The number of degrees of freedom is large enough to preserve the flexibility needed in representing a wide spectrum of differently shaped distributions.

A NL-LS procedure may be applied to fit (10.17), or (10.19), to the given set of n pairs of quantile values, $\{z_i, \log(y_i)\}$, where

$$P_i = F_W[\log(y_i)] = \Phi(z_i) , \; i= 1,2,..,n, \quad (10.20)$$

and F_W is the true (exact) CDF of $W= \log(Y)$, namely,

$$F_W[\log(y)] = G(y). \quad (10.21)$$

Derivation of the set of quantiles, to be used in the fitting procedure, is described by the following path:

$$\text{Select } y_P \rightarrow \text{Find CDF value } P=G(y_P) \rightarrow \text{Identify } z_P \quad (10.22)$$

where z_P is the P-th quantile of the standard normal distribution.

Section 10.5 demonstrates implementation of the two fitting methods to a gamma distribution. In Chapter 19 the two examples given here are expanded to a wider sample of distributions.

10.5. Two Numerical Examples

10.5.1. A moment-matching example

The moment-matching fitting procedure, expounded in Section 10.3, is used to fit the RMM error distribution to the gamma(3,2) distribution. The latter has, for the first four moments (mean, variance, skewness and kurtosis) and the log median (LMed)

$$\mu = 6.00,\ \sigma^2 = 12.00,\ Sk = 1.155,\ Ku = 2.00,\ LMed = 1.67675.$$

Note that the skewness and kurtosis measures are defined by

$$Sk = \mu_3 / \sigma^3,\ Ku = \mu_4 / \sigma^4 - 3. \tag{10.23}$$

For the log-transformed gamma, $\log(Y)$, we obtain by numerical integration the exact first six standardized moments, given in Table 10.1.

Implementing the minimization routine (10.13), the following parameters' values are derived:

$$\alpha = 31.115,\ \lambda_0 = \lambda\sigma_{\varepsilon1} = -0.233097,\ \mu_2 = 1.6882;$$

$$\sigma_{\varepsilon1} = 0.018305,\ \sigma_{\varepsilon2} = 0.25888,\ \rho = -0.10503. \tag{10.24}$$

(The λ in (10.7) was replaced by $\lambda_0/\sigma_{\varepsilon1}$ in order to ensure that small values will be obtained.)

It is of interest to note ρ. Its value was found in the bounded range $(-1,1)$, where its value is expected to be, even though no restriction was imposed on the search routine.

Table 10.1 compares the first six standardized moments of the log of the gamma variable (exact) with those calculated from the fitted quantile function (10.7). Deviations do not exceed half a percent even though very high-degree moments are compared.

Table 10.1. First six standardized moments of log gamma with parameters $\alpha=3$, $\beta=2$.

Degree of Moment	μ_k (WS) (exact)	M_k (WS) (fitted)	Deviation (%)
k=1	0	-0.0010	--
k=2	1	0.99511	-0.49
k=3	-0.62077	-0.62069	-0.012
k=4	3.7607	3.7805	0.52
k=5	-7.5823	-7.5881	0.077
k=6	33.477	33.388	-0.27

10.5.2. A quantile-matching example

The quantile-based fitting procedure, expounded in Section 10.4, is used to fit the gamma(3,2) distribution. The moments of this distribution were given in the previous example.

To carry out the fitting procedure, a sample of quantile values is selected from the standard normal distribution

$$\{z_i\} = \{-4+(0.1)i\}, \; i = 1,2,.., 71,$$

with corresponding quantile values from the gamma distribution [This routine goes in the opposite direction to that given in (10.22); Both directions are equally valid].

Imposing $\mu_2 = \log(M) = LMed = 1.6767$, and applying a NL-LS procedure [using either of the equivalent forms (10.17) or (10.19)], the parameters obtained are

$$\alpha = 131.6, \; \lambda_0 = -0.2563, \; \sigma_{\varepsilon 1} = 0.00350, \; (\sigma_{\varepsilon 2})\rho = 0.1468 \quad (10.25)$$

The residual variance from this fitting is $\sigma_{res.}^2 = 2.17582(10)^{-6}$. Equating the residual variance to the variance of ε_2 [refer to (10.7)]

$$(\sigma_{\varepsilon 2})^2(1-\rho^2) = 2.17582(10)^{-6}, \quad\quad\quad (10.26)$$

we obtain from the last expression in (10.25) and from (10.26):

$$\rho = 0.99995, \; \sigma_{\varepsilon 2} = 0.1485 \quad\quad\quad (10.27)$$

Note that in the fitted quantile function we have assumed $\rho = \pm 1$. It comes therefore as no surprise that this is the value obtained from the fitting procedure. Also note the small value of the residual variance [$2.2(10)^{-6}$]. This is a typical situation encountered with many distributions, as the sample of distributions, fitted by quantile-matching in Chapter 19, attests.

Employing these values, it is now possible to calculate moments, the d.f. and quantile values, and compare with exact values, either in the original scale or in the transformed scale. Plots of the d.f. and the residuals (on the original scale), and of the quantile errors (on the original scale, standardized), where fitting was performed on the log-transformed scale, are given below in Figures 10.1 and 10.2, respectively.

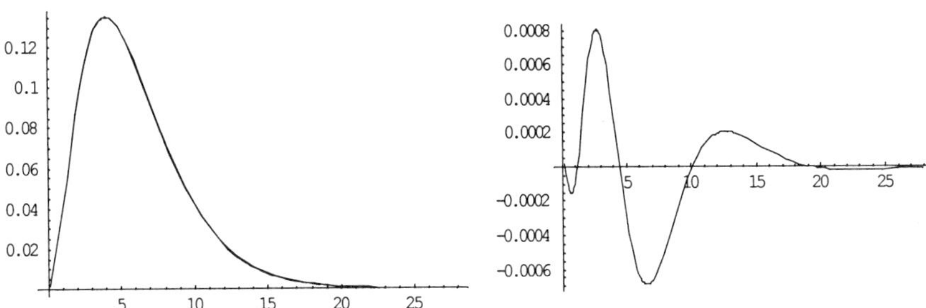

Figure 10.1. Plots of the density function, exact and fitted (left) and of the error for Gamma (3,2) (based on quantile-matching).

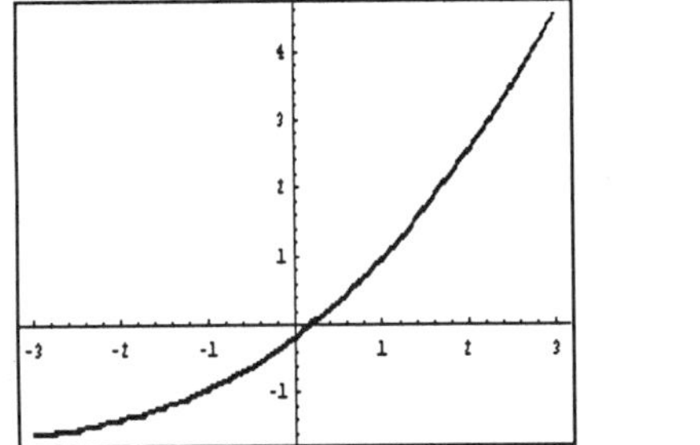
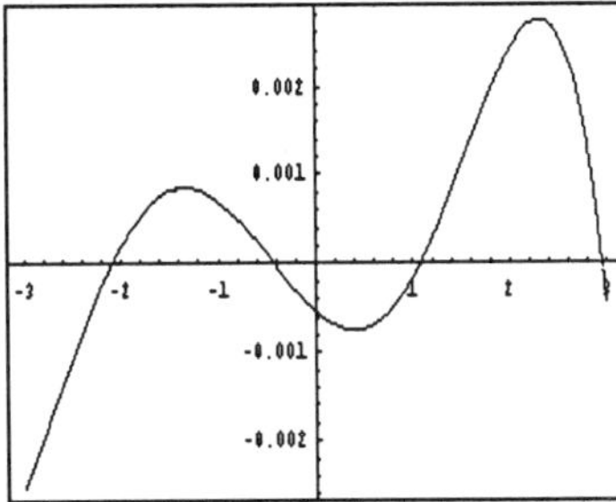

Figure 10.2. Plots of the quantile function (approximate and exact, left) and of the deviation (approximate minus exact, right) for the Gamma (3,2). Horizontal axis is the standard normal quantile (z). Vertical axis (left plot) is the standardized gamma quantile.

References

[1] Burnham, K. P., Anderson, D. R. (2002). *Model Selection and Multimodel Inference-A Practical Information-theoretic Approach.* Springer-Verlag. NY.

[2] Chou, Y-M, Polansky, A. M., Mason, R. L. (1998). Transforming non-normal data to normality in statistical process control. *Journal of Quality Technology,* 30(2), 133-141.

[3] Clements, J. A. (1989). Process capability calculations for non-normal distributions. *Quality Progress,* 22(2), 49-55.

[4] Karian, Z. A., Dudewicz, E. J. (2000). *Fitting Statistical Distributions: The Generalized Lambda Distribution and Generalized Bootstrap Methods.* CRC Press, Boca Raton, Florida, USA.

[5] Pearson, E. S., Johnson, N. L., Burr, I. W. (1979). Comparisons of the percentage points of distributions with the same first four moments, chosen from eight different systems of frequency curves. *Communications in Statistics (Simulation and Computation),* B8(3), 191-229.

[6] Shore, H. (1986). General approximate solutions for some common inventory models. *Journal of the Operational Research Society,* 37(6), 6, 619-629.

[7] Warick, J. (2004). A data-based method for selecting tuning parameters in minimum distance estimators. *Computational Statistics and Data Analysis.* In press.

Chapter 11

Estimating the Error Distribution

11.1. Introduction

RMM provides platforms for relational modeling and for modeling random variation. The former provides models which incorporate both transmitted variation (systematic and random) and self-generated variation (random). In Chapter 8 estimation procedures for this type of models were developed.

To model random variation, the component of systematic variation in the RMM model is assumed to vanish [the linear predictor (LP) is constant], and the resulting RMM error distribution is then employed as a general platform to model diversely-shaped distributions. Estimation procedures for the parameters of tne RMM error distribution are developed in this chapter.

Setting the LP arbitrarily to one, the four variations of the quantile function of the RMM error distribution, corresponding to the four variations in Section 7.2.5 (see also Section 9.4), are

$$Y = (M) \exp\{(\alpha/\lambda)[(1+\varepsilon_1)^\lambda - 1] + \varepsilon_2\}, \qquad (11.1a)$$

$$Y = (M) \exp\{(\alpha/\lambda)[(1+\varepsilon_1)^\lambda - 1]\} \, (1 + \varepsilon_2), \qquad (11.1b)$$

$$Y = (M) \exp\{(\alpha/\lambda)[\exp(\lambda\varepsilon_1) - 1] + \varepsilon_2\}, \qquad (11.1c)$$

$$Y = (M) \exp\{(\alpha/\lambda)[\exp(\lambda\varepsilon_1) - 1]\}(1 + \varepsilon_2), \qquad (11.1d)$$

where (11.1a) corresponds to the original RMM model, derived axiomatically in Chapter 7.

Assume that the errors, $\varepsilon_1 = \sigma_{\varepsilon 1} Z_1$ and $\varepsilon_2 = \sigma_{\varepsilon 2} Z_2$, originate in a bi-variate normal distribution with correlation ρ, or, equivalently, that Z_1 and Z_2 are from a bi-variate standard normal distribution with correlation ρ. By conditioning $Z_2 \mid Z_1 = z_1$, the errors can be re-written as

$$\varepsilon_1 = \sigma_{\varepsilon 1} Z_1,$$

$$\varepsilon_2 = \sigma_{\varepsilon 2}[\rho Z_1 + (1-\rho^2)^{(1/2)} Z_2], \tag{11.2a}$$

where Z_1 and Z_2 are now independent standard normal variables.

Similarly, by conditioning $Z_1 \mid Z_2 = z_2$, we have

$$\varepsilon_2 = \sigma_{\varepsilon 2} Z_2,$$

$$\varepsilon_1 = \sigma_{\varepsilon 1}[\rho Z_2 + (1-\rho^2)^{(1/2)} Z_1], \tag{11.2b}$$

where Z_1 and Z_2 are independent standard normal variables. Equations (11.1) and (11.2) generate altogether 8 possible variations for the quantile function of the RMM error distribution (refer to Section 9.4).

Estimation procedures for two versions of the RMM quantile function are developed: The original model (11.1a), and the variation given by (11.1c). There are three reasons for selecting (11.1c) in addition to (11.1a). First, this model has one less parameter that needs estimation. Secondly, in a numerical search for estimates of the parameters, the term $(1+\sigma_{\varepsilon 1} Z_1)^\lambda$, which appears in (11.1a), may occasionally assume imaginary values (the expression in brackets becomes negative). This problem does not exist when a numerical search is conducted for the parameters that appear in (11.1c). Thirdly, if the normal errors in (11.c) are assumed to be independent ($\rho=0$), $\log(Y)$ is the sum of independent exponential and normal variables. This is a convenient form for the RMM quantile function, which has certain distributional advantages (for example, moments are easy to calculate).

Introducing the errors, in terms of the independent Z_1 and Z_2, from (11.2) into (11.1a), we obtain the two equivalent forms

$$W = \log(Y)$$

$$= \log(M) + (\alpha/\lambda)[(1+\sigma_{\varepsilon 1} Z_1)^\lambda - 1] + \sigma_{\varepsilon 2}[\rho Z_1 + (1-\rho^2)^{(1/2)} Z_2], \tag{11.3a}$$

$$W = \log(M) + (\alpha/\lambda)\{\{1+\sigma_{\varepsilon 1}[\rho Z_2 + (1-\rho^2)^{(1/2)} Z_1]\}^\lambda - 1\} + \sigma_{\varepsilon 2} Z_2. \tag{11.3b}$$

Similarly, for (11.1c) the two equivalent forms are

$$W = \log(M) + (\alpha/\lambda)[\, \exp(\lambda\sigma_{\varepsilon1}Z_1) - 1] + \sigma_{\varepsilon2}[\rho Z_1 + (1-\rho^2)^{(1/2)}\, Z_2], \quad (11.3c)$$

$$W = \log(M) + (\alpha/\lambda)\{\, \exp\{\lambda\sigma_{\varepsilon1}[\rho Z_2 + (1-\rho^2)^{(1/2)}\, Z_1]\} - 1\} + \sigma_{\varepsilon2}Z_2. \quad (11.3d)$$

Note that "equivalent forms" imply that in estimating (11.1a) based on sample data, we cannot tell whether we fit (11.3a) or (11.3b). To realize that, introduce for ρ the three values, $\{-1,\ 0\ 1\}$. The same expressions are obtained, though possibly having different parameters' values. Similarly, if one chose to fit (11.1c), it cannot be determined in advance whether (11.3c) or (11.3d) were fitted.

Given the large number of parameters included in the quantile functions of (11.1) (at most six parameters), one degree of freedom can be given up without losing much in the flexibility of (11.1) to depict differently-shaped distributions. Setting $\rho=\pm1$, (11.3a-11.3d) become, respectively,

$$W = \log(M) + (\alpha/\lambda)[(1+\sigma_{\varepsilon1}Z)^{\lambda} - 1] + \sigma_{\varepsilon2}\rho Z, \qquad (11.4a)$$

$$W = \log(M) + (\alpha/\lambda)[(1+\sigma_{\varepsilon1}\rho Z)^{\lambda} - 1] + \sigma_{\varepsilon2}Z, \qquad (11.4b)$$

$$W = \log(M) + (\alpha/\lambda)[\exp(\lambda\sigma_{\varepsilon1}Z) - 1] + \sigma_{\varepsilon2}\rho Z, \qquad (11.4c)$$

$$W = \log(M) + (\alpha/\lambda)\,[\exp(\lambda\sigma_{\varepsilon1}\rho Z) - 1] + \sigma_{\varepsilon2}Z, \qquad (11.4d)$$

where Z is a standard normal random variable.

Re-written with different parameters, (11.4a) and (11.4b) may be combined to give

$$W = \log(Y) = \log(M) + (A)[(1+BZ)^{C} - 1] + DZ, \qquad (11.5)$$

where $\{M, A, B, C, D\}$ are parameters that need to be estimated. Note that either B or D can assume negative values [but not both, refer to (11.4a,b)].

Similarly, from (11.4c) and (11.4d):

$$W = \log(Y) = \log(M) + (A)[\exp(BZ) - 1] + DZ, \qquad (11.6)$$

and in fitting (11.6) to sample data, provided $\lambda>0$, either B or D can assume negative values [but not both, refer to (11.4c,d)].

For $Z=0$, both (11.5) and (11.6) give $W= \log(M)$, where M is the response median (the median of Y).

Estimating procedures applicable for the parameters in (11.5) and in (11.6) will be developed. We address two types of estimation approaches: Estimation based on percentile-matching (Section 11.2) and estimation based on moment-matching (Section 11.3). Maximum likelihood estimation has been developed for the general RMM model in Chapter 8, and it may be easily adapted for estimation of the RMM error distribution.

In the following sections, it is assumed that a sample of n observations, $\{y_i\}$, is available, and that the observations are expressed also in terms of the log-transformed values, $\{w_i\}$.

11.2. Percentile-Based Estimation

11.2.1. The estimation procedure

A percentile-based matching method is used to estimate the parameters of (11.5) and (11.6). This involves estimating response quantiles (from the given sample), identifying corresponding standard normal quantiles, and then using non-linear least-squares (NL-LS) for fitting.

Two approaches can be pursued:

(A) Select a set of CDF values. Then *identify* (calculate) the associated standard normal quantiles, $\{z_i\}$, and *estimate* (from the sample) the corresponding response quantiles, $\{y_i\}$.
(B) From the sample's ordered observations, *estimate* the corresponding CDF values. Calculate the respective standard normal quantiles.

The first approach requires to determine arbitrarily a set of CDF values $\{p_i\}$, where $p = F(y)$ and $F(y)$ is the CDF of the response (Y). From these values, the corresponding standard normal quantiles are identified. Finally, sample estimates for $\{y_p\}$ are derived, namely, estimates for quantiles of Y such that $Pr(Y \leq y_p) = F(y_p) = p$.

The second approach requires to *estimate* from the sample the value of $p_i = F(y_{(i)})$, where $y_{(i)}$ is the i-th order statistic (the sample value for which there are i-1 smaller values; We will refer to this definition and notation shortly). From this estimate, the respective standard normal quantile, z_i, is calculated.

For both approaches, the final sample for implementation of NL-LS is comprised of the ordered sample set: $\{(z_1, y_1). (z_2, y_2), .., (z_n, y_n)\}$. The two approaches are now expounded in detail.

Method A. Selecting a sample according to a pre-determined set of CDF values

Suppose that we have a predetermined CDF value of p. What is the sample estimate of the respective quantile, namely y_p such that $F(y_p) = p$?

Unfortunately, there is no "consensus" in the statistical literature regarding the best method to estimate quantiles from sample data. Perhaps the best demonstration of this is the STATISTICA Electronic Manual (STATISTICA is a widely-used statistical analysis software package). This manual introduces no less than six (!) methods to estimate percentiles. However, no guidance is given as to the relative benefits of each, or how to select the estimation procedure. We will introduce here one of these estimation methods, described in the above STATISTICA manual. This method is also endorsed by recently published book and paper by Karian and Dudewicz (2000, Section 4.1, and 2003, respectively), and advocated in a personal communication (Dudewicz, 2004).

Calculating a percentile by "The weighted average at Y(n+1)p"

This method estimates the percentile as a weighted average centered at $Y(n+1)p$. Suppose that we have a random sample of n observation $\{y_1, y_2, .., y_n\}$. Let the order statistics from this sample (the observations arranged in an ascending order) be $\{y_{(1)}, y_{(2)}, .., y_{(n)}\}$ [we use notation for order statistics that appears in Ord and Staurt (1987), though other notations exist]. Note that $y_{(1)}$ is the smallest observation, and $y_{(n)}$ is the largest.

Suppose that an estimate of the p-th quantile is desired (namely, the value of the response for which the probability of not exceeding it is p). Estimating comprises two stages:

(1) Express $(n+1)p$ by $r+g$, where r is an integer and $0 \leq g < 1$.

For example, let n=20, p=0.67. We have: (n+1)p= 21*0.67= 14+0.07 (r=14, g=0.07).

(2) Estimate the p-th quantile (denote it by Q_p):

$$Q_p = (1-g)\, y_{(r)} + g\, y_{(r+1)}\,, \quad r<n. \tag{11.7}$$

An example. Consider the sample of ordered observations (n=7): {2,5,6,13,17,22,30}. For p=0.2, we have: (n+1)p=(8)(0.2)= 1+ 0.6. Therefore:

$$Q_{0.2} = 0.4[y_{(1)}] + 0.6[y_{(2)}] = (0.4)(2)+(0.6)(5) = 3.8$$

Note, that for r=n, $y_{(n+1)}$ is replaced by $y_{(n)}$, which implies $Q_p= y_{(n)}$. Also note that this method is based on a simple linear interpolation between the two order statistics adjacent to the required percentile.

The respective standard normal quantile is $\Phi^{-1}(0.2)= -0.841621$. [$\Phi^{-1}(p_j)$ is the standard normal inverse CDF at p_j.]

Method B. Estimating the CDF values associated with the sample order statistics

Using this approach, the given order statistics are included in the sample for the NL-LS procedure. This implies that no interpolation is required, and only the available observations are used. Estimates for the CDF values associated with these observations are derived and thereof the respective standard normal quantiles.

To determine the p value associated with a given order statistic, say, $y_{(j)}$, most statistical packages use the simple approximating expression based on what is known as the "Median Rank" [approximated by (11.8)], or use White's plotting position (11.9):

$$p_j = (j-0.3)/(n+0.4), \tag{11.8}$$

$$p_j = (j-3/8) / (n+1/4). \tag{11.9}$$

The difference between the two approximations will most often be negligible. Find details for both expressions in Dodson (1994).

An example. For the sample of observations given earlier (with n=7), the first order statistic (smallest observation) is $y_{(1)}= 2$. The corresponding

CDF is not the biased value of p= 0.1429 (=1/7), but rather the unbiased value of [refer to (11.9)] p_1= (1-3/8)/(7+1/4)= 0.08620. If the approximate "Median Rank" was used, the CDF value obtained from (11.8) is p_1= (1-0.3)/(7.4)= 0.09459. The "Median Rank" takes account of the sampling distribution of the order statistic, and (11.8) indeed estimates (approximately) the median of this distribution (the exact median rank can be shown to be p_1= 0.09430).

The respective standard normal quantile is $\Phi^{-1}(0.08620)$= -1.36453.

Preparing the Sample for NL-LS Estimating

In the implementation of the NL-LS procedure, we can define a sample of CDF values and estimate (by Method A) the corresponding percentile values based on the available sample. Alternatively, we may confine ourselves only to the given order statistics (Method B) to determine the points which will be included in the fitting procedure.

We pursue the latter option, and the following stages need to be followed to prepare the sample for the NL-LS estimation:

- Arrange the observations in an ascending order (from smallest to largest)
- For the order statistic, $y_{(j)}$ (j=1,2,..,n), calculate p_j from (11.9)
- Find the corresponding standard normal quantile $z_{Pj}= \Phi^{-1}(p_j)$
- Generate the sample for the NL-LS fitting: $\{z_p, y_p\}$, where z_p and y_p are standard normal and response quantiles, respectively.

A detailed numerical example (to prepare a sample for NL-LS). Consider again the sample of ordered observations (n=7): {2,5,6,13,17,22,30}. Using (11.9), we have the p values

$$p = \{0.08620, 0.2241, 0.3621, 0.5000, 0.6379, 0.7759, 0.9138\}.$$

The corresponding z values are

$$z = \{-1.3645, -0.7583, -0.3529, 0, 0.3529, 0.7583, 1.3645\}$$

The sample for the NL-LS fitting is

$$\{(z_p,y_p)\} = \{(-1.3645,2), (-0.7583,5), (-0.3529,6),(0,13),$$
$$(0.3529,17), (0.7583,22),(1.3645,30\}.$$

Once all values of $\{z_p, y_p\}$ are available, NL-LS may be implemented, using for fitting either (11.5) or (11.6). Using the sample median as an estimate for the parameter M [$p=1/2$ in (11.7)], modeling with (11.5) requires estimating four parameters, and three if (11.6) is used.

Other RMM variations may similarly be estimated.

Comment. A relevant issue is the response to employ in the estimation procedure. Either Y or W=log(Y) may be used. It seems that computational convenience should be the over-riding concern, and with this in mind it is recommended that W be used as the response. In that case, the input sample is $\{z_p, w_p\}$, where w_p is the estimate of the p-th quantile of W.

The percentile-based estimating procedure is now demonstrated with two examples from Karian and Dudewicz (2000, henceforth KD), where the generalized Lambda distribution was used for fitting. The data for the examples are reproduced here with permission.

11.2.2. Two numerical examples (percentile-based estimation)

Example 1. Birth weights of twins - The Indiana Twin Study

This example uses data from the Indiana Twin Study (find details in KD, p. 100). The data consist of birth weights of twins, $\{Y_1, Y_2\}$, appearing in Table 11.1 in an ascending order of Y_1 (read lines).

KD fitted the generalized Lambda distribution [GLD, refer to Chapter 5 and to (11.10) below], using moment matching. We will analyze the distribution of Twin 1 (Y_1), using (11.6) as the RMM quantile function. For the Y_1 data, the first four moments are

$$\mu = E(Y_1) = 5.4858, \; \sigma^2 = Var(Y_1) = 1.3189,$$

$$Sk = \sqrt{\beta_1} = E[(Y_1-\mu)^3] / \sigma^3 = -0.04608,$$

$$Ku = E[(Y_1-\mu)^4] / \sigma^4 - 3 = -0.2668.$$

Table 11.1. Example 1 - Birth Weights of Twins

{Y$_1$,	Y$_2$}	{Y$_1$,	Y$_2$}	{Y$_1$,	Y$_2$}	{Y$_1$,	Y$_2$}
2.44	2.81	3	3.78	3.15	2.93	3.17	4.13
3.63	3.19	3.68	5.38	3.69	3.56	3.74	3.24
3.75	3.16	3.83	3.83	3.91	3.81	3.91	4.6
4	3.66	4	4.28	4.1	5	4.12	4.75
4.12	6.31	4.19	4.31	4.2	4.75	4.28	4.5
4.31	3.66	4.31	3.88	4.31	4.69	4.38	3.4
4.38	5	4.44	5.13	4.49	4.15	4.53	3.83
4.56	4.31	4.56	5.38	4.63	4.12	4.69	4.63
4.75	4.56	4.75	4.63	4.81	4.38	4.81	4.44
4.91	5.13	4.94	4.78	4.95	4.22	5	5.38
5	6.16	5.03	4.94	5.13	3.81	5.15	5
5.16	5.75	5.16	6.69	5.19	4.94	5.22	4.75
5.25	5.38	5.25	5.63	5.25	5.81	5.25	6.1
5.25	6.25	5.31	4.69	5.38	5.69	5.38	6.81
5.41	5.69	5.44	5.75	5.47	5.13	5.47	6.75
5.5	5.38	5.56	4.44	5.56	5.48	5.56	6.31
5.59	6.22	5.61	6.18	5.63	4.69	5.63	4.88
5.63	4.97	5.66	5.38	5.72	5.06	5.75	5.63
5.81	5.19	5.81	5.94	5.84	5.56	5.88	4.88
5.88	5.69	5.88	5.88	5.88	5.88	5.91	5.5
5.94	4.81	5.94	5.41	5.94	5.75	5.94	6.31
5.97	5.63	6.06	5.31	6.1	5.19	6.1	6.13
6.16	5.85	6.19	4.44	6.19	5.5	6.31	5.81
6.31	6.1	6.33	8.14	6.38	5.19	6.38	6.05
6.56	6.38	6.59	6.16	6.6	6.53	6.63	6.19
6.63	6.19	6.63	6.38	6.66	7.1	6.69	5.81
6.69	6.13	6.75	6.56	6.78	6.22	6.81	6.19
6.81	6.6	6.81	7.41	6.88	6.06	6.88	6.63
6.94	5.5	6.95	5.72	7.06	7.31	7.25	8
7.31	4.58	7.31	7.31	7.46	7.22	7.69	7.25
7.72	6.44	8.13	7.75	8.44	6.31		

The proximity of Sk and Ku to normal-distribution values (Sk=Ku=0) indicates that the data may be normally distributed. Another indication is obtained by modeling with (11.6). For the normal distribution, (11.6) always gives for the shape-parameter, B, a value of -(1/3). This implies that if a confidence interval for an estimate of B covers this value, Y_1 is probably normal (note that this is not a formal statistical hypothesis testing but rather an indication that such a test needs to be conducted).

The first stage in the analysis is to prepare the sample for a NL-LS run. This comprises arranging values of Y_1 in an ascending order, finding estimates for CDF values associated with the order statistics [use (11.9)], and then transforming the CDF estimates into standard normal values. Results for the lower and upper four order statistics are shown in Table 11.2.

Implementing NL-LS, using for M the sample median (5.56), the results in Table 11.3 are obtained. Note that since non-linear regression was used, the confidence intervals are only asymptotically exact. Figure 11.1 displays the relative errors (in %).

Calculating numerically the first four moments from the fitted (11.6), we obtain (in brackets are the sample values):

$$\mu = 5.5236 \ (5.4858), \ \sigma^2 = 1.3375 \ (1.3189),$$

$$Sk = -0.1598 \ (-0.04608), \ Ku = -0.09168 \ (-0.2668).$$

Table 11.2. Example 1 - Results for the first four and the last four order statistics of the birth weights (Y_1) of Twins (n=123, median is 5.56)

Sample Value:	i=1	i=2	i=3	i=4	i=120	i=121	i=122	i=123
y_i	2.44	3	3.15	3.17	7.69	7.72	8.13	8.44
p_i	.005071	.01318	.02130	.02941	.9706	.9787	.9868	.9949
z_i	-2.5709	-2.2207	-2.0276	-1.8895	1.8895	2.0276	2.2207	2.5709

Table 11.3. Example 1 - Estimates and 95% Confidence Intervals for the RMM Parameters (11.6) derived from NL-LS

Parameter	Estimate	Asymptotic SE	95% Confid. Interval
A	-0.3929	0.1746	{-0.7387,-0.04715}
B	-0.3884	0.08448	{-0.5556,-0.2211}
D	0.05813	0.03272	{-0.0066,0.1229}

The average of the absolute values of the errors in Figure 1.1 is 0.7082%. Good fit is achieved.

The GLD inverse distribution function is

$$Y_1 = \lambda_1 + (1/\lambda_2) \, [P^{\lambda_3} - (1-P)^{\lambda_4}], \ 0 \leq P \leq 1. \qquad (11.10)$$

Figure 11.2 displays the relative errors obtained from fitting GLD by moment matching, with parameters' estimates (KD, 2000, p.101)

$$\lambda_1 = 5.5872, \ \lambda_2 = 0.2266, \ \lambda_3 = 0.2089, \ \lambda_4 = 0.1762.$$

The average of the absolute values of the errors is 1.0425% (vs. 0.7082% for the RMM model). The two models seem to deliver comparable goodness-of-fit.

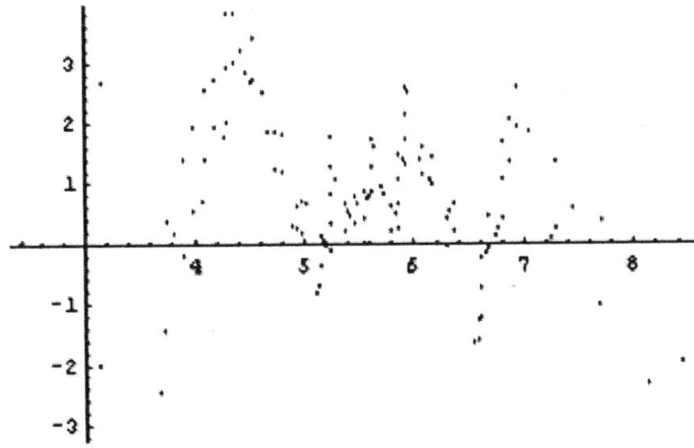

Figure 11.1. Example 1- Relative errors (in %) of the fitted (11.6) vs. actual birth weights

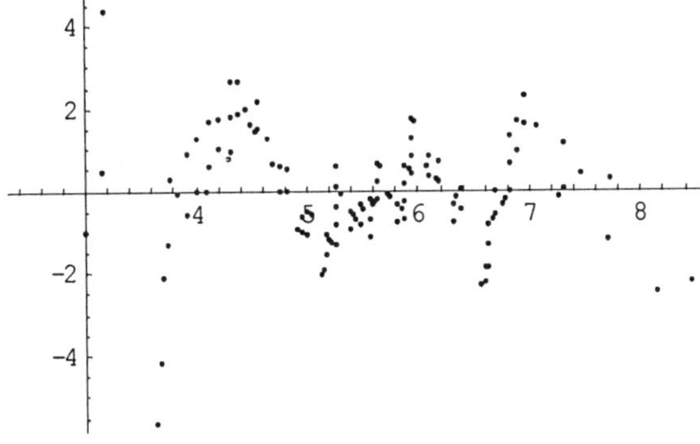

Figure 11.2. Example 1- Relative errors (in %) of the fitted GLD *vs.* actual birth weights

Example 2. Distribution of Intra-Galactic velocities

This example appears in KD (2000, p. 208), and is reproduced here with permission. In astronomy, the cluster named A1775 is believed to consist in fact of two clusters that are in close proximity. Oegerle, Hill and Fitchett (1995) gave velocity observations (in kilometers per second) from this galaxy. Some of these are given in Table 11.4, in an ascending order (read lines).

Proceeding as in Example 1, a first run of NL-LS shows the parameters (AB) and D in (11.6) to be close in their absolute values. Therefore the following modified two-parameter model (apart from the median) was tried [refer to (11.6)]:

$$W = \log(Y) = \log(M) + A_1\{(1/B_1)[\exp(B_1Z)-1] + Z\}, \quad (11.11)$$

where M is the median (22417). Note that a subscript was added to the parameters to distinguish them from those in the original model (11.6).

Re-running NL-LS for the modified model, the results in Table 11.5 are obtained. An analysis of variance delivers a model F-ratio value of $(1.05)10^6$. With 2 and 49 degrees-of-freedom for the mean-squared-errors of the model and the error, respectively, the model is highly significant.

Table 11.4. Example 2- Inter-galactic velocities (km per second)

18499	18792	18933	19026	19111	19111	19130
19179	19225	19404	19408	19595	19595	19619
19673	19740	19807	19866	20210	20210	20875
21911	21993	22192	22193	22417	22417	22426
22513	22625	22647	22682	22738	22738	22744
22779	22781	22796	22809	22922	22922	23017
23059	23121	23220	23261	23303	23303	23408
23432	24909					

Table 11.5. Example 2- Estimates and 95% Confidence Intervals for the RMM Parameters derived from NL-LS

Parameter	Estimate	Asymptotic SE	Confid. Interval
A_1	0.03657	0.004260	{0.02801, 0.04513}
B_1	-0.6838	0.1875	{-1.0606,-0.3070}

Figure 11.3 displays the relative errors (in %). The mean absolute value of the errors is 0.3888%.

Calculating the first four moments of W from the fitted (11.11), we find out that the mean and the variance are well preserved by the fitted equation:

$$\text{Calculated } \{\mu, \sigma^2\}: \{10.0, 0.007439\}$$

$$\text{Sample } \{\mu, \sigma^2\}: \{9.970, 0.006929\}.$$

The sample skewness and kurtosis are not well preserved. This is to be expected given the small sample size and the well-known high MSE of direct sample estimates of skewness and kurtosis.

Consider the KD solution. The authors concluded that the generalized beta distribution (GBD) may model well this dataset (therein, p. 208).

The four-parameter d.f and CDF of GBD is given by (therein, p. 119)

$$f(y) = (y-\beta_1)^{\beta_3}(\beta_1+\beta_2 y)^{\beta_4} / [\beta(\beta_3+1, \beta_4+1)\beta_2^{(\beta_3+\beta_4+1)}],$$

$$\beta_1 \leq y \leq \beta_1+\beta_2, \qquad (11.12)$$

$$F(y) = \int_{\beta_1}^{y} f(u)du, \qquad (11.13)$$

where β is the beta function: $\beta(a,b) = \Gamma(a)\, \Gamma(b) / \Gamma(a+b)$.

For the above data, KD obtain the fitted GLB parameters

$$\beta_1 = 18675.3862, \beta_2 = 4880.6411, \beta_3 = -0.4789, \beta_4 = -0.6045.$$

Figure 11.4 displays the relative errors (in %). An error is defined as the difference between the quantile, y_i, calculated from solving $F(y_i) = p_i$, and the actual i-th order statistic [p_i is the CDF value for the i-th order statistic, calculated from (11.9)]. We realize that the fit is about the same as that obtained via (11.11) (refer to Figure 11.3). The mean absolute value of the relative errors is 0.28% (*vs.* 0.39% for the RMM model).

11.3. Moment-Based Estimation

11.3.1. Introduction

Throughout this book reservation has been expressed about estimating a distribution's parameters via four-moment matching. Various sources had been quoted, including our own, where it was asserted that due to the

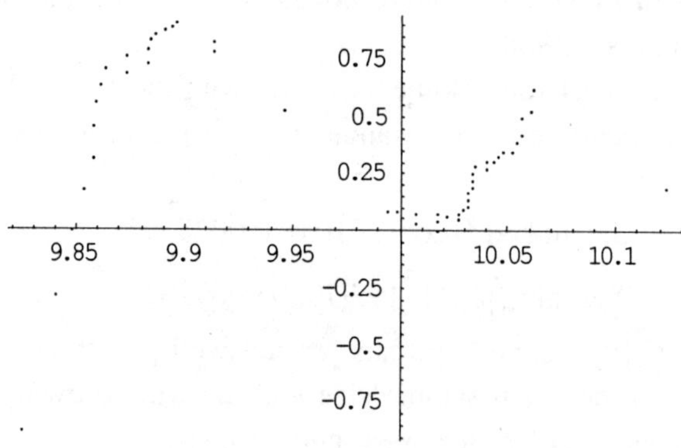

Figure 11.3. Example 2- Relative errors (in %) of the fitted (11.11) vs. actual log of the galaxies velocities (RMM model)

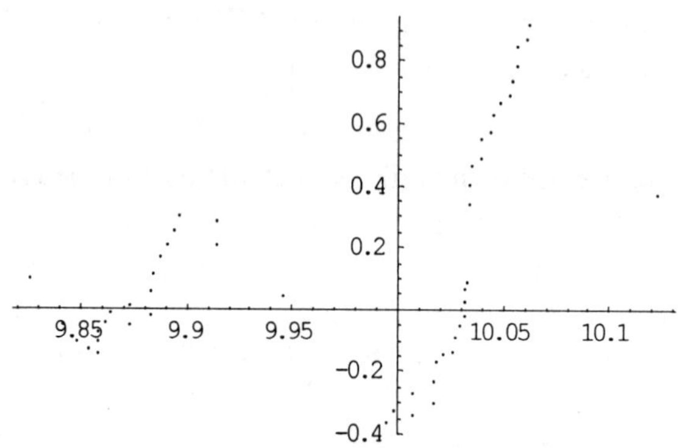

Figure 11.4. Example 2- Relative errors (in %) in the predicted log quantiles [bases on the fitted (11.12)] vs. actual log of the intra-galactic velocities (KD model, based on GBD)

large sampling errors associated with sample estimates of moments of high-degree (third degree and higher), the use of these estimates in a fitting procedure was not recommended.

A solution to this problem has been addressed in this book several times. In essence, the solution uses complete *and* partial moments of low degree in the matching procedure, where partial moments refer to moments calculated for only half the distribution. This moment-matching approach ensures that the adverse effect of the large standard errors, associated with estimates of high-degree moments, is averted. We have alluded to this approach in Chapter 5 and will do that again in Chapters 20 and 21 (refer also to Shore, 2004).

To allow estimation of the parameters of the RMM error distribution by moment-matching with low-degree moments, the error distribution needs to have a manageable number of parameters. For example, if we wish to estimate the parameters using partial moments of second degree at most, this implies that at most four parameters need to be estimated (two for each half of the distribution).

Alternatively, a combination of approaches may be pursued. For example, we may use a sample estimate of the response median to estimate M. Also, we may require moment-parity not only for moments of the response in the original scale (Y) but also for moments in the log-transformed scale (W). It has been demonstrated (Shore, 2000) that this combination of approaches results in good fitting for inverse normalizing transformations (refer to Chapter 20 for details).

Indeed, such marriage of approaches in a single estimation procedure is not at all rare in the statistics literature. For example, in fitting the three-parameter Weibull distribution, Cohen, Whitten and Ding (1984) use both the first two moments (mean and variance) and the smallest order statistic (a percentile) to estimate the Weibull parameters.

Likewise, the Johnson family is often fitted based on moment matching (refer to a review in Staurt and Ord, 1987, section 6.27-6.36), percentile matching (for example, Chou, Polansky and Mason, 1998) or some combination thereof.

In this section, estimation procedures for (11.1c), re-parameterized in (11.6), are developed. Earlier we related to the advantages of (11.6) for modeling random variation (Section 11.1).

There are four parameters in (11.6), $\{M, A, B, D\}$. If M is estimated by the sample median then there are three parameters to estimate. Two moment-based estimation procedures will be developed. They resemble moment-fitting procedures already developed in Chapter 10. The reader, however, should bear in mind a major difference between the procedures of Chapter 10 and those developed here: Considerations of large sampling errors were not valid for the fitting procedures developed in Chapter 10. Therein we have assumed that the RMM error distribution is fitted to *known* distributions, with well-defined and known moments of any order. Conversely, here it is assumed that moments are not known but need to be estimated. This implies that the problem of sampling errors needs to be considered, in other words, sample estimates of high-degree moments (third degree and higher) should be avoided.

Consider the rescaled response in (11.6):

$$Y/M = \exp\{(A)[\exp(BZ) - 1] + DZ\}. \qquad (11.14)$$

Finding values for the parameters requires a search routine coupled with numerical integration (to obtain the moments). Since (11.14) involves double exponential terms, numerical integration can easily deteriorate to states of underflow or overflow with possibly a serious loss of precision. Furthermore, the search routine may take too long to identify the correct parameters.

A simple remedy may be found in the use of a Taylor series expansion, which would eliminate the need for numerical integration. Expanding (11.14) in terms of Z around zero, and taking expectation of the resulting expression, we obtain an expression for the response mean in terms of expected values of power terms of Z. Since all odd-order moments of Z are identically zero, by taking the first six terms of the expected value of the expansion we end up with only three terms.

The resulting expression is $[E(Z^2)=1, E(Z^4)=3]$

$$E(Y/M) \cong 1 + (1/2)[AB^2 + (AB+D)^2] + (3/24)\{AB^4(1 + 7A + 6A^2 + A^3)$$

$$+ 4AB^3D(1 + 3A + A^2) + 6AB^2D^2(1 + A) + 4ABD^3 + D^4\}. \quad (11.15)$$

Consider (11.14) again. It is obvious that $(Y/M)^r$ has the same expression as (Y/M), with A and D replaced by rA, and rD, respectively.

This implies that the approximate expression for E(Y/M) in (11.15) can also be used to approximate $E[(Y/M)^r]$, with the appropriate substitution for B and D, namely,

$$E(Y^r) \cong (M)^r \{ 1 + (1/2)[rAB^2 + r^2(AB+D)^2] + (3/24)\{rAB^4(1 + 7rA$$

$$+ 6r^2A^2 + r^3A^3) + 4r^2AB^3D(1 + 3rA + r^2A^2)$$

$$+ 6r^3AB^2D^2(1 + rA) + 4r^4ABD^3 + r^4D^4\}\}. \tag{11.16}$$

The reader may easily verify that (11.16) is a highly accurate approximation for the non-central moments of Y. To realize that, calculate for several values of the parameters the approximate moments, given by (11.16), and compare with the exact moments (to derive the latter, multiply Y^r, as given by the RHS of (11.14), by the standard normal d.f and integrate).

Since (11.16) provides an explicit expression for the r-th non-central moment, no numerical integration is needed during the search routine, and the parameters that deliver moment-matching may be easily identified (either via root-finding or by a minimization procedure).

It was earlier explained that the moment-based estimation procedures involve matching of partial moments. Consider approximating the partial moments of Y in (11.14). We may wish to use the same approach that led to the derivation of an approximation to the complete moments (11.16). However, odd-order *partial* moments of the standard normal variable do not reduce to zero (as complete moments do). Therefore, into the expression derived from the Taylor expansion, we need to introduce for the odd-order power terms of Z partial moments of the standard normal variable.

Denoting the r-th upper partial moment of Z by $M_U(Z^r)$, namely,

$$M_U(Z^r) = \int_0^\infty z^r \phi(z)dz ,$$

the first six partial moments are

$$M_U(Z) = (2\pi)^{-1/2} , \; M_U(Z^2) = 1/2 , \; M_U(Z^3) = (2/\pi)^{1/2} , \; M_U(Z^4) = 3/2,$$

$$M_U(Z^5) = 4(2/\pi)^{1/2} , \; M_U(Z^6) = 15/2. \tag{11.17}$$

These partial moments may be used to derive approximate explicit expressions for the r-th partial moment of Y, similarly with the development detailed above for the complete moment.

Two estimation procedures based on moment-matching will be derived:

- **Sub-section 11.3.2:** Estimation based on matching of the mean of W (the log-transformed response) and the partial means of the response in the original scale (Y). M will be estimated from the sample median.

- **Sub-section 11.3.3:** Estimation based on matching of the mean and the variance of W, and matching the partial means of the response in the original scale (Y).

For both procedures, it will be demonstrated, based on a sample of commonly applied and differently shaped distributions, that the first four moments of Y are well preserved.

Other moment-based estimation procedures may be conceived. For example, we may opt to develop a procedure that preserves both the mean and the variance of the response in the original scale (Y) and in the log-transformed scale (W). The allied procedure and some other moment-based estimating procedures can be easily developed (refer for details to Shore, 2000).

The two numerical examples of Sub-section 11.2.2 are re-analyzed in Sub-section 11.3.4.

11.3.2. Procedure I: Matching the mean of W =log(Y) and the partial means of Y; M is estimated by the sample median

This estimation procedure uses sample moments of first-degree only. Since the latter, like the median, have the smallest MSEs (relative to MSEs of estimates of higher-degree moments), it is expected that the estimated quantile function will have moments with MSEs smaller than those associated with corresponding direct sample statistics. (This point has been discussed in Chapter 5 and will be discussed again in Section 20.2; Also refer to Shore, 2004, and references therein.)

From (11.14), we have for the mean of W

$$E(W) = \log(M) + (A)[\exp(B^2/2) - 1].$$ (11.18)

Introducing for "A" from (11.18) into (11.14) we obtain

$$W = \log(M)+[E(W)-\log(M)][\exp(BZ) - 1]/[\exp(B^2/2) - 1]+DZ$$ (11.19)

This expression describes W in terms of $\log(M)$, $E(W)$, and two unknown parameters, B and D. Denote by $M_U(Y)$ the response upper partial mean, namely,

$$M_U(Y) = \int_M^\infty yf(y)dy,$$ (11.20)

where $f(y)$ is the d.f. of Y. The lower partial mean, $M_L(Y)$, is equal to $E(Y)-M_U(Y)$. Estimates of both partial moments are obtained from the available sample of observations. M is set to be the sample median.

From (11.19) we have

$$Y = \exp\{\log(M)+[E(W)-\log(M)][\exp(BZ)-1]/[\exp(B^2/2)-1]+DZ\}$$ (11.21)

The upper partial moment of (11.21) is obtained from

$$M_U(Y) = \int_M^\infty (M)\exp\{[E(W)-\log(M)][\exp(Bz)-1]/[\exp(B^2/2)-1]$$

$$+Dz\}\phi(z)dz,$$ (11.22)

where $\phi(z)$ is the standard normal d.f. The lower partial moment is similarly defined. Equating the partial moments to the corresponding sample values, the parameters {B,D} may be obtained by any root-finding routine. If one wishes to eliminate the numerical integration from the search routine, Taylor-based approximations may be developed, as explained earlier.

Table 11.6 shows, for a sample of distributions, the parameters' values and the first four moments obtained with Procedure I. The first four moments are well preserved, even though only first-degree moments took part in the fitting procedure. Coupled with the low sampling errors associated with sample estimates of first-degree moments, this estimation procedure is expected to deliver good sampling accuracy.

Table 11.6. Parameters' values and the first four moments of Y, obtained by fitting (11.14) to a sample of distributions. Exact moments are upper entries in bold. Middle entries refer to Procedure I, and bottom entries to Procedure II. M is the median.

Distribution	A	B	D	Mean	Var.	Sk	Ku
Normal				**14.00**	**4.00**	**0**	**0**
$\mu = M = 14$	-.1342	-.3887	.0905	14.00	4.04	.0090	.0501
$\sigma = 2$	-.1365	-.3612	.0935	14.00	4.03	-.0609	.0473
Gamma				**14.00**	**28.00**	**.7559**	**.8571**
$\alpha = 7, \beta = 2$	-1.444	-.1844	.1193	14.00	28.00	.7555	.8539
M = 13.34	-1.026	-.2158	.1636	14.00	28.09	.7737	.9287
Gamma				**6.000**	**12.00**	**1.155**	**2.000**
$\alpha = 3, \beta = 2$	-1.675	-.2670	.1599	6.000	12.06	1.153	1.974
M = 5.348	-1.419	-.2885	.1953	6.000	12.03	1.166	2.059
Weibull				**4.790**	**18.19**	**1.676**	**4.040**
$\alpha = 1.125, \beta = 5$	-1.933	-.4300	.1951	4.790	18.06	1.635	3.720
M = 3.610	-1.506	-.4769	.2992	4.790	18.56	1.766	4.686
Weibull				**5.000**	**25.00**	**2.000**	**6.000**
$\alpha = 1, \beta = 5$	-2.155	-.4318	.2237	5.000	24.78	1.949	5.546
M = 3.466	-1.700	-.4767	.3342	5.000	25.55	2.105	6.909
Weibull				**5.665**	**50.99**	**2.815**	**12.74**
$\alpha = 0.8, \beta = 5$	-2.649	-.4353	.2893	5.665	50.41	2.737	11.77
M = 3.162	-2.136	-.4764	.4122	5.665	52.39	2.970	14.69
Weibull				**16.62**	**2723**	**11.35**	**287.6**
$\alpha = 0.4, \beta = 5$	-5.059	-.4450	.6314	16.62	2654	10.76	238.8
M = 2.000	-4.360	-.4741	.7890	16.62	2888	12.37	336.1
Rayleigh				**2.507**	**1.717**	**.6311**	**.2451**
$\sigma = 2$	-1.134	-.4214	.09978	2.507	1.709	.6118	.1636
M = 2.355	-.8555	-.4730	.1689	2.507	1.734	.6811	.4046
ExtremeValue				**3.577**	**1.645**	**1.139**	**2.400**
$\alpha = 3, \beta = 1$	0	.7990	.3460	3.577	1.625	1.115	2.282
M = 3.366	-.0129	-.6089	.3385	3.576	1.621	1.070	2.113

11.3.3. Procedure II: Matching the mean and the variance of W, and those of Y (or the partial means of Y); no estimate of median

This estimation procedure uses sample statistics of first and second degree only. A major shortcoming of Procedure I is that the median has to be estimated. Since the median appears as a scale factor in (11.14), any sampling error associated with the sample median will be linearly transmitted to quantile values calculated from the estimated quantile function. A traditional sample-estimate of the medain is based on a single observation (or two at most). This implies high sampling inaccuracy. It is therefore highly desirable that the need to specify the median be eliminated in any estimation procedure. This is a guiding principle in developing Procedure II.

From (11.18) we obtain

$$\log(M) = \mu(W) - A[\exp(B^2 / 2) - 1], \qquad (11.23)$$

where $\mu(W)$ is the mean of W [earlier denoted by E(W)]. Introducing back into (11.14) we obtain

$$Y = \exp\{\log(M)+A[\exp(Bz) - 1]+Dz\} =$$

$$\exp\{\mu(W) + A[\exp(Bz)-\exp(B^2/2)]+Dz\}. \qquad (11.24)$$

From (11.24), by taking log of both sides,

$$W - \mu(W) = A\{\exp(Bz)- E[\exp(BZ)]\}+Dz. \qquad (11.25)$$

Squaring both sides and taking expectations we have

$$Var(W) = A^2 \, Var[\exp(BZ)]+2(A)(B)(D)\exp(B^2/2)+D^2 Var(Z), \qquad (11.26)$$

where $Var(.)$ is the variance operator, and $(B)\exp(B^2/2)$ is the expected value of $(Z)\exp(BZ)$. It is easy to show that

$$Var[\exp(BZ)]= \exp(2B^2) - \exp(B^2). \qquad (11.27)$$

Introducing back into (11.26) we obtain for D (>0)

$$D= -(A)(B)\exp(B^2/2)+\{A^2[(B^2+1)\exp(B^2)-\exp(2B^2)]+Var(W)\}^{(1/2)}. \qquad (11.28)$$

Introducing for D into (11.24), we obtain a two-parameter quantile function, expressed in terms of the mean and the variance of W. The parameters A and B may now be identified either by equating the mean and the variance with those of Y (given by sample estimates), or by equating the partial means. In the former case, the resulting quantile function (11.14) will preserve the means and the variances of both W and Y. In the latter case, the resulting quantile function will preserve the mean and the variance of W and the partial means of Y.

Using the latter case, (11.14) is fitted to the sample of distributions, used earlier, to obtain the results displayed in Table 11.6. Good preservation of all first four moments is attained.

In the following numerical examples, we use the former case, namely, matching the mean and the variance of Y (in addition to those of W).

11.3.4. Two numerical examples (moment-based estimation)

Example 1. Birth weights of twins - The Indiana Twin Study

We will use the Twins Study data of Example 1 in Sub-section 11.2.2. For this example Procedure 1 will be implemented (Sub-section 11.3.2). This procedure requires sample estimates of the median, the mean of W and the partial means of Y_1. From the data in Table 11.1 we obtain

$$\text{Median}(Y_1) = 5.56,\ \mu(W_1) = 1.6788,$$

$$M_U(Y_1) = 3.2025,\ M_L(Y_1) = 2.2833.$$

Note, that to derive an estimate for the upper partial moment, we sum all observations larger than the median, add half the value of the median (since the sample size is odd, n=123) and then divide by n (not n/2). The estimate for the lower partial mean may be similarly derived, or simply calculated by subtracting the upper partial mean from the complete mean (5.4858).

Implementing Procedure I, we obtain the following parameter estimates [refer to (11.21) and (11.22)]:

$$B = -0.3289,\ D = -0.006446,$$

and from (11.18):

$$A = [E(W)\text{-}\log(M)] / [\exp(B^2/2)\text{--}1] = -0.6588$$

The first four moments, calculated numerically from the fitted quantile function, are (with the sample estimates in brackets):

$$\mu(Y) = 5.4858\ (5.4858),\ \text{Var}(Y) = 1.3062\ (1.3189),$$

$$\text{Sk} = -0.3060\ (-0.04608),\ \text{Ku} = -0.1240\ (-0.26678).$$

We realize that even though sample estimates of only first-degree moments took part in the fitting procedure, higher moments are also well preserved. Note that the value of B is close to that obtained earlier by percentile fitting (-0.3884), and both are close to that obtained if (11.6) is fitted to a normal distribution (about -1/3). In fact, the 95% confidence interval for B (refer to Table 11.3) does include -(1/3), implying that the data probably originated in a normal distribution.

Example 2. Distribution of Intra-Galactic velocities

The galaxy velocity observations of Example 2 (Section 11.2.2) are used. Both Procedure I (Section 11.3.2) and Procedure II (Section 11.3.3) will be demonstrated.

Procedure I

This procedure requires sample estimates of the median, the mean of W and the partial means of Y. From the data in Section 11.2.2 we obtain

$\text{Median}(Y) = 22417$, $M_U(Y) = 11486$, $M_L(Y) = 9971$, $\mu(W) = 9.9704$.

Implementing Procedure I we have the following parameter estimates:

$$B = -0.09408, D = -0.9097,$$

$$A = [E(W)-\log(M)]/[\exp(B^2/2)-1] = -10.631.$$

Like in the earlier analysis of this example (Section 11.2.2), it is easy to realize that the parameters (AB) and D are very close in their absolute values, so the following corrected model was tried:

$$Y = (M)\exp\{A_1 \{(1/B_1)[\exp(B_1Z)-1]+Z\}\}, \qquad (11.29)$$

where M is the sample median (22417). This model now has only two parameters (besides the median) so it might be a good idea to switch to matching the first two moments, namely, the mean and the variance.

From (11.29), using a Taylor series-expansion approximation [similar to the derivation of (11.16)], we obtain the approximate explicit expression for the r-th complete moment of Y:

$$E(Y^r) = (M)^r\{1 + (1/2)(rB_1A_1 + 4r^2A_1^2) + (3/4)[rB_1^3A_1/6$$
$$+ B_1^2r^2A_1^2 + (1/2)rB_1A_1(rB_1A_1 + 4r^2A_1^2)$$
$$+ (2/3)rA_1(B_1^2rA_1/2 + 2B_1r^2A_1^2 + rA_1(rB_1A_1 + 4r^2A_1^2))]\}. \quad (11.30)$$

Using this approximate highly accurate expression, we obtain, by matching the means and the variances, the parameters' estimates

$$A_1 = -0.02394, B_1 = 1.9814.$$

The first four moments of the fitted quantile function are (with the sample moments in brackets):

$$M = 22417, \mu(Y) = 21456 \ (21456),$$

$$\mathrm{Var}(Y) = (3.099)10^6 \ [(3.099)10^6],$$

$$\mathrm{Sk}(Y) = 0.8370 \ (1.525), \mathrm{Ku}(Y) = -2.633 \ (-1.453)$$

Given the small sample and the relatively high sampling errors associated with sample estimates of skewness and kurtosis, the moments calculated from the fitted (11.29) are close to the sample values. The reader should again be reminded that only first- and second-degree moments participated in the fitting procedure.

Procedure II

This procedure requires matching the mean and the variance of $W=\log(Y)$ and the partial means of Y. No Estimate of the median is needed. The sample estimates are

$$\mu(W) = 9.970, \mathrm{Var}(W) = 0.006929$$

$$\mathrm{Sk}(W) = 1.547, \mathrm{Ku}(W) = -1.475.$$

To perform moment-matching for W, we may use direct numerical integration since the problem of the double exponential, associated with (11.14), disappears. However, Procedure II requires also matching of the mean and the variance of Y. This implies that use of the Taylor approximation for the moments of Y, given in (11.16), is desirable.

Introducing for D, in terms of A and B, from (11.28) into (11.24), we obtain an expression for Y in terms of the mean and the variance of W, and in terms of the unknown parameters A and B. Equating with the *sample* mean and variance of Y, we obtain for the parameters

$$A = -16.31, B = 0.02562, D = 0.50093 \ [\text{from } (11.28)].$$

The fit is not very good in the sense that the mean and the variance of Y are not preserved to an acceptable degree of accuracy. We realize (like in all previous attempts at modeling the galaxy's velocity data) that B is close to zero and that the parameters (AB) and D are very close in their absolute values. This implies that the correct model is again the one given by (11.29). Furthermore, the earlier solution, where these parameters were determined by matching of the response means and variances, provides the best estimates for the unknown parameters.

We conclude with a comment regarding empirical modeling, as the latter materialized for the galaxy's velocity data (Example 2). Since B consistently turned out to be close to zero, it was realized that a different model was suggested by the data, and whatever solution procedure was implemented the results always pointed to the same model (11.29).

A good lesson may be learned from this example. If the fit achieved is not satisfactory, take a good look at the values of the parameters' estimates. In most cases, they tell a story, and may direct you to the best way to modify your model. In our example, the fact that B was consistently close to zero (in all solution procedures attempted) immediately suggested that instead of modeling this parameter by the term exp(BZ)-1, perhaps the term: [exp(Bz)-1]/B would provide better goodness-of-fit. Likewise, the proximity in the absolute values of (AB) and (D) immediately suggested that perhaps one parameter is superfluous, a fact that would probably manifest itself in high correlations, if estimates' correlations were calculated.

Looking carefully at the results of the statistical analysis, learning from them, and then modifying the model, in an iterative fashion, until the best fit is achieved, is always a good time-honored practice for empirical modeling.

References

[1] Chou, Y-M, Polansky, A. M., Mason, R. L. (1998). Transforming non-normal data to normality in statistical process control. *Journal of Quality Technology*, 30(2), 133-141.

[2] Cohen, A. C., Whitten, B. J., Ding, Y. (1984). Modified moment estimation for the three-parameter Weibull distribution. *Journal of Quality Technology*, 16, 159-167.

[3] Dodson, B. (1994). *Weibull Analysis*. ASQ Press, Milwaukee, WI.

[4] Karian, Z. A., Dudewicz, E. J. (2000). *Fitting Statistical Distributions: The Generalized Lambda Distribution and Generalized Bootstrap Methods*. CRC Press, Boca Raton, Florida, USA.

[5] Karian, Z. A., Dudewicz, E. J. (2003). Comparison of GLD fitting methods: Superiority of percentile fits to moments in L^2 norm. *Journal of the Iranian Statistical Society,* 2(2), 171-187.

[6] Oegerle, W. R., Hill, J. M., Fitchett, M. J. (1995). Observations of high dispersion clusters of galaxies: Constraints on cold dark matter. *The Astronomical Journal,* 110(1), p. 32.

[7] Stuart, A., Ord, K. (1987). *Kendall's Advanced Theory of Statistics. V. 1: Distribution Theory.* Charles Griffin & Company, Ltd., London.

[8] Shore, H. (2000). General control charts for variables. *International Journal of Production Research,* 38(8), 1875-1897.

[9] Shore, H. (2004). Non-normal populations in quality applications- A revisited perspective. *Quality and Reliability Engineering International,* 20(4), 375-382.

Chapter 12

Special Cases of the RMM Model

The RMM model was derived axiomatically in Chapter 7. It has been emphasized that RMM represents well relational models of monotone convex relationships, which one can find in a myriad of current engineering and scientific disciplines. Furthermore, the RMM error distribution represents well existing models of random variation, namely, known statistical distributions.

In this chapter, it is shown that RMM may in fact deliver *exact* representation to various models in engineering and the sciences, as well as to statistical distributions that one often encounters in the application of statistical models. Demonstrating that known models are exact special cases of RMM would confer further validity on the new methodology as a general platform for empirical modeling. We conduct this study pursuing the dual distinction that has been made earlier in the book between "Systematic Variation" (Section 12.1) and "Random Variation" (Section 12.2).

12.1. Current Relational Models as Special Cases of RMM

In Chapter 2 current mainstream relational models in various branches of science and engineering have been introduced. By examining more closely three particular examples (Chapter 3), we have studied commonly shared properties. These properties indeed formed the basis that led to the development of the RMM model. In this section, we re-examine a subset of the models in Chapter 2, and show how these may be derived as special cases of the RMM model.

Four different subject areas are addressed: Chemistry and chemical engineering, physics, electric engineering and non-linear growth modeling. Since all related models have already been introduced in Chapter 2, the focus here is on how assigning particular values to the parameters of the RMM model result in these models.

The reader may wish to complement the examples given here by attempting to derive from RMM the rest of the models in Chapter 2.

For convenience, we re-introduce the four variations of the RMM model (Chapter 11), so that it will be easy to relate to them when various models are introduced:

$$Y = (M)\ \exp\{(\alpha/\lambda)[(\eta+\varepsilon_1)^\lambda - 1] + \varepsilon_2]\}, \tag{12.1a}$$

$$Y = (M)\ \exp\{(\alpha/\lambda)[(\eta+\varepsilon_1)^\lambda - 1]\}\ (1 + \varepsilon_2), \tag{12.1b}$$

$$Y = (M)\ \exp\{(\alpha/\lambda)[\ \eta^\lambda \exp(\lambda\varepsilon_1/\eta) - 1] + \varepsilon_2\}, \tag{12.1c}$$

$$Y = (M)\ \exp\{(\alpha/\lambda)[\ \eta^\lambda \exp(\lambda\varepsilon_1/\eta) - 1]\}(1 + \varepsilon_2). \tag{12.1d}$$

Note, that occasionally M in these models will be introduced in the exponential as μ_2 [$=\log(M)$].

Literature related to the models of this section has been given in Chapter 2. Accordingly, only new references that are not model-related are provided in the references list at the end of this chapter.

12.1.1. Chemistry and chemical engineering

We start with temperature (T) dependence of vapor pressure (P). Referring to (2.1) (Section 2.2):

$$\log(P) = A + B/(\mu_T+\varepsilon_1+C) + \varepsilon_2,\ B < 0, \tag{12.2}$$

where T$= \mu_T+\varepsilon_1$ and μ_T is the expected value. It is easily recognized that this equation is derivable from the RMM model (12.1a), with parameters

$$\eta = \mu_T+C,\ \alpha = -B,\ \lambda = -1,\ \mu_2 = A+B. \tag{12.2a}$$

Equation (2.2) is a polynomial, and will be addressed separately.

For surface tension, S [Eq. (2.3)], re-written here with the errors, we have

$$S = A[1-(\mu_T + \varepsilon_1)/T_c]^C\ (1+ \varepsilon_2). \tag{12.3}$$

It is easy to verify that (12.3) can be derived from (12.1b) with parameters

$$\eta = 1\text{-}\mu_T/T_c, \ \alpha = C, \ \lambda = 0, \ M = A. \qquad (12.3a)$$

Relating next to (2.4), we realize that the upper two equations are special cases of the RMM model (expressing the response as exponential-power and power relationships, respectively). The third equation is a polynomial, which will be addressed shortly.

For solid density, Daubert notes (refer to Chapter 2) that "…a linear equation is normally adequate … In a few cases … a simple quadratic polynomial was recommended. For most situations use of a single exponential value over the entire temperature range is sufficient". All these cases are "steps" on the "Ladder", special cases of the RMM model.

For liquid density, ρ_L, Rackett's equation (2.5) may be re-expressed with the errors by

$$\log(\rho_L) = A + B[1\text{-}(\mu_T+\varepsilon_1)/C]^D + \varepsilon_2, \qquad (12.4)$$

where A, B, C and D are parameters, that need to be determined. It is easy to see that this equation is derivable from (12.1a) with

$$\eta = 1\text{-} \mu_T/C, \ \alpha = BD, \ \lambda = D, \ \mu_2 = A+B. \qquad (12.4a)$$

For solid heat capacity, Daubert recommends a simple polynomial in temperature, where "most data can be fitted with a linear equation, with a quadratic necessary for a few systems". Both the linear and the quadratic are special cases of RMM.

For liquid heat capacity, Daubert recommends using a polynomial in t [Equation (2.6)], where $t = 1\text{-} T_r$. Polynomials will be addressed shortly.

For ideal gas heat capacity, C_P^o, Daubert recommends the use of an exponential [Equation (2.7)]. Re-writing it with the random errors, we obtain

$$C_P^o = A + \{B \exp[-C/(\mu_T+\varepsilon_1)^D]\}(1 + \varepsilon_2). \qquad (12.5)$$

It is easy to see that this is (12.1b), corrected for location, with

$$\eta = \mu_T, \ \alpha = (-D)(-C), \ \lambda = -D, \ \mu_2 = \log(B) - C. \qquad (12.5a)$$

For temperature dependency of liquid viscosity, Daubert recommends a polynomial [Equation (2.8)] to be addressed shortly.

For thermal conductivity, Daubert examines both liquid and low-pressure vapor thermal conductivity, all of which are either polynomials in T or in $t=1-T_r= 1-T/T_c$.

The Arrhenius equation (2.9) has the same structure as (12.5), with location parameter equal zero, and therefore is easily recognized as a special case of the RMM model.

We now consider polynomial models. Polynomials are a linear combination of various powers of the "linear predictor" (η), and therefore constitute a linear combination of various power terms, derivable from the RMM model with $\lambda=0$. Since an implicit objective of using polynomials is often to capture various degrees of "convexity", one can reasonably expect the RMM model to provide viable substitute to the use of polynomials.

A demonstration of this is given by viscosity modeling, delivered in Daubert (1998) by two expressions [Equations (21) and (22), therein], and presented here in a somewhat different form:

$$\mu_L = \exp[A + B/T + C\log(T) + DT^E], \qquad (12.6)$$

$$\mu_L = \exp(AF) - C, \qquad (12.7)$$

where $F= T^B$. Daubert states that (12.6) and (12.7) are nearly equivalent in the degree of accuracy they deliver. Yet (12.6) is a polynomial, while (12.7) is a special case of the RMM model (up to a linear transformation).

Our own experience with modeling by polynomials confirms that for uniformly convex (concave) relationships, RMM regularly delivers good substitute for polynomial representations. Furthermore, the "temptation" to increase the number of parameters does not exist with RMM. Conversely, modeling by a polynomial, one always run the risk that by increasing the number of parameters in the model in order to enhance goodness-of-fit, unstable parameters' estimates can ensue [refer to Shore, Brauner and Shacham (2002) and Shore (2003)]. The point is further discussed in Chapter 17.

The issue of whether a polynomial should be used has to be carefully balanced against the alternative, offered by the RMM approach.

12.1.2. Physics

Introducing kinetic energy (E_k) as a function of velocity (V) with the errors added, we obtain from (2.10)

$$E_k = M(\mu_V+\varepsilon_1)^2 / 2 + \varepsilon_2, \qquad (12.8)$$

with V= $\mu_V+\varepsilon_1$. Since this is a power relationship it is a special case of the RMM model.

For Newton's law of cooling (2.12), we obtain, with the errors added,

$$\Delta T = \Delta T_0 \exp[-a\,(\mu_t+\varepsilon_1) + \varepsilon_2], \qquad (12.9)$$

where measured time is given by t= $\mu_t+\varepsilon_1$. This is an RMM model with parameters $\lambda=1,\alpha= -a$.

For the law (2.13), which binds together saturation vapor pressure, P_{vapor}, with temperature T= $\mu_T + \varepsilon_1$, we have, with the errors added,

$$P_{vapor} \sim \exp\{-E_B / [\kappa_B(\mu_T+\varepsilon_1)] + \varepsilon_2\}, \qquad (12.10)$$

where κ_B is the Boltzmann factor, T is on a Kelvin scale, and E_b is a particle binding energy. This model may be derived from RMM with parameters $\lambda=-1$, $\alpha= E_b/\kappa_B$.

The "Barometric formula" is [Eq. (2.14), with the errors added]

$$\rho(\mu_K) = \rho(0) \exp\{-(Mg/R)(\mu_K+\varepsilon_1) + \varepsilon_2\}. \qquad (12.11)$$

This model is (12.1a) with parameters $\lambda=1$, $\alpha= -(Mg/R)$.

From Einstein's Special Theory of Relativity (2.15), the formula for total energy carried by a moving object is (without errors)

$$E(L) = M_0C^2 (1- L)^{-1/2}, \qquad (12.12)$$

with L=$(v/C)^2$. This power relationship can be derived from RMM with parameters $\eta= 1-L$, $\lambda=0$, $\alpha= -(1/2)$.

Planck's radiation law is [Eq. (2.16), with the error, ε_1, added]

$$S(\mu_{T0}) = (2\pi c^2h/\lambda_0^5) \{\exp[(hc/\lambda_0k)/(\mu_{T0}+\varepsilon_1)]-1\}^{-1}, \qquad (12.13)$$

where S is spectral radiancy, λ_0 is the wavelength at temperature $T_0= \mu_{T0}+\varepsilon_1$, and h is the Planck constant. The response: $(2\pi c^2h/\lambda_0^5)/S(\mu_{T0}) + 1$ (a linear transformation of 1/S) is an exponential-power relationship, derivable from RMM with parameters $\lambda=-1$, $\alpha= -(hc/\lambda_0k)$.

The energy levels of the stationary states of the hydrogen atom are [Eq. (2.17), re-introduced here for convenience]

$$E(n) = - 13.6 \text{ eV} / n^2, \; n=1,2,3,.., \qquad (12.14)$$

where n is the quantum number, and the numerical coefficient is the electron ground-state energy, measured in electron-volts (eV). This model can be derived from RMM with parameters $\eta = n$, $\lambda = 0$, $\alpha = -2$. Since this is a theoretical model and n is an integer, no errors are added.

Radioactive decay rate, R, is given by [Eq. (2.18) with the errors added]

$$R(\mu_t) = \{\, R_0 \exp[-k(\mu_t + \varepsilon_1)]\}(1 + \varepsilon_2), \qquad (12.15)$$

with time $t = \mu_t + \varepsilon_1$. This is an RMM model with parameters $\lambda = 1$, $\alpha = -k$.

The power output at a nuclear reactor at time t, P(t), is given by [Eq. (2.19) with the errors added]

$$P(\mu_t) = P_0 \exp[(k/t_{gen})(\mu_t + \varepsilon_1) + \varepsilon_2], \qquad (12.16)$$

where $t = \mu_t + \varepsilon_1$. This is RMM with parameters $\lambda = 1$, $\alpha = k/t_{gen}$.

12.1.3. Electrical engineering

Coulomb's law is [Eq. (2.20), with the errors added]

$$F(\mu_r) = \{k(q_1 q_2)/(\mu_r + \varepsilon_1)^2\}(1 + \varepsilon_2), \qquad (12.17)$$

where $r = \mu_r + \varepsilon_1$ is the distance. This model is a power law, an RMM model with parameters $\lambda = 0$, $\alpha = -2$.

For a charging capacitor, connected to a charging source having emf E, the voltage across the capacitor at time t, as measured by a resistor connected in parallel, is [Eq. (2.21), with the errors added]

$$V_R(\mu_t) = E \exp[-(\mu_t + \varepsilon_1)/RC + \varepsilon_2], \qquad (12.18)$$

with $t = \mu_t + \varepsilon_1$.

Similarly, a discharging capacitor, with an initial (full) charge of q_0, will have at time t a charge of [Eq. (2.22), with the errors added]

$$q(\mu_t) = q_0 \exp[-(\mu_t + \varepsilon_1)/RC + \varepsilon_2]. \qquad (12.19)$$

Both (12.18) and (12.19) are RMM models, governed by an exponential relationship.

For an electricity current of intensity I that runs through a loop of radius R (area S), the magnetic field is [Eq. (2.23), with the errors added]

$$B(\mu_d) = [(\mu_0/2\pi)(IS)/(\mu_d + \varepsilon_1)^3]\,(1 + \varepsilon_2)$$

$$= [(\mu_0/2)(IR^2)/(\mu_d + \varepsilon_1)^3]\,(1 + \varepsilon_2), \qquad (12.20)$$

This RMM model has parameters $\lambda=0$, $\alpha=-3$.

12.1.4. Growth models

Relating to the growth models in Section 2.7, we have for the logistic growth model (2.32) a power-exponential relationship, for Gompertz (2.33) an exponential-exponential relationship, and for Weibull (2.34) an exponential-power relationship. All are special cases of the RMM model, occasionally requiring introduction of additional parameters to the basic model (as explained in Chapter 7).

12.2. Current Models of Random Variation as RMM Models

In this section the RMM error distribution is shown to comprise as special cases some known statistical distributions, response transformations and distributional approximations. In expounding these cases, we are assisted mainly by Stuart and Ord (1987, henceforth SO) and Johnson, Kotz and Balakrishnan (1994, V. 1, henceforth JKB). Much of the material in this section has appeared in Shore (2004).

12.2.1. The Johnson families of distributions

Johnson (1949, 1954) and Tadikamalla and Johnson (1982) described sets of transformations of a random variable, where the transformations result in a unit normal variable, Z. The most well-known of these appear in Johnson (1949). These are (JKB, p. 34):
The S_B family

$$Z = \gamma + \delta \log [Y/(1-Y)], \, 0 \le Y \le 1, \qquad (12.21)$$

and the S_U family

$$Z = \gamma + \delta \sinh^{-1}Y = \gamma + \delta \log [Y+(Y^2+1)^{1/2}], \, -\infty \le Y \le \infty, \quad (12.22)$$

where $Y= (X-\xi)/\lambda$, X is the original random variable that we wish to model via a member of the Johnson family, and ξ and λ (>0) are location and scale parameters, respectively.

Consider first the S_B family. Introducing $U= (Z-\gamma)/\delta$, we obtain from (12.21)

$$U = \log[Y/(1-Y)], \tag{12.23}$$

or

$$Y = \exp[\ \log(1+e^U)^{-1} + U\ ]. \tag{12.24}$$

If in the RMM model, with $\eta=1$ [relate to (12.1a)], we assume that $\varepsilon_1 = \exp(U)$, $\varepsilon_2 = U$, we obtain

$$Y = \exp\{(\alpha/\lambda)[(1+e^U)^\lambda - 1] + \mu_2 + U\}. \tag{12.25}$$

This is equivalent to (12.24) with $\lambda=0$, $\alpha=-1$, $\mu_2=0$. Note, that in deriving (12.25) we have modified the assumption about the distribution of the error ε_1 (assuming log-normality instead of normality). Given the approximate equivalence of these two distributions, as representations of the errors' distribution (we have alluded to this earlier, in Chapter 7), the modified assumption does not violate the basic equivalence between the RMM error distribution and the S_B family.

For the S_U system we have from (12.22)

$$Y = [\exp(U)-\exp(-U)]/2 = (1/2)[\exp(2U)-1]\ /\ \exp(U), \tag{12.26}$$

or

$$Y = \exp\{\log[-1+\exp(2U)] - \log(2) - U\}. \tag{12.27}$$

It is easily recognized that (12.27) is an RMM model. From (12.1a), introducing $\eta=-1$, we obtain

$$Y = \exp\{(\alpha/\lambda)[(-1+\varepsilon_1)^\lambda - 1] + \mu_2 + \varepsilon_2]\}. \tag{12.28}$$

This is (12.27), with $\alpha=1$, $\lambda=0$, $\mu_2 = -\log(2)$, and errors: $\varepsilon_1 = \exp(2U)$, $\varepsilon_2 = -U$. Note that again ε_1 is log-normally distributed, and thus deviate from our initial assumption (Chapter 7) that both errors are normally distributed. However, for $2U<<1$: $\exp(2U) \approx 1 + 2U$, and the initial assumption about the approximate normality of both errors is restored for this case.

A third transformation that belongs to the Johnson family results in Y having a log-normal distribution, again an RMM model ($\alpha=0$).

12.2.2. Tukey g- and h-systems of distributions

In Hoaglin, Mosteller and Tukey (1985), Hoaglin introduces an approach, originally introduced by Tukey (1977), to model random variables in terms of transformations of the unit normal variable. This approach results in Tukey's *g*- and *h*- systems, which were suggested to model "skewed variables" and "elongated" symmetric variables, respectively. Combining the two systems to represent both skewness and elongation result in the gh system.

The three systems of distributions are, respectively (refer to Hoaglin *et al.*, 1985, Chapter 11),

$$Y_g(Z) = [\exp(gZ) - 1]/g, \; g \neq 0, \tag{12.29}$$

$$Y_h(Z) = (Z) \exp(hZ^2/2), \tag{12.30}$$

$$Y_{g,h}(Z) = (1/g) [\exp(gZ) - 1] \exp(hZ^2/2), \; g \neq 0, \tag{12.31}$$

where Z is a unit normal variable.

These systems may be derived as special cases from the RMM error distribution. Referring first to (12.29), since $Y_g(Z)$ is a log-normal variable, adjusted for location and scale in a particular manner, it is a special case of the RMM error distribution.

To obtain (12.30), we have from the RMM error distribution [refer to (12.1b), with $\alpha=h$, $\lambda=2$, $\log(M)= h/2$]

$$Y = \exp\{(h/2)(1+\varepsilon_1)^2\} (1 + \varepsilon_2). \tag{12.32}$$

Assume that in (12.32): $1+\varepsilon_1= 1+\varepsilon_2= Z$, namely, the errors are normally distributed with mean -1 and unit variance. We obtain from the RMM error distribution the *h*-system (12.30). Again note that to obtain the *h*-system, the basic RMM model is preserved, and only the assumptions about the distributions of the errors have to be slightly modified, relative to the original formulation (Chapter 7).

The *gh* system (12.31) may be considered as a compilation of (12.29) and (12.30). To realize that put g=0 in (12.31) to obtain (12.30). Evidently, (12.31) is also a special case of the RMM model.

12.2.3. Fisher's transformation of the sample correlation

Fisher (1921) introduced the following transformation of the sample correlation, r:

$$X = (1/2) \log[(1+r)/(1-r)]. \tag{12.33}$$

It may be shown that

$$X' = X - (1/2) \log[(1+\rho)/(1-\rho)], \tag{12.34}$$

where ρ is the population correlation, converges to normality more rapidly then r itself. Furthermore, to $O(n^{-2})$ the mean of X' is $(1/2)\rho/(n-1)$ and the variance is $1/(n-1)$, where n is the sample size. The skewness measure, Sk, and the kurtosis measure, Ku, are (SO, V.1, p. 533, and Kendall and Stuart, 1976, p. 95)

$$Sk = \rho^3/(n-1)^{(3/2)} + 0(n^{-3/2}), \quad Ku = 2/(n-1) + O(n^{-2}). \qquad (12.35)$$

This implies that although Fisher's transformation intends to stabilize the variance, it also tends to normalize since the skewness measures approach that of the normal distribution more rapidly than r $[O(n^{-3/2})$ compared to $O(n^{-1/2})$, respectively].

Assume that

$$Z = (n-1)^{1/2} [X' - (1/2)\rho/(n-1)] = \gamma + \delta X \qquad (12.36)$$

is approximately a unit normal variable. Re-writing as $U = (Z-\gamma)/\delta = X$, we obtain from (12.33)

$$r = [1 - \exp(2U)] / [1 + \exp(2U)], \qquad (12.37)$$

where U=0 corresponds to r=0.

Introducing into (12.1b): $\varepsilon_1 = \exp(2U)$, $\varepsilon_2 = -\varepsilon_1$, we obtain thereof:

$$Y = (M) \exp\{(\alpha/\lambda)[(1+ \exp(2U))^\lambda - 1]\} [1 - \exp(2U)]. \qquad (12.38)$$

This is equivalent to (12.37) with M=1, λ=0, α=-1. Again only slight changes in the assumptions about the distributions of the errors were needed to obtain the Fisher transformation from the RMM error quantile function.

12.2.4. Haldane power-transformation and Wilson-Hilferty approximation to χ^2

Wilson and Hilferty (1931) suggested a cube-root transformation of the χ^2 variable (with ν degrees of freedom) in order to normalize it:

$$Z = \{ (\chi^2/\nu)^{1/3} + 2/(9\nu) - 1\}(9\nu/2)^{1/2}, \qquad (12.39)$$

where Z is approximately a unit normal variable. Haldane (1938) generalized this result, and derived two transformations for a random variable, X, which has all its cumulants of order n, where n is the sample size (for example, the binomial has this property).

Let κ_j be the j-th cumulant of X. Defining

$$Y = (X/\kappa_1)^h = [1 + (X-\kappa_1)/\kappa_1]^h, \qquad (12.40)$$

Haldane (1938) has shown that if we put: $h = 1-\kappa_1\kappa_3/(3\kappa_2^2)$, Y then has skewness of order $n^{-3/2}$ (as compared to $n^{-1/2}$ for X). More rapid convergence to normality is achieved. Furthermore, by adding a constant to X to make its mean equal to g, and choosing

$$g = (12\kappa_3\kappa_2^2) / (20\kappa_3^2 - 9\kappa_4\kappa_2), \ h = (16\kappa_3^2 - 9\kappa_4\kappa_2) / (20\kappa_3^2 - 9\kappa_4\kappa_2), \qquad (12.41)$$

the re-defined variable

$$Y = [1 + (X-\kappa_1)/g]^h \qquad (12.42)$$

has skewness of order $n^{-3/2}$ (as before), and kurtosis of order n^{-2} (compared to n^{-1} for X). Find details in Kendall and Stuart (1976, p. 119).

Let us re-write the power transformed variable as

$$\eta + \delta Z = (\xi + \beta X)^h, \qquad (12.43)$$

where Z is standard normal, and all the other terms are real-valued parameters. Then, Haldane's transformation implies

$$X = (-\xi/\beta) + (1/\beta)(\eta + \delta Z)^{1/h}. \qquad (12.44)$$

This is a power transformation of a normal variable, a linear transformation of a special case of RMM ($\lambda=0$, $\alpha = 1/h$).

12.2.5. Box-Cox normalizing transformation

This well-known transformation (Box and Cox, 1964) is widely used in empirical relational modeling, when the response is non-normal. The transformation is commonly applied to the response so that the normal scenario is revoked and linear regression analysis can be applied to the transformed data. In this sub-section, we address the Box-Cox (BC) normalizing transformation as an empirical model for random variation.

Denote by $Y^{(\lambda)}$ the BC transformed variable:

$$Y^{(\lambda)} = \begin{cases} (Y^{\lambda'} - 1)/\lambda', & \lambda' \neq 0, \\ \log(Y), & \lambda' = 0. \end{cases} \qquad (12.45)$$

Note that the parameter of this transformation is λ' to distinguish it from the RMM parameter, λ.

Re-writing $Y^{(\lambda)} = \eta' + \delta Z$, where Z is a unit normal variable, we obtain from (12.45)

$$Y = [\lambda'(\eta' + \delta Z) + 1]^{1/\lambda'} = (1 + \lambda'\eta' + \lambda'\delta Z)^{1/\lambda'}. \tag{12.46}$$

This implies that the BC transformation tacitly assume that Y can be expressed as a power transformation of a normal variable.

The inverse BC transformation (12.46) is derivable from the RMM model (12.1a), with parameters

$$\eta = 1 + \lambda'\eta', \ \sigma_{\varepsilon 1} = \lambda'\delta, \ \lambda = 0, \ \alpha = 1/\lambda', \ \mu_2 = 0, \ \sigma_{\varepsilon 2} = 0. \tag{12.47}$$

For $\lambda'=0$, the inverse BC transformation results in a log-normal variable, also a special case of the RMM model.

12.2.6. Cauchy distribution

Let U and V be independent standard normal variables. The ratio

$$Y = U / V \tag{12.48}$$

has a standard Cauchy distribution (JKB, p. 318). Let us re-write Y as

$$Y = \delta\exp[\log (\delta V)^{-1}] \ U, \tag{12.49}$$

where

$$\delta = \begin{cases} -1, \ V < 0, \\ 1, \ V > 0, \end{cases} \tag{12.50}$$

is introduced to allow for negative values of V in (12.49).

From the RMM error quantile function [Eq. (12.1b) with $\eta=0$], introducing M=1, $\varepsilon_1 = V$, $1 + \varepsilon_2 = U$ (namely, ε_1 is standard normal and ε_2 is normal with mean -1 and unit variance), we obtain

$$Y = \exp\{(\alpha/\lambda)[(V)^\lambda - 1]\}U. \tag{12.51}$$

For $\lambda=0$, $\alpha=-1$, the RMM model (12.51), with the introduction of δ, is equivalent to the Cauchy distributed Y (12.49).

12.2.7. Generalized Inverse Gaussian distribution and the Levy distribution

Let D be the distance traveled by a particle in time interval T, when it moves along a line with a uniform velocity V_0. If the particle is now subject to a linear Brownian motion (namely, the velocity, V, is random), it is well known that for a fixed time $T=t_0$, the distance over which the particle travels is a random variable having normal distribution with mean $V_0 t_0$, and variance βt_0, where β is a diffusion constant (find details in JKB, p. 260). Conversely, if the distance is constant, D_0, then the time, T, required to cover this distance, follows an Inverse Gaussian distribution, with density function (JKB, p. 260)

$$f(t) = (2\pi\beta t^3)^{-1/2} D_0 \exp\{-(D_0-V_0 t)^2/(2\beta t)\}, \ t > 0. \qquad (12.52)$$

The standard ("canonical") two-parameter density function is (one of several possible forms, refer to JKB, p. 262, or SO, p. 382)

$$f(y) = (2\pi y^3/\lambda)^{-1/2} \exp\{-(\lambda/2\mu^2)(y-\mu)^2 / y\}, \ y, \lambda, \mu > 0. \quad (12.53)$$

Writing the exponent as $-(1/2)S$, it may be shown (SO, p. 383, based on Folks and Chhikara, 1978), that S is distributed as χ^2 with one degree of freedom, namely,

$$S = (\lambda/\mu^2)(Y-\mu)^2 / Y = Z^2, \qquad (12.54)$$

where Z is a standard normal variable. It is easy to see that Z^2 is a member of the RMM error distribution (put in the RMM model: $\eta=0$, $\lambda=0$, $\alpha=2$, assume that ε_1 is standard normal and that $\mu_2=0$, $\varepsilon_2=0$). Thus, the function of the Inverse Gaussian variable, Y, given by S, is a member of the RMM error distribution. Furthermore, $1/S$ is distributed like $1/Z^2$. The latter has density function

$$f(y) = (2\pi)^{-1/2} \exp[-1/(2y)] \ y^{-3/2} , \ 0 \le y < \infty, \qquad (12.55)$$

which is the Levy distribution (refer for details to SO, Exc. 11.25, p. 382). Since $1/Z^2$ is a member of the RMM family of models (insert as above with the replaced value of $\alpha= -2$), the Levy distribution is also a special case of the RMM error distribution.

Robert (1991) has suggested a particular form of the generalized inverse Gaussian distribution, with density function

$$f(x) = \{K(\alpha,\mu,\tau)/|z|^{\alpha}\} \exp\{-[(1/x)-\mu]^2 / (2\tau^2)\}, \ \alpha > 1, \tau > 0, \ (12.56)$$

where K is a normalizing transformation, expressed in terms of a confluent hyper-geometric function. For $\alpha=2$, $\tau=\sigma$, X of (12.56) follows the distribution of the inverse of U, namely, $X=1/U$, where U is $N(0, \sigma^2)$. It is easy to verify that X is a member of the RMM family of models.

Thus, for this special case, the RMM provides also a member of the generalized inverse normal distribution, as the latter has been defined by Robert (1991). Find details in JKB (p. 288).

12.2.8. Generalized gamma distributions

Suppose that $Y = [(X-\gamma)/\beta]^c$ ($c>0$) has the standard gamma distribution, with parameter α and density function

$$f(y) = y^{\alpha-1}\exp(-y)/\Gamma(\alpha), \; y \geq 0. \tag{12.57}$$

Then X has a generalized gamma distribution, as defined by Stacy (1962). The density function is

$$f(x) = \{c/[\beta^{c\alpha}\Gamma(\alpha)]\} \, (x-\gamma)^{c\alpha-1} \exp\{-[(x-\gamma)/\beta]^c\}, \, x \geq \gamma. \tag{12.58}$$

Special cases of the generalized gamma variable include Weibull ($\alpha=1$), half-normal distributions ($\alpha=1/2$, $c=2$, $\gamma=0$) and the ordinary gamma distribution ($c=1$). It can be shown that while for the gamma variable, Y, the skewness and kurtosis measures are, respectively, $Sk(Y) = 2/\alpha^{1/2}$, $Ku(Y) = 6/\alpha$ (JKB, p. 339), for $W = \log(Y)$ the respective measures are (JKB, p. 383): $Sk(W) \approx -1/(\alpha-1/2)^{1/2}$, $Ku(W) \approx 2/(\alpha-1/2)$.

Thus W is closer to normality than the standard gamma. Re-writing W in terms of the standard normal variable, Z, we have, in terms of the generalized gamma, X,

$$\eta + \delta Z = \log[(X-\gamma)/\beta]^C, \tag{12.59}$$

or

$$X = \gamma + \exp[\log(\beta) + (1/c)(\eta + \delta Z)]. \tag{12.60}$$

We realize that the distribution of a generalized gamma variable may be approximated by a log-normal distribution (see also respective comment in JKB, p. 344). Therefore the generalized gamma is also approximately a member of the RMM error distribution. Note, however, that when the quantile function of the RMM error distribution is fitted to members of the generalized gamma distribution (like Weibull and gamma), the resulting approximation has appreciably better accuracy than obtained by the above log-normal approximation. Furthermore, the derived value of the parameter α (in the RMM model) is far from the value $\alpha=0$, suggested by approximation (12.59) (refer to Shore, 2000, 2001, for details, and also to Chapters 10, 19 and 20).

References

Comment: For model-related references, refer to Chapter 2

[1] Box, G. E. P., Cox, D. R. (1964). An analysis of transformations. *Journal of the Royal Statistical Society, Series B,* 26, 211-243.

[2] Fisher, R. A. (1921) On the probable error of a coefficient of correlation deduced from a small sample. *Metron,* 1(4), 1.

[3] Folks, J. L., Chhikara, R. S. (1978). The inverse Gaussian distribution and its statsitcial application - a review. *Journal of the Royal Statistical Society, Series B,* 40, 263.

[4] Haldane, J. B. S. (1938). The approximate normalization of a class of frequency distributions. *Biometrika,* 29, 392.

[5] Hoaglin, D. C., Mosteller, F., Tukey, J. W. (1985). *Explorative Data, Tables, Trends, and Shapes.* John Wiley & Sons.

[6] Johnson, N. L. (1949). Systems of frequency curves generated by methods of translation. *Biometrika,* 36, 149-176.

[7] Johnson, N. L. (1954). Systems of frequency curves derived from the first law of Laplace. *Trabajos de Estadistica,* 5, 283-291.

[8] Johnson, N. L., Kotz, S., Balakrishnan, N. (1994). *Continuous Univariate Distributions — Volume 1.* 2nd Edition. John Wiley & Sons.

[9] Kendall, M., Stuart, A. (1976). *The Advanced Theory of Statistics. V. 3: Design and Analysis, and Time-series.* Charles Griffin & Co, London.

[10] Robert, C. (1991). Generalized inverse normal distributions. *Statistics & Probability Letter,* 11, 37-41.

[11] Shore, H. (2000). General control charts for variables. *International Journal of Production Research,* 38(8), 1875-1897.

[12] Shore, H. (2001). Modeling a non-normal response for quality improvement. *International Journal of Production Research,* 39 (17), 4049-4063.

[13] Shore, H. (2003). Response Modeling Methodology (RMM)- A new approach to model a chemo-response for a monotone convex/concave relationship. *Computers and Chemical Engineering,* 27(5), 715-726.

[14] Shore, H. (2004). Response Modeling Methodology (RMM)- Current distributions, transformations and approximations as special cases of the RMM error distribution. *Communications in Statistics (Theory and Methods),* 33(7), 1491-1510.

[15] Shore, H., Brauner, N., Shacham, M. (2002). Modeling physical and thermodynamic properties via inverse normalizing transformations. *Industrial and Engineering Chemistry Research,* 41, 651-656.

[16] Stacy, E. W. (1962). A generalization of the gamma distribution. *Annals of Mathematical statistics,* 33, 1187-1192.

[17] Stuart, A., Ord, J. K. (1987). *Kendall's Advanced Theory of Statistics, V. 1: Distribution Theory.* Charles Griffin & Co. Ltd. London.

[18] Tadikamalla, P. R., and Johnson, N. L. (1982). Systems of frequency curves generated by transformations of logistic variables. *Biometrika,* 69, 461-465.

[19] Tukey, J. W. (1977). Modern techniques in data analysis. *NSF-sponsored regional research conference at Southeastern Massachusetts University, North Darmouth, MA.*

[20] Wilson, E. B., Hilferty, M. M. (1931). The distribution of chi-square. *Proceedings of the National Academy of Sciences*, 17, 684-688.

Chapter 13

Evaluating RMM for Compliance

13.1. Introduction

Ten requirements that an empirical modeling methodology should satisfy have been expounded, explained and demonstrated in Chapter 6. Current major methodologies for empirical modeling were then assessed for compliance with these requirements. In particular, existing major methodologies for *relational* modeling, namely, linear regression, data normalization and generalized linear models (GLM) were addressed. We have realized that these methodologies fall short of satisfying the desirable requirements.

In previous chapters of this part of the book (Chapters 7-12), the RMM approach has been developed, the model, its variations and fitting and estimation procedures. We have also demonstrated that numerous current models of both systematic variation and random variation are indeed special cases of the RMM model.

It is now appropriate to subject the RMM approach to the same scrutiny that current methodologies have undergone in Chapter 6.

Evaluating RMM for compliance to requirements is partitioned into two, similarly to earlier evaluations: Compliance to requirements relevant to modeling systematic variation (Section 13.2) and compliance to those relevant to modeling random variation (Section 13.3).

13.2. Compliance for Modeling Systematic Variation

We compare in this section the RMM properties relative to the two major

219

alternatives for modeling systematic variation, namely, the BC normalizing transformation and GLM. Linear regression is a special case of both and therefore will not be addressed separately.

Evaluation of RMM relative to the requirements, as expounded in Chapter 6, is displayed in a tabular form in Table 13.1. This is an expansion of Table 6.1, which now includes RMM. Evaluation of compliance ranges from low ("-") to medium ("+") to high ("++").
The comments that follow intend to focus on some properties that seem to set RMM apart from current methodologies, and discuss these properties. Some further comparisons are detailed in the "Example" sections that follow the introduction of each requirement in Chapter 6.

Consider first the most important Requirement 1. The latter requires that the modeling methodology allows for a monotone convex

Table 13.1. Comparison of compliance with requirements of major approaches for modeling variation

Requirement	Methodology			
	Linear regression	BC Transformation	GLM	RMM
1 (Convexity)	-	+	+	++
2 (All Data Driven)	-	++	-	++
3 (Incl. Linear)	++	++	++	++
4 (Dual-error)	+	-	-	++
5 (Changing error dist.)	-	-	-	++
6 (Moments preserved)	-	-	-	++
7 (Representability)	-	-	+	++
8 (Parsimony)	++	++	++	++
9 (Compat. with current method.)	++	++	++	++
10 (Easy to apply)	++	+	+	+

relationship between the linear predictor and the response. The richer the spectrum of monotone convex relationships the methodology accommodates, the more universal its status as a general platform for empirical modeling, and the more compliant it is with Requirement 1.

The "richness" of RMM in representing various convex relationships has been demonstrated in Chapter 12, where current engineering, scientific and statistical models have been shown to be special cases of the RMM model (refer also to Shore, 2004ab).

In Chapter 6 we have demonstrated, via the inverse function, that the BC transformation implicitly assumes two types of monotone convex relationships: Power and exponential. The transition from one type of relationship to another is continuous, via the parameter λ, just as with RMM. By contrast, in GLM a link function has to be determined in advance. This eliminates the possibility of a continuous transition from one form of convex relationship to another, as afforded when data analysis determines the final structure of the model (refer to the discussion of Requirement 2 below). From a different perspective, GLM is more flexible than the BC normalizing transformation since unlike the latter, which is confined to a power data transformation, GLM has no restrictions with respect to the permissible link functions.

Finally, with RMM excepted, no existing methodology allows convexity to be modeled as a continuous spectrum, which is unbounded from above (relate to the "Ladder", introduced in Chapter 3).

It could therefore be asserted that RMM, with its richness of possible convex relationships, is the most compliant with Requirement 1, followed by the normalizing approach and GLM. The latter two are graded equally in Table 13.1 with respect to Requirement 1.

Requirement 2 asks for all features of the model to be determined by the data analysis ("Data driven"). Consider GLM. This methodology requires three different decisions in the modeling process: Determining the link function (transformation of the mean to achieve additivity of effects), the error distribution, or, equivalently, the relationship between the mean and the variance and the composition of the linear predictor. Only the third decision is determined by data analysis, when a stepwise procedure is followed. This should be contrasted with BC transformation, where the parameter λ, estimated from data analysis, is a shape parameter that determines the structure of the final model.

In RMM, two parameters dominate in determining the model's structure, α and λ (for extremely convex relationships, more than two parameters may be needed, as explained in Chapter 7). Availability of two parameters to determine the shape of the model (in addition to a scale parameter, μ_2) is what confers on RMM its flexibility in accommodating such a wide spectrum of monotone convex relationships. If empirical modeling aspires to construct models which require no a-priori assumptions, RMM is the most compliant, followed by the BC transformation and finally by GLM.

Requirement 4 for a dual-error structure is explicitly embedded in the RMM model. In terms of compliance, RMM is the only empirical modeling methodology that explicitly satisfy this requirement (relate however to the discussion following presentation of Requirement 4 in Chapter 6, where GLMM seems to come close to fulfilling this requirement).

Requirement 5 demands that we take account of the empirical fact that systematic variation that spans several orders of magnitude may modify the underlying error distribution in a major way (namely, not by merely changes in location and/or scale). Let us examine, therefore, how each of the examined methodologies accommodates this requirement.

For the BC transformation, it is assumed that whatever the value of the linear predictor, the transformed variable is normally distributed. Expressed in terms of the original variable, Y [refer, for example, to (12.46)], the inverse BC transformation implies that proximity to normality of the modeled response is achieved not as a result of changes in the linear predictor, but when the parameter λ approaches unity. Thus, approximate normality for the error distribution as a result of large increase in the linear predictor is not rendered by this modeling approach. Neither is decoupling of the variance from the mean implied by it.

Implementation of GLM requires that the error distribution is specified in advance, and, consequently, so is the relationship between the variance and the mean. Since the latter relationship, needed to determine weights in the iteratively re-weighted least-squares estimation routine, is specified in advance, no decoupling of these distributional parameters is indeed possible. GLM is far from satisfying Requirement 5.

RMM differs in a major way. Since in RMM both α and λ determine the structure of the model, and since the linear predictor is associated with a separate error, ε_1, the relative size of this error (relative to η) will determine whether changes in the systematic variation component (changes in η) would affect the shape of the error distribution. As η becomes larger, ε_1 tends to be negligible, and the variance of the error distribution will tend to stabilize. The error distribution will then be determined mainly by the other error term, ε_2, which has a stable variance. As a consequence, the log-transformed response will also have a stable variance.

Requirements 6 and 7, though relevant to modeling of systematic variation (since they relate to the error distribution of the *relational* model), are addressed in the next section.

13.3. Compliance in Modeling Random Variation

To discuss the relative merits of the RMM error distribution as a general platform for modeling random variation, let us recall variations of the basic model that result from assuming various error structures. All these variations are developed in Chapter 9 and repeated here for convenience.

The basic model, with correlated errors, is

$$Y = \exp\{(\alpha/\lambda)[(\eta+\varepsilon_1)^\lambda - 1] + \mu_2 + \varepsilon_2\}$$

$$= \exp\{(\alpha/\lambda)[(\eta+\sigma_{\varepsilon1}Z_1)^\lambda - 1] + \mu_2 + \sigma_{\varepsilon2}[\rho Z_1 + (1-\rho^2)^{(1/2)} Z_2]\},$$

$$W = (\alpha/\lambda)[(\eta+\sigma_{\varepsilon1}Z_1)^\lambda - 1] + \mu_2 + \sigma_{\varepsilon2}[\rho Z_1 + (1-\rho^2)^{(1/2)} Z_2], \qquad (13.1)$$

where ε_1 and ε_2 are errors derived from a bi-variate normal distribution with correlation ρ. By conditioning $Z_2 \mid Z_1{=}z_1$, we have obtained for the errors

$$\varepsilon_1 = \sigma_{\varepsilon1}Z_1, \quad \varepsilon_2 = \sigma_{\varepsilon2}[\rho Z_1 + (1-\rho^2)^{(1/2)} Z_2], \qquad (13.2)$$

where Z_1 and Z_2 are *independent* standard normal variables.

Assuming that the linear predictor is constant, arbitrarily setting it to one, one obtains from (13.1) and (13.2)

$$Y = \exp\{(\alpha/\lambda)[(1+\varepsilon_1)^\lambda - 1] + \mu_2 + \varepsilon_2\}$$

$$= \exp\{(\alpha/\lambda)[(1+\sigma_{\varepsilon1}Z_1)^\lambda - 1] + \mu_2 + \sigma_{\varepsilon2}[\rho Z_1 + (1-\rho^2)^{(1/2)} Z_2]\}$$

$$W = (\alpha/\lambda)[(1+\sigma_{\varepsilon1}Z_1)^\lambda - 1] + \mu_2 + \sigma_{\varepsilon2}[\rho Z_1 + (1-\rho^2)^{(1/2)} Z_2]. \qquad (13.3)$$

Variations of the basic model, assuming different error structures, are (refer to Chapter 9 for details)

$$Y = \exp\{(\alpha/\lambda)[(1+\varepsilon_1)^\lambda - 1]+\mu_2\} \, (1+\varepsilon_2)$$

$$= \exp\{(\alpha/\lambda)[(1+\sigma_{\varepsilon1}Z_1)^\lambda - 1] + \mu_2\} \, \{1+\sigma_{\varepsilon2}[\rho Z_1 + (1-\rho^2)^{(1/2)} Z_2]\}, \quad (13.4)$$

$$Y = \exp\{(\alpha/\lambda)[\exp(\lambda\varepsilon_1) - 1] + \mu_2 + \varepsilon_2\}$$

$$= \exp\{(\alpha/\lambda)[\exp(\lambda\sigma_{\varepsilon1}Z_1) - 1] + \mu_2 + \sigma_{\varepsilon2}[\rho Z_1 + (1-\rho^2)^{(1/2)} Z_2]\}, \quad (13.5)$$

$$Y = \exp\{(\alpha/\lambda)[\exp(\lambda\varepsilon_1) - 1] + \mu_2\} \, (1+\varepsilon_2)$$

$$= \exp\{(\alpha/\lambda)[\exp(\lambda\sigma_{\varepsilon1}Z_1) - 1] + \mu_2\} \, \{1+\sigma_{\varepsilon2}[\rho Z_1 + (1-\rho^2)^{(1/2)} Z_2]\}. \quad (13.6)$$

By conditioning $Z_1 \mid Z_2=z_2$, four equivalent forms of (13.3)-(13.6) can similarly be developed (find details in Chapter 9).

Let us examine how well the RMM error distribution, with variations given by (13.3)-(13.6), fulfills the requirements that relate to modeling random variation (Requirements 6 and 7).

Requirement 6 asserts that at least the first three or four moments of the underlying error distribution be preserved in the empirically fitted model. As explained earlier (Chapter 6), this requires two features. First, that there are enough parameters, apart from those associated with the linear predictor (assumed now to be constant). This would allow flexibility in reflecting variously-shaped distributions. Secondly, that the allied estimation procedure guarantees that the estimated quantile function preserves well (low MSEs) at least the first three of four moments of the true error distribution.

The new RMM model attains both features. With respect to the first feature, the three parameters, $\{\alpha, \lambda, \mu_2\}$, provide the flexibility needed to accommodate variously-shaped error distributions. This has been demonstrated for a large set of distributions, with different shape characteristics, in Shore (2000), for the inverse normalizing transformations (INTs), and in Shore (2002), for the general RMM model. Chapter 19 is dedicated to demonstrating this feature.

With respect to the second property (low MSEs for the moments of the fitted error distribution), moment matching procedures, based on moments of second degree at most (partial and complete), have been repeatedly shown to result in low MSEs (Shore, 1998, 2000, 2004c, and references therein). Estimation procedures for the RMM error distribution, developed in Chapters 11 and 20, are all confined to moments (partial and complete) of second-degree at most. Thus, the requirement for small MSEs is well attained.

Addressing Requirement 7, this calls for good coverage of the $(\sqrt{\beta_1}, \beta_2)$ plane. The issue of coverage was addressed in Shore (2002) and in Chapter 9. While unable, for computational reasons, to derive a comprehensive "map" of areas in the $(\sqrt{\beta_1}, \beta_2)$ plane occupied by the RMM error distribution, we have demonstrated, using different values for the parameters of this distribution, that a large spectrum of $(\sqrt{\beta_1}, \beta_2)$ values are obtainable.

The RMM error distribution is based on a transformation of a symmetric (normal) variable (Z_1) that spans monotone convexity ranging, in a continuous manner, from linear to power to exponential and beyond. Therefore the capability of the error distribution to represent distributions with extremely varied shapes is similar to other commonly applied parameter-rich families of distributions, like Burr, Pearson or Johnson. With respect to the latter, we have demonstrated that the Johnson family is a special case of the RMM error distribution (Chapter 12 and also Shore, 2004b). RMM therefore covers at least areas in the $(\sqrt{\beta_1}, \beta_2)$ plane occupied by the Johnson family of distributions.

Contrasted with how current approaches to model random variation comply with Requirements 6 and 7 (refer to an earlier discussion in Chapters 5 and 6), it seems that adopting the RMM error distribution as a platform for modeling random variation can provide benefits not widely entertained by current methodologies.

References

[1] Shore, H. (1998). A new approach to analysing non-normal quality data with application to process capability analysis. *International Journal of Production Research (IJPR)*, 36(7), 1917-1933.

[2] Shore, H. (2000). General control charts for variables. *International Journal of Production Research,* 38(8), 1875-1897.

[3] Shore, H. (2002). Response Modeling Methodology (RMM)- Exploring the implied error distribution. *Communications in Statistics (Theory and Methods),* 31(12), 2225-2249.

[4] Shore, H. (2004a). Response Modeling Methodology (RMM)- Validating evidence from engineering and the sciences. *Quality and Reliability Engineering International,* 20, 61-79.

[5] Shore, H. (2004b). Response Modeling Methodology (RMM)- Current distributions, transformations and approximations as special cases of the RMM error distribution. *Communications in Statistics (Theory and Methods),* 33(7), 1491-1510.

[6] Shore, H. (2004c). Non-normal populations in quality applications- A revisited perspective. *Quality and Reliability Engineering International,* 20(4), 375-382.

Part III

MODELING SYSTEMATIC VARIATION - APPLICATIONS

Chapter 14

Comparative Solutions for Relational Models

14.1. Introduction

This is the first in a series of chapters that demonstrate application of RMM to modeling systematic variation (Chapters 14-18). In this chapter we address data-sets that have appeared in the literature and analyzed by some existing methodology for empirical relational modeling. RMM solutions derived for a sample of problems are compared to solutions obtained by current approaches.

Since detailed RMM solutions, with respective major summary results extracted from the computer's runs, have already been demonstrated in Chapter 8, the results displayed here are confined to main results and goodness-of-fit statistics. Showing only essential statistics relevant to assess the quality of the derived solutions, one may find it easier to compare the effectiveness of the solutions attained by the different methodologies.

A sample of four problems is analyzed. The first two problems, related to experimental data and to field data, are entirely new to the book. RMM solutions are developed and compared to solutions derived by other methodologies. This is the subject of Section 14.2. In the next Section 14.3 we relate to two problems, already solved by RMM in Chapter 8. The two solutions are compared to those obtained by other current methodologies.

For all four problems, we compare the linear predictor (LP), obtained by RMM, to LPs obtained by linear regression (applied to the original response), by a Box-Cox (BC) normalizing transformation and via GLM.

Section 14.4 is a brief overview of current commonly used criteria to compare goodness-of-fit of models.

14.2. Two New Problems

14.2.1. Example 1 - The Windshield Molding Slugging Experiment (logistic regression)

This example addresses a 2^{4-1} fractional factorial experiment, conducted at an ITT Thompson plant in a stamping process manufacturing windshield molding. During the stamping operation, debris carried into the die appears as dents or slugs in the product. Martin, Parker and Zenick (1987) originally presented and analyzed the data from this experiment. The response is the number of "Good parts" (conforming parts) out of a lot of 1000 produced parts. Four factors were studied, each at two levels: (A) Poly-film thickness (0.0025, 0.00175), (B) Oil mixture (1:20, 1:10), (C) Gloves (cotton, nylon), (D) Metal blanks (dry underside, oily underside).

The experimental design, with the observations (number of good parts), are given in Table 14.1. The original researchers used a standard linear regression analysis on the original data, ignoring the binomial

Table 14.1. Example 1- Experimental Design and Observations

Run	Factors				Y
	A	B	C	D	("good parts")
1	+	+	+	+	338
2	+	+	-	-	826
3	+	-	+	+	350
4	+	-	-	-	647
5	-	+	+	-	917
6	-	+	-	+	977
7	-	-	+	-	953
8	-	-	-	+	972

nature of the response. They concluded that the main effects A, C and D are significant.

The data were re-analyzed by Hamada and Nelder (HN, 1997) in the framework of GLM, using a logit link function, and assuming a binomial distribution. They derived the following model:

$$\text{logit}(P) = \log[P/(1\text{-}P)]$$

$$= -0.605 + 3.127(1/2)(1\text{-}A) + 1.585(1/2)(1\text{-}C), \quad (14.1)$$

with P as the probability of manufacturing a good (conforming) part. Note, that this model does not appear in HN, who had introduced their design coded to (1, 2) and their estimates based on coding to (0,1), respectively. Nowadays the more conventional coding is (1,-1), respectively. The latter coding is used in Table 14.1 and in the model (14.1), based on the estimates given in HN.

Lewis, Montgomery and Myers (2001) re-analyzed the data, and, based on residual analysis, concluded that perhaps including in the model also the factor D and interactions AB and BC would deliver a better model, as judged also by the lengths of the confidence intervals for the means (the actual final model is not specified). Note, that AC and D are aliased (cannot be distinguished in the present design), so that an alternative model may be concluded.

Consider an RMM solution for this problem. First, define as the response the proportion of conforming items (Y/1000). This is a departure from the conventional practice in logistic regression where logit (the log of the odds ratio) is commonly defined as the response [refer to the LHS of (14.1)].

Starting with Phase 1, the number of terms to include in the response Taylor expansion needs to be determined. Given that we have four main effects and only eight observations, the option of using a fourth degree expansion is not possible (we would obtain infinity for the first eigenvalue in the CCA). Therefore, we examine only polynomial Taylor expansions of a second degree and a third degree.

The results for these two CCA runs are given in Table 14.2. For both options, the first canonical eigenvalue accounts for about 95% of the variation in the data. Given the small number of available observations,

Table 14.2. Example 1- Results from CCA with different number of terms in the Taylor approximation to G(Y) (optimal choice is starred). "YT$_i$" means terms up to and including YT$_i$

Can. Correlations	Taylor expansion up to term:	
(Significance)	YT$_2$*	YT$_3$
First Can. Cor.	.9976	.9992
(Significance)	(<.0037)	(<.0314)
Second Can. Cor.	.9581	.9809
(Significance)	(<.0389)	(<.0862)

and since no particular alternative is dominant in terms of the criteria expounded earlier (refer to Chapter 8), we opt for an expansion with the smaller number of terms, to obtain the following first canonical variable:

$$W_1 = -0.1156 \, YT_1 + 1.8384 \, YT_2. \tag{14.2}$$

Standardizing W_1 to have zero mean and unit variance, it becomes the response for a multiple stepwise regression, to yield the following model for the LP:

$$LP = 0.9192 + 0.393A + 0.543C + 0.649AC, \tag{14.3}$$

with root-mean-squared-error of 0.09447, R^2-adj.=0.9911 and model F-ratio of 260.1 (p<0.0001). Since AC is aliased (confounded) with main effect D, the latter may replace the former in the model. Given the significance of the main effects that also comprise the interaction, model (14.3) seems plausible.

However, inclusion of D (instead of AC) should not be ruled out. It is here where the expertise of the researcher and her/his know-how, and those of the colleagues, is called for to decide between alternative models.

While (14.3) supports Lewis *et al.*'s conclusion (2001, p. 270) that main effect D is also significant and should be included in the model, it does not justify inclusion of the other interactions (as suggested therein). Thus, (14.3) is more in line with the original model, analyzed by Martin *et al.* (1987). The significance-levels of the results in (14.3), however, by far exceed the levels obtained by repeating Martins *et al.*'s analysis.

For example, the model F-ratio for the latter analysis is only 39.7 (p<0.002) compared to 260.1 (p<0.0001) for the current model. Similarly, the significance level for AC is 0.0182, compared to p<0.0001 in our model. In fact, all effects in (14.3) have significance level of the same order of magnitude as that for AC (namely, p<0.0001).

Given the LP in (14.3), ML estimates for the RMM parameters can now be obtained. The estimated RMM model would provide the non-linear relationship between the LP and the response.

The latter may be the proportion, P (as used in deriving the LP), or the actual frequency, 1000P (as given in Table 14.1). We proceed with P.

From (14.2) and (14.3) we have

$$-0.1156 \ YT_1 + 1.8384 \ YT_2$$

$$= 0.9192 + 0.393A + 0.543C + 0.649AC + \varepsilon. \qquad (14.4)$$

Recall from Chapter 8 that in estimating an RMM model, two sets of parameters need to be estimated by two different procedures. We use weighted non-linear least-squares (W-NL-LS) to estimate the "non-linear" parameters, $\{\alpha, \lambda, \mu_2\}$ (Set II), and we use a log-likelihood function, based on the density function (d.f.) of the transformed response, $W = \log(Y)$, to estimate the "error" parameters, $\{\rho, \sigma_{\varepsilon 1}, \sigma_{\varepsilon 2}\}$.

Given the small number of available observations for this example (8), one may find it difficult to estimate the complete RMM model (after the LP was estimated). This is due to the fact that a complete RMM estimation requires estimating another six (!) parameters (three of Set I and three of Set II). Furthermore, the high value of R^2-adj.= 0.991, obtained from linear regression in Phase 1 of the analysis, indicates that the best course of action is to regard (14.4) as the final model and not attempt fitting an RMM model to the data.

From (14.4), predicted values for P may be derived. Introducing into (14.4) for YT_1 and for YT_2 in terms of P, we obtain

$$\eta = -0.1156 \ [(P-\mu_P)/\sigma_P] + 1.8384 \ [(P-\mu_P)/\sigma_P \ ]^2 / 2$$

$$= 0.9192 + 0.393A + 0.543C + 0.649AC, \qquad (14.5)$$

with $\mu_P = 0.7475$ and $\sigma_P = 0.2715$ being the mean and the standard deviation of P. Values of the predicted P for individual observations may be extracted from (14.5) either via a root-finding routine, or by solving the quadratic equation explicitly for P. Pursuing the latter option, two solutions will be obtained for each observation. Only one of these provides a satisfactory predicted value.

For the HN model, P is extracted from [relate to (14.1)]

$$P = [1+\exp(-\eta)]^{-1}, \qquad (14.6)$$

where η is given by the RHS of (14.1).

As a third modeling effort, we opted to use logit as a *transformed response* in stepwise linear regression to obtain the following model:

$$\log [P/(1-P)] = 1.697 - 1.497A - 0.667C. \qquad (14.7)$$

This model has R^2-adj.= 0.939, with F-ratio of 54.47. With 2 and 5 degrees of freedom for the model and the error, respectively, this result is highly significant (p=0.0004).

Comparing the residual (error) standard deviations for all three models, we obtain:

For the GLM model [(14.1) and (14.6)]: $\sigma_{res.}$ = 0.0499.

For the RMM model (14.5): $\sigma_{res.}$ = 0.0254.

For the data-transformed "logit model" (14.7): $\sigma_{res.}$ = 0.0570.

For the last two models, both of which are based on linear regression, judging by either the R^2-adj. values or by the residual standard deviations, the RMM estimation approach delivers better goodness-of-fit.

Comment. For the RMM approach, two "off-line" estimates were derived from the data (the coefficients of the canonical variable). These were not included as lost degrees of freedom in assessing the significance of the results of the subsequent linear regression analysis. A similar practice prevails in applications of GLM, where "off-line" pre-analysis decisions (like selecting a link function and the error distribution appropriate for the data) are commonly not accounted for as lost degrees of freedom. The same practice is also commonly pursued in regard to data transformations. For example, in estimating the parameter λ of the Box-Cox transformation, this is commonly regarded as "off-line", not considered in the determination of the model's degrees of freedom when linear regression is subsequently applied.

For a small sample size, this commonly accepted ignoring of "off-line" decisions and estimation may need to be given a higher weight (for both RMM, GLM and the BC transformation) than currently practiced.

Finally, note that although (14.5) had been derived as part of the estimation procedure of the RMM model, only Phase 1 was implemented. This means that no particular features of the RMM model affected the estimation process. The goodness-of-fit that has been achieved should be attributed, in this case, to the merits of the new approach to derive the LP.

14.2.2. Example 2 - The Economist Big Mac Parity Index

This example is based on real-life data. The example appears in Cook and Weisberg (1999, henceforth CW), and is reproduced here with permission. The data may be found in the file *big-mac.lsp* at the website, associated with the above reference:
http://www.stat.umn.edu/arc/index.html.

The file includes economic data on 45 world cities from the period 1990-1991. The Big Mac hamburger is a simple commodity that is virtually identical worldwide. One might expect that the price of a Big Mac be the same everywhere, but obviously it is not. The *Economist* magazine publishes a Big Mac Parity Index, which compares the costs of a Big Mac in various places, as a measure of inefficiency in currency exchange. In this example, economic indicators are used to describe each city. These indicators have been employed to model the price of a Big Mac. The data in the file are described in Table 14.3.

We will now apply the RMM estimation procedure. Starting with Phase 1, several CCAs are attempted, starting with no transformation at all (only YT_1 included in the Taylor expansion) and concluding with fourth-degree expansion. The results are displayed in Table 14.4.

Several measures for the quality of the results may be employed to determine the final structure of the Taylor approximation to the transformed response (relate to a discussion in Chapter 8). We use here two indicators: The increase in the canonical correlation, as more terms are added to the expansion, and the significance level of the first root extracted, relative to the significance level obtained when this root is removed. One wishes the significance level for the first root to be small and for the rest of the roots (with the first one removed) to be high.

Table 14.3. Example 2 - Variables for the Big Mac Parity Index

Variable Name	Description
BigMac (B-M) (Response)	Minutes of labor required by an average worker to buy a BigMac and French fries
Bread (B)	Minutes of labor required to buy one kilogram of bread
BusFare (BF)	The lowest cost of a ten-kilometer bus, tram or subway ticket, in U.S. dollars
EngSal (ES)	The average annual salary of an electric engineer, in thousands of U.S. dollars
EngTax (ET)	The average tax rate paid by engineers
Service (S)	Annual cost of 19 services, primarily relevant to Europe and North America
TeachSal (TS)	The average annual salary of a primary school teacher, in thousands of U.S. dollars
TeachTax (TT)	The average tax rate paid by primary teachers
VacDays (VD)	Average days of vacation per year
WorkHrs (WH)	Average hours worked per year
City	Name of city

Table 14.4. Example 2- Results from CCA with a different number of terms in the Taylor approximation to G(Y) (the optimal choice is starred). "YT_i" means terms up to and including YT_i

Statistic	YT_1	YT_2*	YT_3	YT_4
R^2 (Canon. Corr. squared)	.5727	.8008	.8022	.8258
Significance	.0001490	.00000	.00000	.00000
Sig. (1st root removed)		.2840	.02281	.07883
Sig. (2nd removed)			.2712	.5039
Sig. (3rd removed)				.9027

It is obvious from Table 14.4 that selecting terms up to and including YT_2 (namely a quadratic equation) delivers the best Taylor approximation for the transformed response.

The associated expression for the canonical scores [the approximation for G(Y)] is

$$W_1 = -1.6320 \, YT_1 + 1.0624 \, YT_2. \tag{14.8}$$

The canonical scores calculated from (14.8) for individual observations are now incorporated in a stepwise multiple linear regression analysis. The initial model for effects to be included in the analysis is factorial of second degree.

Diagnostic checking of the resulting model shows that out of the 45 cities that participate in the analysis, seven cities have studentized residuals that exceed about ±2. These are (order number, name, studentized residual):

{(#2, Athens, 1.94), (#6, Buenos Aires, -2.32),
(#22, Luxembourg, -2.08), (#23, Madrid, -2.05),
(#25, Mexico City, 2.0), (#35, Sao Paulo, -2.10), (#42, Tokyo, 1.98)}.
(A negative residual implies that the response is smaller than the model suggests.)

Excluding the above from pursuing analyses, we obtain the final model for LP (significance level appears below each effect, in brackets):

$$LP = -0.003876(B) + 0.7000(BF) + 0.01900(B-26.4)(BF-1.0095)$$
$$(p=0.0674) \quad (p<0.0001) \quad (p=0.0005)$$
$$- 0.002987(S) - 0.007796(S-245.5)(BF-1.0095)$$
$$(p<0.0001) \quad (p<0.0001)$$
$$+ 0.03222(TS). \tag{14.9}$$
$$(p<0.0001)$$

Note, that the intercept is zeroed since it was not significant.

To assess in an adequate manner the influence of the various predictors (which are measured on different scales), the LP is presented with scaled estimates (scaling is performed by subtracting the mean and dividing by half the range in the data), to give

$$LP = 0.5306 - 0.4108(B) + 0.8995(BF) + 2.588(B-26.4)(BF-1.0095)$$

$$- 0.5525(S) - 1.8533(S-245.5)(BF-1.0095) + 0.9230(TS). \quad (14.10)$$

One can learn from (14.10) that the interactions play important role in modeling Big Mac. Also, given the main effects and the interactions that turned out as significant, this model makes sense on intuitive grounds.

Regarding goodness-of-fit statistics, the model has R^2-adj. of 0.915, with residual standard deviation of 0.291. The model F-ratio is 66.2, which, with model d.f. of $v_1=6$, and error d.f. of $v_2=32$, is highly significant (p of the order of 10^{-16}).

We proceed to Phase 2 of the analysis and estimate the parameters of the RMM model. The latter will relate the LP to the response, Big Mac (Table 14.3).

To avoid negative values in the LP, the minimal value (-2.17) is first added to each observation (a justification for this was given in Chapter 8). Adopting the basic RMM model, and assuming for simplicity that the errors are independent ($\rho=0$), an un-weighted non-linear least-squares (NL-LS) is first applied, with the mean, E(W), given by (8.11a). An estimate for λ is obtained very near 1 ($\lambda=0.987$), and an estimate for $\sigma_{\varepsilon 1}$ very near zero. It is therefore assumed that $\lambda= 1$ and $\sigma_{\varepsilon 1}=0$. Note that the small value of $\sigma_{\varepsilon 1}$ implies that a log-transformed response has a constant variance, and therefore no weighting in the NL-LS is warranted. Additionally, the value of λ implies that a log transformation of the response would deliver additive effects, and therefore there is no need to proceed to the second stage of the analysis (where the RMM error parameters are estimated),

A re-run of un-weighted NL-LS with the above values gives

$$\alpha = -0.6172, \; \mu_2 = 4.385, \; \sigma^2_{res.} = 0.02691.$$

Some goodness-of-fit statistics relating to these estimates are given in Table 14.5. Given the assumption that $\sigma_{\varepsilon 1}=0$, one can consider the residual variance, $\sigma^2_{res.}$, obtained in the NL-LS procedure, as an estimate of the parameter $\sigma_{\varepsilon 2}^2$.

Figure 14.1 displays a scatter plot of the relative errors, where the latter are calculated on the original scale of the response (Big-Mac). We realize that for most of the data, the relative error does not exceed about $\pm 20\%$.

To appreciate the quality of the modeling effort developed above, we relate to the solution given in CW. They advocate (therein, p. 326) transforming both the response and the predictors, (ES) and (TS).

Specifically, the suggested model is

$$(BM)^{(-1/3)} = \beta_0 + \beta_1(ES)^{(1/3)} + \beta_2(TS)^{(1/3)} + error, \qquad (14.11)$$

where the error is expected to conform to the normal scenario, namely, linear regression analysis may be applied. Implementing linear regression, we realize that $(ES)^{(1/3)}$ is non-significant ($p=0.432$). Furthermore, there are three cities with outlying studentized residuals: (#23, Madrid, -1.932), (#38, Stockholm, -2.56), (#39, Sydney, 2.27).

Re-running the model with a single predictor, $(TS)^{(1/3)}$, we realize that two additional cities have exceeding residuals: (#14, Helsinki, -2.089), (#25, Mexico City, -2.319). Note that two of the cities, (#23, Madrid) and (#25, Mexico City), appear as outliers in the RMM model too.

The final model (n= 40) is

$$(B\text{-}M)^{-(1/3)} = 0.1366 + 0.06571(TS)^{(1/3)}. \qquad (14.12)$$

This model has R^2-adj. of 0.860, residual standard deviation of 0.0219 and F-ratio of 240.2 (with model d.f. of $v_1=1$, and error d.f. of $v_2=38$). This F value is highly significant ($p<0.0001$).

Figure 14.2 displays a scatter plot of the relative errors, where the latter are calculated in the original scale of the response (Big-Mac).

Table 14.5. Example 2 - Final estimates of the "Structural" parameters, $\{\alpha, \mu_2\}$, with standard errors and 95% confidence intervals

Parameter	Estimate	S.E.	95% C.I.
α	-0.6172	0.0283	{-.674,-.559}
μ_2	4.385	0.0424	{4.30, 4.47}

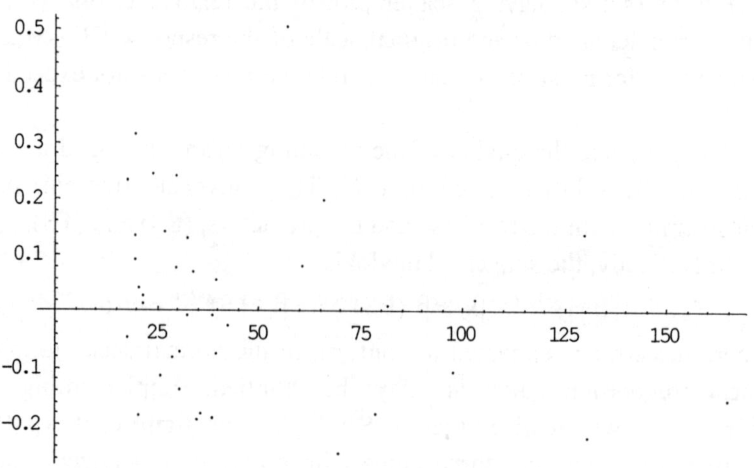

Figure 14.1. Example 2 - A scatter plot of the relative errors (n=38) for the Big Mac example (RMM Analysis). Horizontal axis is Big Mac in the original scale.

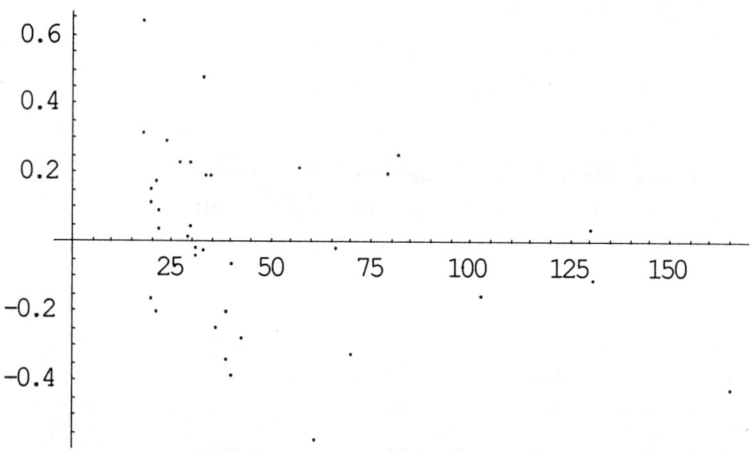

Figure 14.2. Example 2 - A scatter plot of the relative errors (n=38) for the Big Mac example (CW model). Horizontal axis is Big Mac in the original scale.

Comparing the average absolute value of the relative error obtained from the two models (including only the 38 observations analyzed via the RMM), we realize that for the RMM model we obtain a value of 13.2% vs. 20.0% for the CW model.

Which model is more plausible, (14.9) or (14.12), is left open.

14.3. Two Familiar Problems (Cont'd from Chapter 8)

14.3.1. Example 3 - The Wave-Soldering Process (a Poisson model)

This example deals with attaching components to an electronic circuit card assembly by a wave-soldering process. The response is the number of defects in the solder joint. The process involves baking and preheating of the circuit card, and passing it through a solder wave by a conveyor.

Condra (1993) presented the results, and Hamada and Nelder (1997) re-analyzed, using GLM with a variance function equal to the mean (a Poisson model) and a log-link function. Later, Myers *et al.* (2002, p. 145) re-analyzed using GLM. The following presentation of the problem is taken from the latter source.

The factors in this problem are (A) pre-bake condition, (B) flux density, (C) conveyor speed, (D) preheat condition, (E) cooling time, (F) ultrasonic solder agitator, and (G) solder temperature. Only the interactions ab, ac, ad, bc, bd and cd were considered to be important (refer to Hamada and Nelder, 1997), and therefore all analyses are restricted to these interactions.

The experimental design and the observations are given in Section 8.5, where an RMM solution is also developed. The reader is referred to this section for details. Solutions obtained by current approaches, some of which are reported in the literature, are now introduced.

Hamada and Nelder (1997), via GLM, obtained the model

$$\log(\mu) = 3.10 - 0.14A + 0.21B + 0.48C + 0.59D - 0.21E - 0.76G$$

$$+ 0.73AC - 1.19BD. \tag{14.13}$$

Myers *et al.* (2002) derive a different model, using the same variance function and different link functions, however accounting for over-dispersion. For the log link they obtain (therein, p. 148)

$$\log(\mu) = 3.08 + 0.44C - 0.40G + 0.28AC - 0.31BD. \quad (14.14a)$$

For the square-root link they obtain (therein, p. 185)

$$\mu^{1/2} = 5.02 + 0.43A - 0.40B + 1.19C - 0.39E - 1.10G$$

$$+ 0.58AC - 0.96BD. \quad (14.14b)$$

We re-analyze these data, starting with simple linear regression, proceeding with a BC transformation, and concluding by re-introducing results from the RMM solution (Section 8.5).

Applying stepwise linear regression, with an initial factorial model of degree two, the following model is obtained:

$$LP = 29.70 + 6.98A - 6.06B + 14.93C - 6.59E - 14.06G$$

$$+ 8.31(A+0.02)(C+0.02) - 12.76(B-0.02)(D-0.02), \quad (14.15)$$

with R^2-adj.$=0.7782$ and model F-ratio of 24.06. Lack-of-fit test proves to be highly significant ($p<0.0006$), which invalidate this model.

Applying a BC transformation to the response results in

$$Y^{(\lambda)} = (Y^{0.4}-1) / 0.0689. \quad (14.16)$$

(note that this is a normalized BC transformation, where the denominator contains the geometric average of the transformed values; Normalization is standard practice that allows comparison of differently-scaled BC transformations, refer to Chapter 4 for details).

From linear regression applied to the transformed response, we obtain

$$E[Y^{(\lambda)}] = LP = 36.2 + 10.11C - 8.66G$$

$$+ 4.97(A+0.02)(C+0.02) - 8.10(B-0.02)(D-0.02). \quad (14.17)$$

Note, that since the factors are given in standardized (or nominal) values of "-1" and "1", there is no meaning to "centering" by subtracting the "mean", ±0.02. This operation is automatically conducted by the computer in displaying interactions, and we present the results exactly as obtained from the computer output. Also note that estimated effects with significance level of at most 1% were included in both of the above models.

With R^2-adj.$= 0.6821$ and model F-ratio of 25.6, (14.17) introduces very little improvement in goodness-of-fit statistics relative to (14.15).

However, the transformed response results in a simpler model (five parameters are estimated instead of eight). A test of lack-of-fit still proves to be mildly significant ($p<0.0661$).

Proceeding with the RMM approach, the LP derived in Section 8.5 is

$$LP = -0.1786 + 0.164A - 0.144B + 0.526C - 0.157E - 0.495G$$

$$+ 0.224(A+0.02)(C+0.02) - 0.431(B-0.02)(D-0.02), \quad (14.18)$$

with R^2-adj.$=0.848$ (vs. 0.682 for the BC transformation) and model F ratio of 37.7 (vs. 25.6 for the BC transformation). Lack-of-fit test proves to be highly non-significant ($p=.569$). Note the small values of the coefficients of A, B and E, relative to the other coefficients. These effects also prove to be the least significant.

Comparing (14.18) to (14.14b) (the GLM solution with a square-root link), the two models are nearly equivalent, up to a linear transformation. To realize that, compare the relative size of coefficients in the two models. For example:

For A/C:
From (14.18): $0.164/0.526 = 0.31$; From (14.14b): $0.43/1.19 = 0.36$

For (AC)/(BD):
From (14.18): $0.224/(-0.431) = -0.52$; From (14.14b): $0.58/(-.96) = -0.60$.

If only highly significant estimates were included ($p<0.001$), as in Myers *et al.* (2002) with GLM analysis and a log link (14.14a), we would have obtained their model, with R^2-adj.$=0.777$ [vs. 0.848 for (14.18)], and model F ratio of 41.2 (vs. 37.7 for the above model). However, a lack-of-fit test turns out to be mildly significant ($p<0.0187$). Therefore the more extended model (14.18) is adopted, which is compatible with the GLM model assuming a square-root link (14.14b).

The similarity of the models, derived by the new approach and by GLM (assuming either log or square-root links), corroborates Phase 1 of the RMM estimation process as a valid approach for modeling the LP. Furthermore, it demonstrates its unique feature, namely, that even though no explicit transformation has been specified either for the response (like with the BC transformation) or for the response mean (as with GLM), linear regression still delivered a good model with a structure that resembles models derived via GLM.

14.3.2. Example 4 - The resistivity data (a gamma error distribution)

This example originally appeared in Myers and Montgomery (1997), and later re-introduced in Myers *et al.* (2002, therein p. 176). The experiment has an un-replicated 2^4 factorial design, which was run at a certain step in a semiconductor manufacturing process. The response variable, resistivity of the test wafer, is a characteristic well-known to have a non-normal, long right-tailed distribution. The design, with the observations collected, is given in Section 8.5, where a detailed RMM solution is developed.

Analyzing this problem, Myers and Montgomery (1997) first applied a log transformation to the data to obtain the following model:

$$E[\log(Y)] = 2.351 + 0.027X_1 - 0.065X_2 + 0.039X_3. \qquad (14.19)$$

This model was judged to be unsatisfactory, and a GLM model was derived, assuming a gamma error distribution and a log link. The resulting model is (Myers *et al.*, 2002, p.181)

$$\log(\mu) = 5.414 + 0.0613X_1 - 0.150\ X_2 + 0.090X_3 - 0.028X_4$$

$$- 0.040X_1X_3 - 0.044X_2X_3 - 0.046X_3X_4. \qquad (14.20)$$

It is interesting to notice that while for X_4 the significance level is $p=0.0195$, for all other main effects (X_1, X_2 and X_3) the significance level is $p<0.0000$.

Re-analyzing the data, a BC transformation is first applied and then the RMM solution, derived in Section 8.5, is addressed.

A BC transformation, based on a model with the effects $\{X_1, X_2, X_3, X_2X_3, X_3X_4\}$, where significance level of $p<0.05$ is required for inclusion in the model, is

$$Y^{(\lambda)} = (Y^{0.4}-1)/0.01554. \qquad (14.21)$$

Using the transformed Y as a response, we obtain from stepwise linear regression

$$E[Y^{(\lambda)}] = 498.4 + 13.35X_1 - 34.04X_2 + 20.63X_3$$

$$- 11.16X_2X_3 - 10.56X_3X_4. \qquad (14.22)$$

The goodness-of-fit statistic is R^2-adj.= 0.847, with model F-ratio of 17.63. However, only X_2 and X_3 are highly significant ($p<0.001$).

Proceeding with RMM, we obtained in Section 8.5

$$LP = - 0.1451 + 0.293X_1 - 0.707X_2 + 0.409X_3$$

$$- 0.201X_1X_3 - 0.188X_2X_3 - 0.205X_3X_4, \qquad (14.23)$$

with R^2-adj.=0.882 (vs. 0.847 for the BC transformation), and model F-ratio of 19.7 (vs. 17.6 for the BC transformation). With 6 and 9 degrees of freedom for the model and the error, respectively, the F-ratio is highly significant ($p=0.0001$). Also, all main effects are highly significant ($p<0.01$).

As with the previous example, only main effects participated in deriving the *initial* LP (the first stage of Phase 1 of the analysis). This did not prevent significant interactions from entering into the *final* LP, derived in the second stage of Phase 1 of the analysis.

Comparing (14.23) to (14.20), obtained in a GLM framework, we find the two models to be very similar. The coefficients are also numerically close, up to a change in scale. For example:

For X_1/X_2:
Eq. (14.23): 0.293/(-0.707)= -0.414; Eq. (14.20): 0.061/(-0.150)= -.4070

For $(X_1)/(X_1X_3)$:
Eq. (14.23): 0.293/(-0.201)= -1.46; Eq. (14.20): 0.0613/(-.040)= -1.53.

The only difference between the models is the inclusion of X_4 in the GLM model. However, this main effect is appreciably less significant also in the GLM model ($p=0.0195$), and it was probably included in the LP because of the inclusion of the interaction term X_3X_4. (The practice of retaining all main effects taking part in a certain significant interaction, even when some of these main effects are only mildly significant, is advocated by many for some good reasons; Find details, for example, in Draper and Smith, 1998, Section 12.3.)

If X_4 was deleted in (14.20), we would have obtained an LP similar to that in (14.23).

14.4. Comparison of Models

14.4.1. Mallow's C_p

Different statistics may be used for the selection of "the best model". In this sub-section we introduce briefly Mallow's C_p criterion. Akaike's Information Criterion (AIC) will be treated in Sub-section 14.4.2.

Mallow (1973, 1995), has suggested use of the C_p statistic for linear regression:

$$C_p = RSS_p / S^2 - (n-2p), \qquad (14.24)$$

where RSS is the residual sum of squares from a model containing p parameters [including β_0, the intercept (the "constant")], and S^2 is the residual *mean* square from the *complete* model (the model that includes all possible predictor variables). It is assumed that the model with p parameters is a sub-model of the complete model. S^2 is assumed to be a reliable unbiased estimate of the true error variance, σ_ε^2, namely, $E(S^2) = \sigma_\varepsilon^2$. As terms are added to a model, RSS will decrease but C_p will usually increase.

If a model is adequate, that is, does not suffer from lack of fit, then $E(RSS_p) / (n-p) = \sigma_\varepsilon^2$, and we have for an adequate model

$$E(C_p) = E(RSS_p) / E(S^2) - (n-2p)$$

$$= (n-p)\sigma_\varepsilon^2 / \sigma_\varepsilon^2 - (n - 2p) = p. \qquad (14.25)$$

This implies that for "adequate" models we expect C_p to be about equal to p. Equations with considerable lack of fit, that is, biased equations, will give rise to C_p values above the $C_p = p$ line. Note, however, that due to random variation, we may occasionally find also well-fitting equations somewhat below this line. Therefore we would look for a model with low C_p that is *about* equal to p.

Refer to further comments in Draper and Smith (1998, p. 332), and in Cook and Weisberg (1999, p. 275). The latter give an F test, based on C_p, to test the null hypothesis that the sub-model is adequate against the alternative hypothesis that the extended model is the correct one.

14.4.2. Akaike's Information Criterion (AIC)

Another general criterion to compare models (not necessarily linear), with different numbers of parameters, is Akaike's Information Criterion (AIC) (Akaike, 1973, 1983). This criterion may be formulated by different statistics, dependent on the context and the analysis performed on the data. A minimum value of AIC is an indicator of the "best" model.

In the framework of maximum-likelihood estimation (MLE), AIC for a specified model is given by

$$AIC = -(2)LL_p + 2p, \qquad (14.26)$$

where p is the *total* number of parameters that need to be estimated from the data, and LL_p is the value of the maximized log-likelihood function (the value of the log-likelihood function where parameters are replaced by their ML estimates, namely, those that maximize the log-likelihood function). For a non-linear model, p includes both the parameters in the linear predictor and other parameters that need to be estimated [like (α, λ, μ_2) in the RMM model].

In the framework of GLM, the AIC criterion is (refer, for example, to Cook and Weisberg, 1999, p. 536)

$$AIC = Deviance + 2p, \qquad (14.27)$$

where p is the number of parameters estimated, namely, those in the LP plus other parameters that need to be estimated, like that of the dispersion parameter. [Deviance in GLM is equivalent to the sum of squared residuals ("residual sum of squares") used in linear regression.]

In the framework of non-linear regression (namely, assuming that the errors are normally distributed with constant variance), the AIC criterion is (Motulsky and Christopoulos, 2003)

$$AIC = n \log(RSS/n) + 2(p+1), \qquad (14.28)$$

where n is the sample size. Note that the authors claim that since regression is estimating also the sum-of-squares, a value of p+1 should be used in the second term of the formula (instead of p). A more adequate model is indicated by a smaller value of AIC.

For small n (n small relative to p), a corrected AIC_c is

$$AIC_c = AIC + 2(p+1)(p+2)/(n-p-2). \qquad (14.29)$$

To compare two models, say a simple model, A, with p_A parameters, and a more complicated model, B, with a larger number of parameters, $p_B > p_A$, the difference in the AIC values is the meaningful statistic:

$$\Delta AIC = AIC_B - AIC_A = n \log(RSS_B/RSS_A) + 2(p_B - p_A). \quad (14.30)$$

A negative value suggests that model B is better than A, namely, the reduction in RSS achieved by model B ($RSS_B < RSS_A$) more than compensate for the larger number of parameters. A positive value would likewise indicate that the simpler model A is better. Note, that these are all intuitive evaluations with no formal statistical testing.

A good source for the use of AIC to compare models with different numbers of parameters is Burnham and Anderson (2002).

References

[1] Akaike, H. (1973). Information theory and an extension of the maximum likelihood principle. In B. N. Petrov and F. Csaki (Eds.), *Second International Symposium on Information Theory*. Budapest: Akademiai Kiado.

[2] Akaike, H. (1983). Information measures and model selection. *Bulletin of the International Statistical Institute: Proceedings of the 44th Session*, V.1, 277-290.

[3] Burnham, K. P., Anderson, D. R. (2002). *Model Selection and Multi-model Inference- A Practical Information-theoretic Approach.* 2nd edition. Springer.

[4] Condra, L. W. (1993). *Reliability Improvement with Design of Experiments*. Marcel Dekker, NY.

[5] Cook, R. D., Weisberg, S. (1999). *Applied Regression Including Computing and Graphics.* John Wiley & Sons.

[6] Draper, N. R., Smith, H. (1998). *Applied Regression Analysis.* 3rd edition. John Wiley & Sons.

[7] Hamada, M., Nelder, J. A. (1997). Generalized linear models for quality-improvement experiments. *Journal of Quality Technology*, 29(3), 292-304.

[8] Lewis, S. L., Montgomery, D. C., Myers, R. H. (2001). Confidence interval coverage for designed experiments analyzed with GLMs. *Journal of Quality Technology*, 33(3), 279-292.

[9] Mallow, C. P. (1973). Some comments on C_p. *Technometrics*, 15, 661-675.

[10] Mallow, C. P. (1995). More comments on C_p. *Technometrics*, 37, 362-372. Also refer to *Technometrics*, 1997, 39, 115-116.

[11] Martin, B., Parker, D., Zenick, L. (1987). Minimize slugging by optimizing controllable factors on Topaz Windshield Molding. In *Fifth Symposium on Taguchi methods*. American Supplier Institute, Inc., Dearborn, MI, 519-526.

[12] Motulsky, H., Christopoulos, A. (2003). *Fitting Models to Biological Data Using Linear and Nonlinear Regression- A practical guide to curve fitting. GraphPad Software Inc., San Diego CA,* www.graphpad.com.

[13] Myers, R. H., Montgomery, D. C. (1997). A tutorial on generalized linear models. *Journal of Quality Technology*, 29(3), 274-291.

[14] Myers, R. H., Montgomery, D. C., Vining, G. G. (2002). *Generalized Linear Models with Applications in Engineering and the Sciences*. John Wiley & Sons.

Chapter 15

Reliability Engineering (with Censoring)

15.1. Introduction

Characterization of the life-times of parts, or systems, operating in their natural end-user environments is a major subject of reliability-engineering modeling. How life-time is affected by related factors, like stress, is modeled by dedicated relational models. One such model is the Random Fatigue Limit (RFL) model, introduced recently in a series of papers by Pascual and Meeker (henceforth PM) and by Pascual (PM, 1997, 1999; Pascual, 2003, 2004, and references therein). The RFL model is the focus of this chapter. It was first introduced in Section 2.5, and will be addressed again later in the chapter, when a numerical example demonstrates using RMM to analyze fatigue-life data that include censored observations.

Hardware reliability-engineering models and censoring are two inter-linked issues. Censoring relates to observations known to be above or below a certain threshold value, without knowing their exact values. For example, we may observe life-times of a sample of n components for a limited time, say T_0. If at time $T=T_0$, k components are still operative (in a working condition), this implies that for these observations the exact life-times are not known, only that they are larger than T_0. Out of the n observations, k observations are *right-censored*.

Similarly, if a certain observation is known to be in the interval $[0,T_0)$, however we do not know the value of this observation since only values above the threshold T_0 are registered, this observation is *left-censored*. Censoring is a permanent companion to modeling life-times in hardware-reliability studies, hence the strong link between the two subjects.

In this chapter the RMM estimation procedures, introduced in Chapter 8, are modified to incorporate right-censored observations. The modified procedures are developed in Section 15.2. A data-set with right-censored observations, which was formerly analyzed by PM using the RFL model, is re-analyzed by the modified RMM estimation procedure. This analysis is introduced in Section 15.3. The RMM results are compared to PM's results.

15.2. RMM Estimating with Censored Data

In Chapter 8 estimation procedures for the RMM model were developed. It was assumed that the linear predictor (LP), η, had been estimated by a separate procedure so that in estimating the RMM parameters we relate to the LP as a single regressor (independent variable) with given (known) sample values. Once these values are known, an ML procedure is detailed for estimating the RMM parameters.

In this section, we modify the estimation procedure to accommodate right-censored observations. For convenience, the main results from the estimation procedures of Chapter 8 are first re-introduced, and then modified to accommodate right censoring. We assume that in the derivation of the parameters of η, in Phase 1 of the estimation procedure, only complete (uncensored) observations were used. This is a restrictive assumption, which is not necessarily true. However it simplifies appreciably the developed procedure.

The basic RMM model is

$$Y = \exp\{(\alpha/\lambda)[(\eta+\varepsilon_1)^\lambda-1] + \mu_2 + \varepsilon_2\}$$

$$= \exp\{(\alpha/\lambda)[(\eta+\sigma_{\varepsilon 1}Z_1)^\lambda-1] + \mu_2 + \sigma_{\varepsilon 2}[\rho Z_1 + (1-\rho^2)^{(1/2)} Z_2]\}, \qquad (15.1)$$

$$W = \log(Y) = (\alpha/\lambda)[(\eta+\sigma_{\varepsilon 1}Z_1)^\lambda-1] + \mu_2$$
$$+ \sigma_{\varepsilon 2}[\rho Z_1 + (1-\rho^2)^{(1/2)} Z_2], \qquad (15.2)$$

where ε_1 and ε_2 are correlated errors, and Z_1 and Z_2 are two independent (uncorrelated) standard normal variables.

From the expression for the density function (d.f.) of W, $f_W(w;\theta)$ [Equation (9.4)], the likelihood function, L, for a given sample of n complete (uncensored) observations, $\{(\eta_1, w_1), (\eta_2, w_2),\ldots, (\eta_n, w_n)\}$, is

$$L(\theta \mid w) = \prod_{i=1}^{n} f_W(w_i; \eta_i, \theta) = \prod_{i=1}^{n} \int_{-\infty}^{\infty} f_{W \mid \varepsilon 1}(w_i \mid \varepsilon_1; \eta_i, \theta)\, f_{\varepsilon 1}(\varepsilon_1) d\varepsilon_1$$

$$= \prod_{i=1}^{n} \int_{-\infty}^{\infty} [(1-\rho^2)^{1/2}\sigma_{\varepsilon 2}]^{-1}\phi\{[w_i - \mu(W \mid \varepsilon_1; \eta, \theta)] / [\sigma(W \mid \varepsilon_1; \theta)]\}\, f_{\varepsilon 1}(\varepsilon_1) d\varepsilon_1$$

$$(15.3)$$

where

$$f_{\varepsilon 1}(\varepsilon_1) = (2\pi\sigma_{\varepsilon 1}^2)^{-1/2} \exp[-\varepsilon_1^2/(2\sigma_{\varepsilon 1}^2)], \qquad (15.4)$$

$$\mu(W \mid \varepsilon_1; \eta, \theta) = (\alpha/\lambda)[(\eta+\varepsilon_1)^\lambda - 1] + \mu_2 + (\sigma_{\varepsilon 2}\rho/\sigma_{\varepsilon 1})\varepsilon_1, \qquad (15.5)$$

$$\sigma(W \mid \varepsilon_1; \theta) = (1-\rho^2)^{1/2}\, \sigma_{\varepsilon 2}, \qquad (15.6)$$

and $\theta = \{\alpha, \lambda, \mu_2, \rho, \sigma_{\varepsilon 1}, \sigma_{\varepsilon 2}\}$. Refer for further details to Chapter 9. Note that since W is a relational model, the LP (η) is part of the error density function [unlike in (9.4)].

Suppose now that n sample data were collected with some of the observations right-censored (Type I censoring). Denote by LT_0 the log censoring time, $LT_0 = \log(T_0)$, and let $F_W(LT_0; \eta_i, \theta)$ be the cumulative distribution function (CDF) of W for the censored observation i.

Then the likelihood function is

$$L(\theta) = \prod_{i=1}^{n} f_W(w_i; \eta_i, \theta)^{\delta i} [1 - F_W(LT_0; \eta_i, \theta)]^{1-\delta i}, \qquad (15.7)$$

where

$$\delta_i = \begin{cases} 1, & \text{if } w_i \text{ is given (observed)} \\ 0, & \text{if } w_i \text{ is censored (non-observable),} \end{cases} \qquad (15.8)$$

Separating the given data-set into two sub-sets of k_1 observed (complete) observations and k_2 censored observations ($k_1+k_2=n$), the log-likelihood function, LL, is, from (15.7),

$$LL(\theta) = \log[L(\theta)]$$

$$= \sum_{i=1}^{k1} (\delta_i) \log[f_W(w_i; \eta_i, \theta)] + \sum_{i=1}^{k2} (1-\delta_i) \log[1 - F_W(LT_0; \eta_i, \theta) \qquad (15.9)$$

The ML estimates for the parameters in θ are those which maximize (15.9). A detailed estimation procedure is now developed.

From the expressions for the conditional d.f. of W, the marginal (unconditional) expression for the d.f. is (refer to Chapters 9 and 11)

$$f_W(w; \eta, \theta) = \int_{-\infty}^{\infty} [\sigma(z; \theta)]^{-1} \phi\{[w - \mu(z; \eta, \theta)] / [\sigma(\theta)]\} \phi(z)\, dz. \qquad (15.10)$$

Similarly, the marginal CDF of W is

$$F_W(w; \eta, \theta) = \int_{-\infty}^{\infty} \Phi\{[w - \mu(z; \eta, \theta)] / [\sigma(\theta)]\} \phi(z) dz, \qquad (15.11)$$

where $\phi(.)$ is the standard normal d.f., $\Phi(.)$ is the standard normal CDF, and

$$\mu(z; \eta, \theta) = (\alpha/\lambda)[(\eta + \sigma_{\varepsilon 1} z)^{\lambda} - 1] + \mu_2 + \sigma_{\varepsilon 2}\rho z, \qquad (15.12)$$
$$\sigma(\theta) = \sigma_{\varepsilon 2}(1 - \rho^2)^{(1/2)}. \qquad (15.13)$$

Inserting (15.10) and (15.11) into (15.9), an expression explicit in the parameters is obtained. Some approximations may now be applied to facilitate numerical maximization of (15.9).

First, to eliminate the need for integration in $f_W(w; \eta, \theta)$ (15.10), a Taylor-based approximate expansion has been developed in Chapter 8. This approximate expression is given in (9.10) and is re-introduced here for convenience:

$$f_W(w; \eta, \theta) = \int_{-\infty}^{\infty} f_{W|\varepsilon 1}(w \mid \varepsilon_1; \eta, \theta)\, f_{\varepsilon 1}(\varepsilon_1) d\varepsilon_1$$
$$\cong [2\pi(1 - \rho^2)\sigma_{\varepsilon 2}^2]^{-1/2} \exp[-(1/2)Z_f^2]$$
$$\times \{1 + (1/2)\sigma_{\varepsilon 1}^2[(1 - \rho^2)^{1/2}\sigma_{\varepsilon 2}]^{-2}[(\alpha\eta^{\lambda - 1} + \rho\sigma_{\varepsilon 2}/\sigma_{\varepsilon 1})^2 (Z_f^2 - 1)$$
$$+ \alpha\eta^{\lambda - 2}(\lambda - 1)(1 - \rho^2)^{1/2}\sigma_{\varepsilon 2}Z_f]\}, \qquad (15.14)$$

where

$$Z_f = Z_f(w) = [w - (\alpha/\lambda)(\eta^{\lambda} - 1) - \mu_2] / [(1 - \rho^2)\sigma_{\varepsilon 2}^2]^{1/2}. \qquad (15.15)$$

Similarly, we wish to obtain a simple approximate expression for the standard normal CDF, which would eliminate the need for a single numerical integration in obtaining the CDF of W (15.11).

An approximation with such desirable properties has been recently developed in Shore (2004a):

$$\Phi(z) = 1 - \exp\{-\log(2) \exp[B[\exp(Cz) - 1] + Dz]\}, \; z > 0, \qquad (15.16)$$

where $B = -1.81280$; $C = -0.472245$; $D = 0.294549$. For $z > 0$, this simple approximation has a maximum error of ± 0.00002. Note, that the parameters have been determined so that the maximum absolute error for

$z>0$ is minimized. For $z<0$, the absolute error may be somewhat larger (maximally around 0.0007). It is recommended for $z<0$ to replace z by $-z$ and $\Phi(z)$ by $1-\Phi(z)$.

A more accurate approximation (an error of 10^{-7}), though somewhat algebraically more complex, appears in Chapter 19.

Introducing from (15.16) into (15.11), we obtain for the CDF of W at $W=LT_0$:

$$F_W(LT_0;\; \eta,\; \theta) = \int_{-\infty}^{\infty} \Phi\{[LT_0-\mu(z;\eta,\; \theta)]/[\sigma(\theta)]\}\; \phi(z)dz$$

$$= \int_{-\infty}^{0} \Phi(z_F)\phi(z)dz + \int_{0}^{\infty}\Phi(z_F)\phi(z)dz$$

$$\cong \int_{-\infty}^{0} \exp\{-\log(2)\exp[B[\exp(-Cz_F)-1] - Dz_F]\}\; \phi(z)dz$$

$$+ \int_{0}^{\infty} \{1- \exp\{-\log(2)\exp[B[\exp(Cz_F)-1] + Dz_F]\}\}\; \phi(z)dz, \qquad (15.17)$$

where

$$z_F = [LT_0-\mu(z;\; \eta,\; \theta)] / [\sigma(\theta)], \qquad (15.18)$$

and $\mu(z;\; \eta,\theta)]$ and $\sigma(\theta)$ are given by (15.12) and (15.13), respectively.

Introducing from (15.17) for F_W and from (15.14) for f_W, we obtain for (15.9) an approximate expression that needs to be maximized, with a single numerical integration for each censored observation to derive $F_W(LT_0;\eta_i,\theta)$, $(i=1,2,..,k_2)$.

While the new approximate log-likelihood function seems simpler to solve [relative to the original based on (15.10) and (15.11)], maximizing (15.9) is still no easy numerical task (the parameters' vector includes six parameters!). It is therefore suggested that separation of the estimation procedure into two sub-routines, each estimating a different sub-set of the parameters, be followed, similarly to the estimation procedure developed for uncensored observations in Chapter 8.

Specifically, the RMM "structural" parameters of Set II, (α,λ,μ_2), will be estimated by weighted non-linear least-squares (NL-LS), using only complete observations. The error parameters of Set III, $\{\rho,\; \sigma_{\varepsilon 1},\; \sigma_{\varepsilon 2})$, will be estimated via maximizing (15.9), using all available observations (including censored observations). The reader is referred to Section 8.4, where these solution procedures are expounded in detail.

For the NL-LS procedure, the following approximate mean-value can be used for (15.2) (refer for details to Section 8.4.2):

$$E(W) \cong (\alpha/\lambda)[(\eta)^{\lambda} - 1] + (1/2)\alpha(\lambda-1)\sigma_{\varepsilon1}^{2}\eta^{\lambda-2} + \mu_2 \qquad (15.19)$$

Accordingly, the following estimating routine is suggested (recall that it is assumed that values of the linear predictor for individual observations are known (estimated) from Phase 1 of the analysis):

I. Using the residual variance from the linear regression stage of Phase 1 as an estimate for $\sigma_{\varepsilon1}^{2}$, apply NL-LS to maximize S (8.10) in order to find initial estimates for $\{\alpha, \lambda, \mu_2\}$. Use *un-weighted* NL-LS for this step. Use only complete observations.

II. For the derived estimates of $\{\alpha, \lambda, \mu_2\}$, maximize the log-likelihood function (15.9) to find estimates for the error parameters, $\{\rho, \sigma_{\varepsilon1}, \sigma_{\varepsilon2}\}$. Use both complete and censored observations.

III. Calculate variances for individual complete observations, using Method II (detailed in Section 8.4). Use the reciprocal values of the derived variances as weights for the next step.

IV. Derive updated estimates for $\{\alpha, \lambda, \mu_2\}$, using *weighted* NL-LS to maximize S (8.10). Use only complete observations.

V. Test for convergence of the solution (various criteria may be used, for example, significant reduction in the residual variance). If convergence has not yet occurred- go to step II. Otherwise- terminate the estimation procedure.

Comments.

(1) It is always desirable to estimate all the parameters of θ via maximization of (15.9), where all observations are used (including censored ones). If the initial estimates of the parameters $\{\alpha, \lambda, \mu_2\}$, derived in stage I, are used as starting values for a search routine that maximizes (15.9), and estimates for *all* six parameters can be derived thereof, then these ML estimates should be considered as the final ML

estimates for all parameters of the RMM model. Otherwise, some iterations would be needed.

(2) It is always a good idea to reduce the number of parameters in the model by *assuming* that certain parameters have specified values. For example, assuming that the errors are uncorrelated ($\rho=0$) eliminates one parameter that needs estimating. Likewise, assuming $\rho=\pm1$ allows μ_2 to be estimated by the sample log median. This eliminates two parameters that need estimating.

The relatively large number of parameters associated with RMM allows such flexibility without risking loss of accuracy in fitting the RMM model to given data.

The estimation procedure expounded in this section may be modified to accommodate other versions of the RMM model.

A detailed numerical example follows where the RMM model given by (15.2) will be used.

15.3. A Numerical Example - The RFL model

The Random Fatigue Limit (RFL) model expresses fatigue life, Y, as a function of S, the stress level. The model is

$$\log(Y) = \beta_0 + (\beta_1)\log(S-\gamma) + \varepsilon, \, S > \gamma, \qquad (15.20)$$

where β_0 and β_1 are fatigue curve coefficients, γ is the (random) fatigue limit of the specimen, and ε is the error term. PM have assumed that the fatigue limit is random, and they characterize the random variable $V = \log(\gamma)$, after standardization, as following either the standardized smallest extreme value (Sev) distribution or the standardized normal distribution. The distribution of ε is similarly modeled as either Sev or normal.

PM applied the RFL model to the laminate panel data, given by Shimokawa and Hamaguchi (1987; Refer to Section 4 in PM paper). The following exposition of the problem pursues PM, who were also kind enough to provide us with the data. The latter come from 125 specimens in four-point out-of-plane bending tests of carbon eight-harness-satin/epoxy laminate. Fiber fracture and final specimen fracture occurred simultaneously. Thus, fatigue life is defined to be the number of cycles until specimen fracture. The data set includes 10 right-censored observations (known as "run-outs" in fatigue literature). These observations will be treated as missing values in the NL-LS part of the estimation procedure.

The data are displayed in Table 15.1.

The response variances both in the original scale (Y) and in the log transformed scale (W), as well as the coefficient of variation (CV) for W, are given in the bottom three rows. For the log-transformed response, the variance seems to stabilize.

Using the RFL model, PM (1999) obtained the following equation, assuming normal distribution for both V and ε (refer to Table 1 in PM, 1999):

$$\log(Y) = 30.272 + (-5.100)\log(S\text{-}\gamma) + \varepsilon, \ S > \gamma, \qquad (15.21)$$

with parameters

$$E(V) = \mu_\gamma = 5.366, \ \sigma(V) = \sigma_\gamma = 0.031, \ \sigma(\varepsilon) = \sigma = 0.289.$$

We next model these data via the RMM model.

PM describe in their paper (PM, 1999) two properties of fatigue-life data that any modeling approach should consider: "There are two main considerations in modeling the applied stress and fatigue life. First, often the standard deviation of fatigue life decreases as the applied stress increases. Secondly, curvature in fatigue curves suggests the inclusion of a fatigue limit in the statistical model for fatigue life". (Fatigue curves, also referred to as S-N curves, traditionally display stress, S, versus the median fatigue life, N; The latter is often expressed in cycles to failure, characteristic for products like jet engines).

The two requirements that PM address are compatible with the RMM model. First, this model provides typical curvature observed in S-N curves, namely, monotone convex curvature. Secondly, we have demonstrated in Section 7.4 that for the RMM model the coefficient of variation asymptotically tends to zero as the linear predictor becomes large. Since fatigue life *decreases* with increased stress, the associated

Table 15.1. Data for the numerical example. S - Stress (in MPa), Y - Life time (number of cycles to specimen fracture, kilocycles). Censored observations are starred. Statistics in the bottom three rows exclude censored observations. W=log(Y). CV is the coefficient of variation.

S	Y	S	Y	S	Y	S	Y	S	Y
380	34.2	340	125.5	300	954	280	2604.2	270	5163.1
380	37.7	340	156.9	300	959.4	280	2610.7	270	5269.9
380	42	340	173.6	300	1194.6	280	2773.4	270	7863.8
380	42.3	340	176.9	300	1240.5	280	3093.2	270	8188.5
380	48.2	340	179.4	300	1250.4	280	4270.2	270	9488.6
380	52.5	340	188.5	300	1285.5	280	5993.8	270	10171.8
380	55.9	340	195.1	300	1410.5	280	6460.5	270	10794.2
380	58.3	340	208.1	300	1495.1	280	7430	270	11556.2
380	61.7	340	211.9	300	1518.7	280	8105	270	12161.5
380	64.7	340	224.1	300	1544.7	280	8497.6	270	13314
380	65	340	226	300	1551.4	280	8594.7	270	15857.2
380	65.5	340	253	300	1585.9	280	8630.4	270	15996.3
380	70.4	340	255.5	300	1639.1	280	8816	270	17274.7
380	71	340	259	300	1683.7	280	8820.3	270	18804
380	72.4	340	274	300	1926.1	280	9013.4	270	19514.5
380	75.2	340	292	300	2011.3	280	10124.8	270	19736.5
380	77.4	340	300.4	300	2171.8	280	10163.7	270	20354.5
380	77.8	340	302.3	300	2391.5	280	11144	270	20504.7*
380	87.8	340	308.3	300	2569.4	280	12469.8	270	20532.3*
380	93.4	340	406.3	300	2674.9	280	13232.9	270	20532.3*
380	94	340	420.7	300	2921.7	280	13813.4	270	20536.1*
380	97.2	340	428.5	300	3046.5	280	15338.4	270	20538.2*
380	99.6	340	664.8	300	3105.5	280	16040.7	270	20555*
380	116.7	340	776.1	300	3523.2	280	20172.3*	270	20779.4*
380	122.5	340	793.9	300	4311.7	280	20707.1*	270	20916.3*
Var(Y)	551.4		3.3E4		7.4E5		2.5E7		2.5E7
Var(W)	.0218		.0437		.0312		.0593		.0363
CV(W)	.04409		.03512		.01658		.01619		.01145

linear predictor, a linear transformation of stress, is expected to *decrease* with increasing stress. This implies that the log fatigue life modeled by (15.2) would have coefficient of variation that *increases* with S (stress). Indeed, this property is compatible with the given data. Observing

CV(W) in Table 15.1 we realize that it increases with S. For example, for S=270 we have CV(W)= 0.01145, while for larger S=380, we have larger CV(W)= 0.04409. The RMM model is thus compatible with both requirements expounded by PM.

We start the RMM estimation with modeling of the linear predictor. PM model the linear predictor by [refer to (15.21)]

$$\eta + \varepsilon_1 = S - \gamma = \exp(x) - \exp(V), \quad x > V, \tag{15.22}$$

where x=log(S), and V, after standardization, is either standard normal or standard Sev. Alternatively, using the RMM model, the linear predictor with the error may be modeled by

$$\eta + \varepsilon_1 = \beta_0 + \beta_1 S + \varepsilon_1. \tag{15.23}$$

Attempting to implement Phase I of the estimation procedure, based on canonical correlation analysis (refer to Chapter 8 for details), did not provide better fit (in the pursuing stepwise linear regression) than simple application of the latter to the log fatigue-life. This could perhaps be anticipated since canonical correlation analysis is intended to correlate two sets of variables whereas in our case one set comprises a single variable, namely, the regressor S (stress). Applying simple linear regression to W, we obtain for the LP the *initial* estimate

$$E(W) = E[\log(Y)] = \eta = 22.11 - 0.04772S. \tag{15.24}$$

The F-ratio value is 1729 (p<.0000), with an adjusted R-squared of 93.8%, and residual standard deviation of $\sigma_{\varepsilon 1}$= 0.4963. The average of the absolute value of the relative error (relative deviation of the model from the observed value) is 6.53%.

Introducing from (15.24) for η into (15.19) and implementing un-weighted NL-LS to find estimates for the RMM parameters, $\{\alpha, \lambda, \mu_2\}$, inserting for $\sigma_{\varepsilon 1}$ the value estimated from the linear regression analysis, we obtain the parameters' estimates

$$\alpha = 0.1244, \lambda = 2.104, \mu_2 = 3.181. \tag{15.25}$$

The asymptotic F-ratio is huge (1993/0.2046) and the residual standard deviation is 0.4523. The average of the absolute value of the relative error (relative deviation of the model from the observed value) is 5.32%. Figure 15.1 displays the relative errors for various values of W (the log fatigue life).

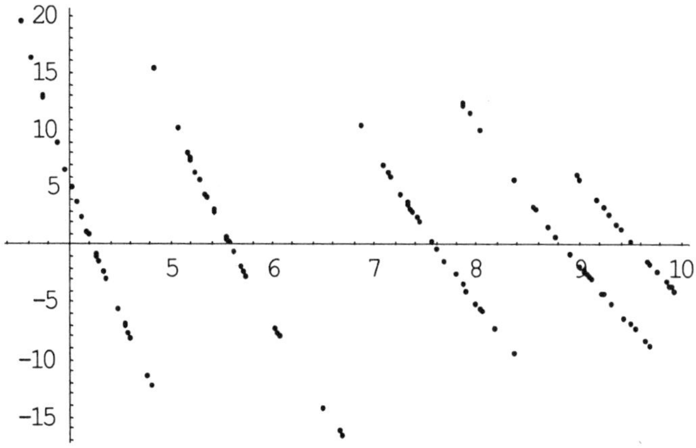

Figure 15.1. Relative errors (in %) for model (15.19) vs. W

Comparing the residual standard deviations obtained from the NL-LS with that obtained from linear regression (0.4523 vs. 0.4963, respectively), it is doubtful that the non-linear model adds much to the goodness-of-fit relative to the linear model. Furthermore, estimates of α and λ are very highly correlated ($\rho=-0.998$). On the other hand, quantile values calculated from the RMM model are appreciably more accurate, as will shortly be demonstrated. Results for both models will be presented. Introducing from (15.25) into (15.9) and maximizing with respect to $\{\sigma_{\varepsilon 1}, \sigma_{\varepsilon 2}\}$ (assuming uncorrelated errors, namely, $\rho=0$), we obtain

$$\sigma_{\varepsilon 1} = 0.3270, \ \sigma_{\varepsilon 2} = 0.3106, \tag{15.26}$$

with maximum log-likelihood of -99.12. Note the proximity between the value of the parameter σ, for the normal-normal case, in PM's model ($\sigma= 0.289$) and the corresponding statistic here ($\sigma_{\varepsilon 2}= 0.3106$).

Iterating the solution procedure, particularly in order to improve the initial estimate of the linear predictor (15.24), does not yield a better solution.

From the *linear* model, given by (15.24), the final model is

$$Y = \exp(22.11 - 0.04772s + 0.4963Z), \tag{15.27}$$

where Z is a standard normal variable.

The *non-linear* fitted RMM model is (15.2), with η given by (15.24), and the RMM parameters given by the estimates in (15.25) and (15.26).

Table 15.2 shows 0.05 quantile values for fatigue life for different values of stress and different models. Values for PM model (the RFL model), under various distributional scenarios, are taken from PM, (1999), Table 1. Values for the log fatigue-life linear model are taken from (15.27), introducing therein Z= -1.64485. Values for the log fatigue life RMM model are calculated numerically by solving

$$F_W(w_{0.05}; \eta, \theta) = 0.05,$$

where w_p is the p-th quantile of W and F_W is obtained numerically from (15.11). One realizes that the simpler linear model (15.27) provides quantile values that are exceptionally deviant relative to the other models (observe values for $S \geq 300$). The RMM model delivers quantile values compatible with those of the RFL model. The non-linear RMM model is the preferred model (relative to the linear model), even though judging only by goodness-of-fit statistics (like the residual standard deviations) does not indicate such preference.

Finally, note that the RFL model has been shown to be a special case of RMM (Shore, 2004b), however the random errors are differently defined in the two models [compare (15.22) and (15.23)]. This is probably the source for the slight discrepancies between the quantile values obtained for the two models in the Normal-Normal case.

Table 15.2. Predicted fatigue-life 0.05 quantile values for different stress levels (S) and different models

Stress level (S)	RFLM (according to the distributions of V and of log(Y), given γ)				RMM model	
	Sev-Sev	Normal-Normal	Sev-Normal	Normal-Sev	Linear Eq. (15.27)	Non-linear Eq. (15.2)
270	4443.0	6136.0	5530.0	6139.0	4533.7	5691.2
280	2319.0	2963.0	2810.0	2899.0	2813.8	2954.4
300	751.0	884.0	888.0	840.0	1083.9	897.6
340	126.0	144.0	150.0	134.0	160.8	133.0
380	32.0	38.0	39.0	35.0	23.86	36.3

References

[1] Pascual, F. G., and Meeker, W. Q. (1997). Analysis of fatigue data with run-outs based on a model with non-constant standard deviation and a fatigue limit parameter. *Journal of Testing and Evaluation*, 25, 292-301.

[2] Pascual, F. G., and Meeker, W. Q. (1999). Estimating fatigue curves with the random fatigue-limit model. *Technometrics,* 41 (4), 277-290.

[3] Pascual, F. G. (2003). A standardized form of the Random Fatigue-Limit Model. *Communications in Statistics (Simulation and Computation)*, 32(4), 1205-1221.

[4] Pascual, F. G. (2004). The random fatigue-limit model in multi-factor experiments. *Journal of statistical Computation and Simulation.* In press.

[5] Shimokawa, T., and Hamaguchi, Y. (1987). Statistical evaluation of fatigue strength in circular-holed notched specimens of a carbon eight-harness-satin/epoxy laminate. In *Statistical Research on Fatigue and Fracture (*current Japanese Materials Research, Vol. 2), Tanaka, T., Nishijima, S., Ichikawa, M., Eds.; Elsevier: London; 159-176.

[6] Shore, H. (2004a). Response Modeling Methodology (RMM)- Current distributions, transformations and approximations as special cases of the RMM error distribution. *Communications in Statistics (Theory and Methods)*, 33(7), 1491-1510.

[7] Shore, H. (2004b). The Random Fatigue Life Model as a special case of the RMM model- A comment on Pascual (2003a). *Communications in Statistics (Simulation and Computation).* 33(2), 537-539.

Chapter 16

Software Reliability-Growth Models

16.1. Introduction

Software reliability-growth models relate to the modeling of the cumulative number (rate) of errors (failures), found in a software product, as a function of the effort invested in detecting these errors. The detection effort is commonly expressed in testing hours, and it is assumed that as more errors are exposed and corrected the reliability of the product tends to grow. In other words, no new errors are injected into the software as a result of the repair activities. A further assumption commonly made is that the errors arrive in a Poisson stream. However the mean rate is not constant but changes as a function of T, the cumulative number of testing hours. Thus, the number of errors that have occurred by any given time, T=t, may be modeled by a non-homogenous Poisson Process (NHPP), with mean value function $\mu(t)$.

Modeling software reliability-growth usually refers to the modeling of $\mu(t)$ as a function of t. In Section 2.6, commonly applied software reliability-growth models (SRGM) were reviewed. In the review questions, the reader is requested to show that these are indeed special cases of the RMM model (up to a linear transformation). Refer also to Shore (2002, 2004), where this subject is addressed.

In this chapter we demonstrate application of RMM to two of Musa's data sets (Musa, 1979). The latter are commonly accepted as industrial-standard for comparing SRGM, and we pursue in that regard previous references that used Musa's data-sets for comparative evaluation of various SRGM (refer, for example, to Arnoux *et al.*, 2000). For the first example (Section 16.2), we use Musa's M_1 data-set (sample size of n=136), with no comparison to current models. For the second example (Section 16.3), we use Musa's M_3 data-set (n=38), and compare the goodness-of-fit of this model to that of current models, estimated via maximum likelihood. The log-likelihood function, assuming Poisson data, is developed in the same Section 16.3.

For a relatively recent good exposition of SRGM, the reader is referred to Xie (2000).

16.2. Example 1 - Musa's M_1 Data-Set

We use for this example Musa's M_1 data-set (Musa, 1979). The data consist of 136 observations, which are displayed (cumulative number of failures *vs.* cumulative number of testing hours) in Figure 16.1, and given in Table 16.1. It is evident From Figure 16.1 that the dispersion of points shows a trend of monotone convexity which qualifies the displayed data to be modeled by RMM.

For an RMM model, the quantile function for the log-transformed response, W=log(Y), and approximate expressions for the mean

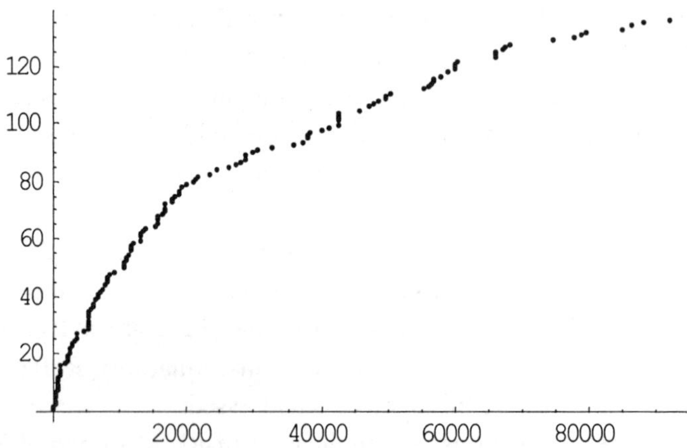

Figure 16.1. Example 1- Number of cumulative detected failures *vs.* number of software testing hours (Musa's M_1 data set)

Table 16.1. Example 1- Musa's M_1 data set. No. is the cumulative number of failures by time t.

No.	t	No.	t	No.	t	No.	t
1	3	35	5324	69	16106	103	42596
2	33	36	5389	70	16485	104	42597
3	146	37	5865	71	16529	105	45706
4	227	38	5923	72	16658	106	46953
5	342	39	6380	73	17468	107	47896
6	351	40	6680	74	17758	108	48596
7	353	41	6777	75	18058	109	49471
8	444	42	7040	76	18587	110	49716
9	556	43	7492	77	18868	111	50445
10	571	44	7747	78	19028	112	55342
11	709	45	7944	79	19856	113	55789
12	759	46	8137	80	20867	114	56175
13	836	47	8143	81	21312	115	56621
14	860	48	8222	82	21608	116	56743
15	968	49	9038	83	23363	117	57733
16	1056	50	10389	84	24427	118	58681
17	1726	51	10537	85	26210	119	59763
18	1846	52	10558	86	27070	120	59785
19	1872	53	10791	87	28053	121	59860
20	1986	54	10925	88	28760	122	60342
21	2311	55	11282	89	28793	123	65851
22	2366	56	11475	90	29661	124	65951
23	2608	57	11711	91	30385	125	65961
24	2676	58	11742	92	32708	126	67032
25	3098	59	12111	93	35638	127	67403
26	3278	60	12859	94	37099	128	68193
27	3288	61	12860	95	37942	129	74343
28	4434	62	13091	96	37954	130	77664
29	5034	63	13421	97	38215	131	78709
30	5049	64	13786	98	40015	132	79357
31	5085	65	15008	99	40880	133	84842
32	5089	66	15551	100	42315	134	86002
33	5090	67	15561	101	42345	135	87866
34	5097	68	15577	102	42488	136	91982

response, E(Y), and for the mean of the log-transformed response, E(W), are, assuming uncorrelated errors (refer to Chapter 7 for details),

$$W = \log(Y) = (\alpha/\lambda)[(\eta+\varepsilon_1)^\lambda - 1] + \mu_2 + \varepsilon_2$$

$$= (\alpha/\lambda)[(\eta+\sigma_{\varepsilon1}Z_1)^\lambda - 1] + \mu_2 + \sigma_{\varepsilon2}Z_2, \tag{16.1}$$

$$E(Y) \cong \exp[(\alpha/\lambda)(\eta^\lambda - 1) + \mu_2 + (\sigma_{\varepsilon2}^2/2)]$$

$$\times \{1 + (\sigma_{\varepsilon1}^2/2)[\eta^{2(\lambda-1)}\alpha^2 + \eta^{\lambda-2}\alpha(\lambda-1)]\}, \tag{16.2}$$

$$E(W) = E\{(\alpha/\lambda)[(\eta+\sigma_{\varepsilon1}Z_1)^\lambda - 1]\} + \mu_2$$

$$\cong (\alpha/\lambda)[(\eta)^\lambda - 1] + (1/2)\alpha(\lambda-1)\sigma_{\varepsilon1}^2\eta^{\lambda-2} + \mu_2. \tag{16.3}$$

These expressions will be used later in the analysis.

The RMM estimation procedure starts with determining the structure of the linear predictor (LP). Consider a combined mean function of both current software reliability-growth models III and V (these are defined in Section 2.6):

$$\mu(t) = (a)[\log(1+bt)]^c, \tag{16.4}$$

where {a,b,c} are parameters that need to be determined. Note that for c=1, we obtain Model III, and for b=1, we obtain Model V.

Model (16.4) is easily recognized as the mean of an RMM model. Assume that $\sigma_{\varepsilon1}^2$ and $\sigma_{\varepsilon2}^2$ in (16.2) are negligible (setting them equal to 0). Then (16.4) is a special case of (16.2) if we introduce in the latter, in terms of the parameters of (16.4),

$$\lambda = 0, \ \alpha = c, \ \mu_2 = \log(a), \ \eta = \log(1+bt). \tag{16.5}$$

Let us replace (16.4) by the more general RMM model. However preserve the LP (η), as given in (16.5). For the response means given by (16.2) and (16.3), neglecting terms that include the errors' variances, we obtain, respectively,

$$E(Y) = \exp[(\alpha/\lambda)(\eta^\lambda-1)+\mu_2] = \exp\{(\alpha/\lambda)\{[\log(1+bt)]^\lambda - 1\} + \mu_2\},$$

$$E(W) = E[\log(Y)] = (\alpha/\lambda)\{[\log(1+bt)]^\lambda - 1\} + \mu_2. \tag{16.6}$$

In the analysis that follows, we do not pre-specify values for α, λ or μ_2, and let the data decide these values. Furthermore, since the LP contains a single unknown parameter (b), we include estimation of this parameter within the non-linear least-squares (NL-LS) procedure. Thus, the need to iterate between NL-LS and another procedure (that estimates the LP) is avoided. However, keep in mind the approximating assumptions made here, which render the estimation procedure numerically simpler: We have assumed that the errors are uncorrelated (ρ=0), and we have assumed that the standard deviations of the errors are small enough to be ignored in modeling the mean ($\sigma_{\varepsilon 1}$= $\sigma_{\varepsilon 2}$= 0). These assumptions reduced the number of parameters that need estimating by 3, allowing us to use only NL-LS to estimate the RMM parameters.

Applying NL-LS to the given data set (with n=136), we obtain for the parameters of the mean function

$$\alpha = 1.079, \lambda = 0.4481, \mu_2 = 2.113, b = 0.00330. \qquad (16.7)$$

While the quality of the allied statistics (like standard errors of the estimates or correlations between estimates) is less than desirable, the parameters' estimates imply that the following may constitute a good model for the data:

$$E(W) = E[\log(Y)] = (2)[\log(1+bt)]^{1/2}. \qquad (16.8)$$

(namely, α= 1, λ= 1/2, μ_2= 2).

As in previous examples, we have modified the model according to results obtained in the estimating procedure, while delaying examination of goodness-of-fit statistics until the final model is selected. More on this at the end of this section.

Refitting (16.8) via the NL-LS procedure, we obtain for the single parameter: b= 0.0046384. The asymptotic standard error of this estimate is of the order of $6(10)^{-5}$ (implying short confidence interval), and the model F-ratio is 2227/0.007 (with 1 and 135 degrees of freedom for the model and for the error, respectively). This F value is highly significant.

Good fit is obtained, as we may also realize graphically from Figures 16.2 and 16.3. The former shows the predicted vs. actual number of failures (on a log scale), the latter shows the errors vs. the actual number of failures (also on a log scale). It is interesting to note that in Figure 16.3 the first six observations are not shown: Their deviations far exceed those of the rest of the observations (absolute deviation of less than 0.3). Apparently, these observations represent a "warm up" period, where the testing hours put in detecting the first few failures do not comply with the pattern shown by succeeding observations.

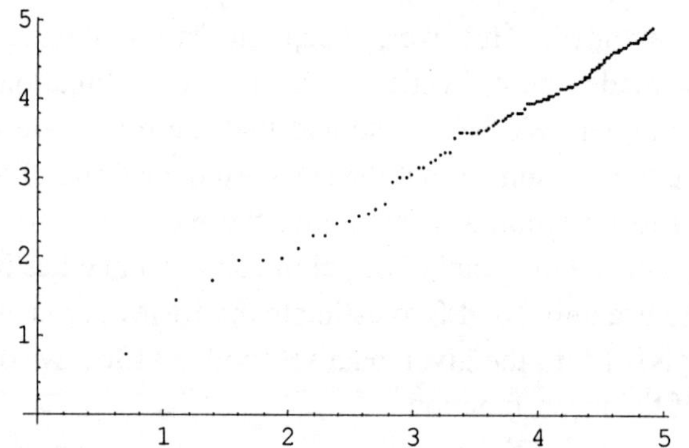

Figure 16.2. Example 1- Predicted *vs.* actual number of failures (both on a log scale)

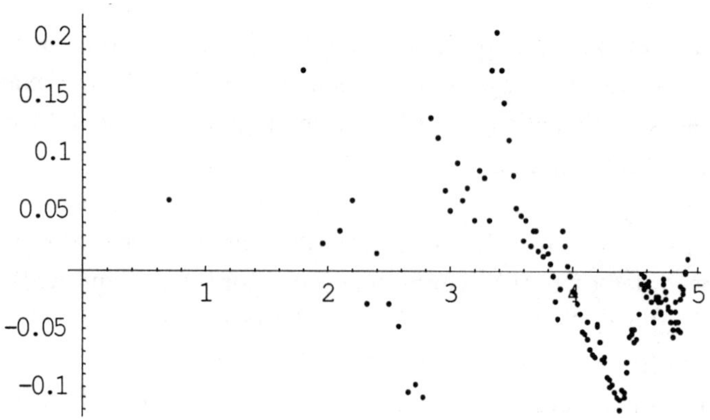

Figure 16.3. Example 1- Errors (actual minus model) for the number of failures *vs.* the actual number of failures (both on a log scale)

Comment. Good empirical modeling is the art of deriving the right model, based on trial-and-error iterations, where in each we learn something about the patterns revealed by the data. In particular, the derived values of the parameters can be informative in deciding on the next tested model. The iterative nature of empirical modeling has already been demonstrated in modeling the galaxies' velocities in Section 11.2.3. There, like here, the model finally selected, which provided the best fit, was different from the model with which we have started the empirical modeling process.

The excellent fit obtained in modeling Musa's M_1 data-set attests to the success of the iterative learning process that we have exercised.

16.3. Example 2 - Musa's M_3 Data-Set

Musa's M_3 data-set is analyzed (Musa, 1979). The data comprise 38 observations, given in Table 16.2. Unlike in the previous example, these data are analyzed both via RMM and by current major models, as these were introduced in Section 2.6. We start by deriving maximum-likelihood estimates (MLE) for models I-V (as depicted in that section). Since the underlying assumption of all these models is that the error-rate is Poisson (with non-homogenous mean), we may derive the MSE easily from the Poisson probability function, $f(n)$, with the assumed structure of the mean function.

Given a data set of k observations, $\{n_1, n_2, ..., n_k\}$, where n_i is the number of failures during time interval $[t_{i-1}, t_i)$, the log likelihood-function (LL), assuming an underlying Poisson law, is

$$LL = \log[\prod_{i=1}^{k} f[n_i \mid \mu(t_i) - \mu(t_{i-1})]$$

$$= -\sum_{i=1}^{k} [\mu(t_i) - \mu(t_{i-1})] + \sum_{i=1}^{k} n_i \log[\mu(t_i) - \mu(t_{i-1})] - \sum_{i=1}^{k} \log(n_i!)$$

$$= -\mu(t_k) + \sum_{i=1}^{k} n_i \log[\mu(t_i) - \mu(t_{i-1})] - \sum_{i=1}^{k} \log(n_i!). \tag{16.9}$$

Table 16.2. Example 2 - Musa's M_3 data set

No.	t	No.	t
1	115	20	6147
2	116	21	6162
3	198	22	6552
4	376	23	8415
5	570	24	9752
6	706	25	14260
7	1783	26	15094
8	1798	27	18494
9	1813	28	18500
10	1905	29	23061
11	1955	30	26247
12	2026	31	36818
13	2632	32	37381
14	3821	33	40151
15	3861	34	40803
16	4649	35	46396
17	4871	36	58092
18	4943	37	64816
19	5558	38	67362

For a given model for the response mean, $\mu(t)$, the parameters' values that maximize (16.9) are MLEs.

Using Musa's M_3 data-set, MLEs for the parameters $\{a,b\}$ of models I-V are identified, together with values of LL at the optimal point (LL*). These are given in Table 16.3 with the associated standard deviations of the residuals (STDres.). It is evident, both from values of LL* and those of STDres., that only Models IV and V provide acceptable accuracy (goodness-of-fit).

Scatter-plots of the residuals from these models are given as the top two plots of Figure 16.4.

Referring to the RMM approach, no specific mean function is assumed. Unlike the models above, we let the data determine its structure. However, the form of the LP needs to be decided.

Table 16.3. Example 2- MLEs and other results from the analysis of Musa's M_3 data-set

Model	a	b	LL*	STDres.
I	189638	0.009551	-69365	10464
II	160952	0.03201	449633	8147
III	338825	0.005169	432702	10154
IV	.9342	3.0667	470961	2988
V	0.02345	11.45	468773	2962

Consider the simplest model for LP

$$\eta(t) = t. \tag{16.10}$$

From a theoretical point of view, this is the preferred choice since this LP embodies the ideal of the new approach, namely, simplicity in the definition of the LP. Only if this choice does not provide results competitive with those obtained by the current models (Models I-V), will we opt for an expanded LP, which will be a linear transformation of the $\eta(t)$ above, namely,

$$\eta(t) = \beta_0 + \beta_1 t. \tag{16.10a}$$

Equation (16.10a) may be an optional model only if the parameters' estimates are significantly different from zero. Therefore we start the modeling process by examining the viability of (16.10a).

From (16.1) and (16.10a), the response quantile is

$$W = (\alpha/\lambda)[(\beta_0 + \beta_1 t + \varepsilon_1)^\lambda - 1] + \mu_2 + \varepsilon_2. \tag{16.11}$$

Introducing for ε_2 its expected value, $E(\varepsilon_2)=0$, we obtain for the i-th observation the transformed response value

$$\eta(t_i) + \varepsilon_1 = \beta_0 + \beta_1 t_i + \varepsilon_1 = [(\lambda/\alpha)(w_i-\mu_2) + 1]^{(1/\lambda)}, \tag{16.12}$$

where ε_1 is the random error associated with the i-th value of the LP and w_i is the i-th logged value of the original response.

The data transformation given by the RHS of (16.12) may serve as response input for linear regression analysis, once initial values for the RMM parameters, (α, λ, μ_2), have been obtained via NL-LS.

We start the first iteration assuming $\sigma_{\varepsilon 1}=0$ in the expression for $E(W)$ (16.3). In subsequent iterations, an estimate of $\sigma_{\varepsilon 1}$ will be available from the most recent application of the linear regression stage of the analysis, and this estimate may be used in the expression for $E(W)$.

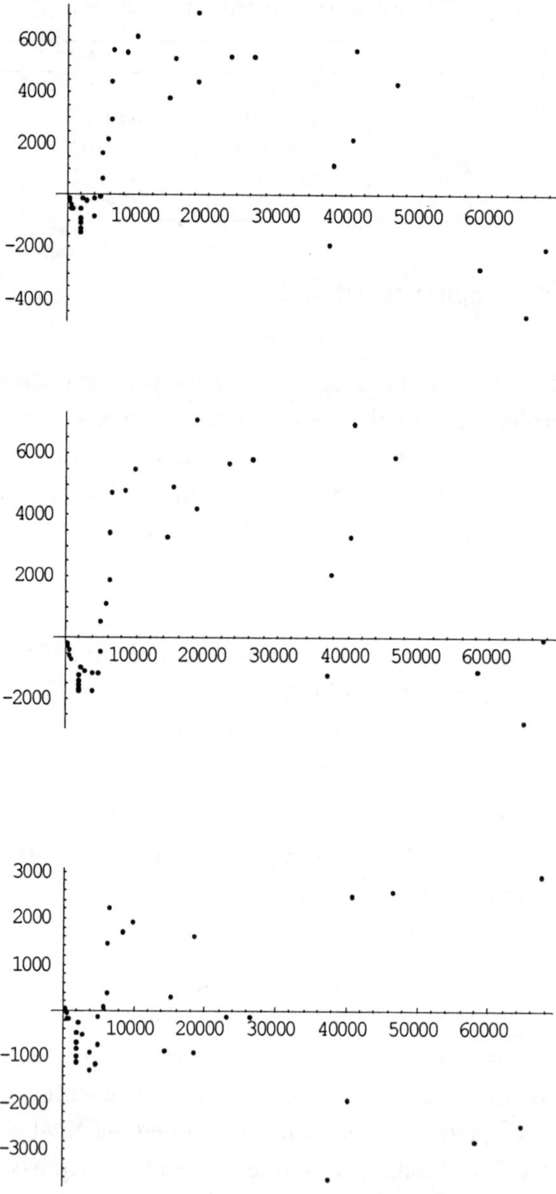

Figure 16.4. Example 2 - Scatter plots of residuals (*vs.* exact value of Y), from fitting (top to bottom) Model IV, Model V and the RMM model [(16.3) and (16.10)]

Also, no weighting is introduced in the first iteration. In subsequent iterations, we may use weighting if the residual variance varies from one observation to the next, that is, $\sigma_{\varepsilon 1}$ is not negligible relative to typical values of η.

A first run of NL-LS, using the approximate procedure, delivers

$$\alpha = 0.7433, \lambda = 0.4809, \mu_2 = 3.8113, \text{STDres.} = 1817.$$

The latter statistic is about 61% of the best of the STDres. in Table 16.3. Observing the residuals we find observation 31 to have an outlying residual of 6256.6. Excluding this observation from the analysis we obtain

$$\alpha = 0.4335, \lambda = 0.6437, \mu_2 = 4.832, \text{STDres.} = 1471. \qquad (16.13)$$

The latter statistic is about 50% of the best of the STDres. in Table 16.3. Approximate confidence intervals for the estimates of the RMM parameters may be derived from the NL-LS procedure, as these are provided as standard output from any statistical package that performs NL-LS analysis.

Turning next to the linear regression part of the analysis, we model the LP as in (16.12), with the above RMM parameters, to obtain

$$\beta_0 = 1.5211, \beta_1 = 0.9488.$$

Goodness-of-fit statistics give an F value of 2064, R^2-adj. of 98.3% and $\sigma_{\varepsilon 1} = 1.391$. Both estimates of β_0 and β_1 are significant, implying that they are not zero. Model (16.10a) is therefore a possible candidate for modeling the LP.

Consider the estimate for β_0. Since t is measured on a scale of 10^4 (namely, $\beta_0 \ll t$) not much accuracy is sacrificed if we select $\beta_0 = 0$, $\beta_1 = 1$ [as initially suggested, refer to (16.10)]. With these values, the new estimate is $\sigma_{\varepsilon 1} = 1.570$. Re-running NL-LS, the final estimates for the RMM model are nearly the same as in (16.13).

A scatter-plot of the residuals from this model is given in Figure 16.4 (bottom plot). The scatter of the residuals for the new model seems more random than displayed by the other two models (no formal statistical analysis of randomness has been performed). Furthermore, while application of the RMM model required estimation of three parameters, $\{\alpha, \lambda, \mu_2\}$, vs. two parameters for the other two models, the error, as evidenced by the value of STDres.= 1471, is nearly halved relative to the best performing current model (Model V).

References

[1] Arnoux, F., Gaudoin, O., Makni, C. (2000). *The generalized power family in software reliability data analysis.* In: MMR'2000: Second International Conference on Mathematical Methods in Reliability- Abstracts' Book, 1, 107-110.

[2] Musa, J. D. (1979). *Software Reliability Data.* Technical Report. Rome Air Development Center.

[3] Shore, H. (2002). Modeling a response with self-generated and externally-generated sources of variation. *Quality Engineering,* 14(4), 563-578.

[4] Shore, H. (2004). Response Modeling Methodology (RMM)- Validating evidence from engineering and the sciences. *Quality and Reliability Engineering International,* 20, 61-79.

[5] Xie, M. (2000). Software reliability models- Past, present and future. Chapter 21 in Limnios and Nikulin (Editors), *Recent Advances in Reliability Theory- Methodology, Practice and Inference.* Birkhauser, Boston, c/o Springer-Verlag, NY.

Chapter 17

Modeling a Chemo-Response

17.1. Introduction

Monotone convex/concave relationships abound in all branches of chemistry and chemical engineering. This has been demonstrated in the survey of Chapter 2, where current models related to chemo-responses, like liquid density, liquid heat capacity or Arrhenius equation, were reviewed. In Chapter 12 we re-addressed these models, and have shown that they may be considered special cases of the RMM model. In this chapter, RMM is applied to model real data-sets in the chemical engineering arena, and typical goodness-of-fit obtainable in such modeling endeavors is demonstrated.

For monotone convex/concave relationships, it is acceptable practice within the chemical-engineering discipline to enhance the goodness-of-fit of empirical models by using high-order polynomial models. For example, the basic Antoine equation, which relates vapor pressure to temperature, is derived from theoretical arguments (refer to Chapter 2 for details). This relationship was enhanced by both Wagner and Riedel (Section 17.2), who had added polynomial terms that resulted in empirical models with more parameters than in Antoine equation. The enhanced models have been shown to attain better goodness-of-fit (Daubert, 1998).

High-order polynomials, however, may result in unstable models. As discussed in Shore, Brauner and Shacham (2002), this may occur when the lengths of confidence intervals for the model's parameters are larger than the parameters' estimates. Furthermore, high-order polynomials tend to introduce inflections, which are undesirable if the actual relationship is known to be monotone.

For all of the above reasons and others, discussed in previous chapters, RMM may produce a good substitute to various polynomial-based current models in chemical engineering.

In this chapter, implementation of the RMM model to data-sets from chemical engineering is demonstrated. We use two sources for the compilation of the numerical examples: Shore (2003) and Shore, Brauner and Shacham (2002). Since the two papers have used RMM models with different error structures (as will be expounded later on), the reader may find it instructive to compare, for those cases where the same data-sets are used in both references, the goodness-of-fit obtained by the different models.

The numerical examples are divided into two sets. In Section 17.2, we use the RMM model as implemented in Shore (2003). This is the original model, introduced in Chapter 7. In Section 17.3, a variation of the basic model is employed, based on inverse normalizing transformations (the latter are developed in Chapter 20). This is the same model as in Shore, Brauner and Shacham (2002).

The numerical examples in the two sections roughly correspond to those in the two sources cited.

17.2. Applying RMM to a Chemo-Response - First Variation

17.2.1. Example 1 - Temperature dependence of vapor pressure

This example is based on data from Wagner (1973) and Wagner *et al.* (1976), related to Argon (n=57), Nitrogen (n=68) and Oxygen (n=183), and data from Osborn and Douslin (1966), related to 1- Heptanethiol (n=18) and 1-Propanethiol (n=15). For Argon and Nitrogen, data range from the triple point to the critical point. For Oxygen, data range from the normal boiling point to the critical point. For most of these vapors (exceptions are detailed shortly), temperature, T, is in deg. K. For pressure, P, measurements are in "bar" for Argon and Nitrogen and in Mpa for Oxygen. For 1-Heptanethiol and 1-Propanethiol, temperature is in deg. C., and pressure in mmHg.

To fit the RMM model, we assume that log(P) has approximately a constant variance. Recall that for the response, Y, and the log-transformed response, W=log(Y), we have, with the associated approximate expressions for the means (refer to Chapter 7),

$$W = (\alpha/\lambda)[(\eta+\varepsilon_1)^\lambda - 1] + \mu_2 + \varepsilon_2$$
$$= (\alpha/\lambda)[(\eta+\sigma_{\varepsilon 1}Z_1)^\lambda - 1] + \mu_2 + \sigma_{\varepsilon 2}Z_2, \qquad (17.1)$$

$$E(Y) \cong \exp[(\alpha/\lambda)(\eta^\lambda - 1) + \mu_2 + (\sigma_{\varepsilon 2}^2/2)]$$
$$x \{1 + (\sigma_{\varepsilon 1}^2/2)[\,\eta^{2(\lambda-1)}\,\alpha^2 + \eta^{\lambda-2}\alpha(\lambda-1)]\}, \qquad (17.2)$$

$$E(W) = E\{(\alpha/\lambda)[(\eta+\sigma_{\varepsilon 1}Z_1)^\lambda - 1]\} + \mu_2$$
$$\cong (\alpha/\lambda)[(\eta)^\lambda - 1] + (1/2)\alpha(\lambda-1)\sigma_{\varepsilon 1}^2\eta^{\lambda-2} + \mu_2, \qquad (17.3)$$

where $\varepsilon_1 = \sigma_{\varepsilon 1}Z_1$ and $\varepsilon_2 = \sigma_{\varepsilon 2}Z_2$ are assumed to be uncorrelated normal errors, Z_1 and Z_2 are independent standard normal variables, and $\theta = \{\alpha, \lambda, \mu_2, \sigma_{\varepsilon 1}, \sigma_{\varepsilon 2}\}$ is the parameters' vector.

Assume that $T=\mu_T+\varepsilon_1$, where μ_T is the expected value of T. For most reported measurements of T, it is reasonable to assume that $|\varepsilon_1|/\mu_T \ll 1$. If that is valid, we can neglect ε_1 ($\sigma_{\varepsilon 1}\cong 0$), and assume that the error variance of W= log(P) is determined solely by ε_2 [refer to (17.1)]. In that case, log(P) may reasonably be assumed to have error variance that is unaffected by the value of T. The assumption of a constant variance for log(P) is corroborated by the common practice of modeling log(P), rather than P, in various non-linear models of vapor pressure (refer, for examples, to Daubert, 1998). This assumption will allow us to use un-weighted non-linear least-squares (NL-LS), which, in the case of a homogenous variance, deliver maximum likelihood (ML) estimates (Myers *et al.*, 2002, Section 3.3.3).

In the analyses that follow, a detailed analysis for the Oxygen data is first carried out. Later, we display the analyses results for all vapors, as specified earlier. A plot of the Oxygen data is given in Figure 17.1. Obviously, it is monotone convex, a basic assumption of RMM.

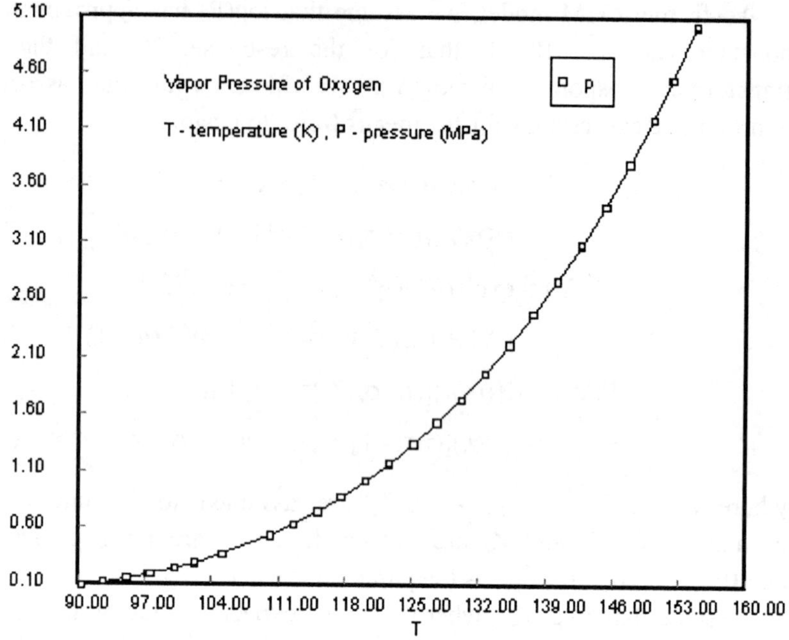

Figure 17.1. Plot of the Oxygen data (pressure vs. temperature)

Introducing $\sigma_{\varepsilon 1}=0$ in E[log(P)] (17.3), we obtain, via NL-LS estimation, initial values of

$$\alpha = 603.1, \lambda = -0.9300, \mu_2 = -640.9.$$

The *estimated* value of λ is near the value given by Antoine equation ($\lambda=-1$).

From (17.1), recalling that $E(\varepsilon_2)= 0$, we obtain explicitly for the linear predictor (LP) of observation i

$$\eta_i + \varepsilon_1 = \{[\log(P_i)+640.9](-0.93/603.1)+1\}^{(-1/0.93)}, \qquad (17.4)$$

where $\log(P_i)$ is an estimate of $E(W)=E[\log(P_i)]$ for observation i.

Note that assuming uncorrelated errors and that $\sigma_{\varepsilon 1}$ is negligible allowed derivation of (17.4). The latter provides a non-linear response transformation from which the coefficients of the LP can be estimated straightforwardly, via stepwise linear regression. This "spares" us the more complex "standard" estimation procedure, expounded in Chapter 8.

To estimate the linear predictor (LP), consider an initial model for observation i:

$$\eta_i + \varepsilon_1 = \beta_0 + \beta_1\mu_{Ti} + \varepsilon_1,$$

where μ_{Ti} is the expected value of temperature reading for observation i.

Applying linear regression to the transformed response values we obtain

$$\eta = 0.2618+0.9953T,$$

where we replace μ_{Ti} by T_i, the *actual* temperature reading for observation i. The associated F-Ratio is 8E6, R^2-adj. is 0.99998, and the standard deviation of the residuals ($\sigma_{\varepsilon 1}$) is 0.09316. While the estimate of β_0 is significantly non-zero (t=5.929, p<0.000), the estimate of β_1 has an appreciably larger t value of 2829. This indeed overshadows the significance of β_0. Furthermore, the latter is negligible relative to the size of T.

The problem with the significant but small intercept estimate, encountered in Example 2 of Section 16.3, is repeated here, and the same determination is made to set $\beta_0=0$. Re-running linear regression, the results obtained are $\beta_1= 0.9973$, with F-Ratio= 2.78E8, R^2-adj.=0.99999 and $\sigma_{\varepsilon 1}=0.1015$.

The small value of $\sigma_{\varepsilon 1}$ (relative to T which is measured in 10^2) shows that the random component delivered to the response via T is indeed small (relative to μ_T), as initially postulated. Implementing the NL-LS routine with no weighting [namely, assuming a constant variance for log(P)] is therefore supported by the analysis results.

Observing studentized residuals, we note that some of the upper-scale observations are outliers (absolute value of studentized residual larger than 2). Deleting observations with outlying deviations (observations 178-183) had not meaningfully improved the fit, and therefore these observations were left in the analysis. The final model is

$$\log(P) = (-603.1/0.93)[(0.9973\mu_T+\varepsilon_1)^{(-0.93)} - 1] - 640.9 + \varepsilon_2. \quad (17.5)$$

Once all parameters have been estimated, expected values for individual observations may be calculated, as well as any desirable quantile.

To assess the goodness-of-fit of the estimated RMM model (17.5), we evaluated E[log(P)] (17.3) for individual observations, neglecting the term that contains the negligible $\sigma_{\varepsilon 1}$. Residuals from this expression are used to calculate "STD Res.", the residual standard deviation, which is displayed in Table 17.1. This table also shows results from the analyses of the other data-sets, relating to the other vapors. These analyses were conducted similarly to the detailed analysis of Oxygen.

Inspecting Table 17.1, it is worth noting that the F-ratio values obtained for all vapors are all measured in over 1E6. This attests to the good fit obtained during the linear regression phase of the analysis. Also, the values of λ for Argon, Nitrogen and Oxygen are compatible with those expected by the Antoine formula (-1). However, no theory-based consideration has been involved in deriving these empirical values.

To appreciate the *relative* accuracy obtained by the new approach, the vapor samples are re-analyzed using other models. It was judged that only Argon, Nitrogen and Oxygen had a large enough sample size to justify accuracy comparisons. All analyses were therefore confined to these vapors. To carry out the comparisons, we re-applied the NL-LS routine to Antoine, the extended-Riedel and the Wagner equations (refer to Daubert, 1998). The first two have been addressed in Chapter 2, and all are re-introduced here for convenience.

Antoine equation:

$$log(P) = A + B / (T+C). \tag{17.6}$$

Table 17.1. Example 1- Results from the analysis of vapor-pressure data ["STD Res." refers to STD of residuals from fitting log(P)]

Vapor	Linear Regression Results			NL-LS Results			STD Res.
	F-Ratio	Excluded Obs.	$\sigma_{\varepsilon 1}$	α	λ	μ_2	
Argon (n=57)	34.4E6	none	.1331	597.5	-.9378	-627.5	.006692
Nitrogen (n=68)	33.4E6	none	.1425	813.2	-1.032	-778.8	.009032
Oxygen (n=183)	278E6	none	.1015	603.1	-.9300	-640.9	.004689
Hept. (n=18)	71.6E6	No. 18	.06945	1.914	.1631	-8.931	.002361
Propt. (n=15)	2.3E6	No. 15	.1569	.1659	.6063	3.377	.006240

Extended Riedel:
$$\log(P) = A + B/T + C\log(T) + DT^2 + E/T^2. \qquad (17.7)$$

Wagner equation:
$$\log(P_r) = (1/T_r)\,(At + Bt^{1.5} + Ct^3 + Dt^6), \qquad (17.8)$$

where $t = 1 - T_r = 1 - T/T_c$, $P_r = P/P_c$, and (T_c, P_c) are measured at the critical point (refer to Section 2.2).

The standard deviations of the residuals (errors), obtained from NL-LS fitting, are exhibited in Table 17.2.

One realizes (expectedly) that the accuracy increases with the number of parameters that need to be estimated. Thus, Antoine and RMM (with three parameters) have error standard deviations of the order of E-3 (10^{-3}), the extended Riedel (five parameters) has standard deviations of the order of E-4, and Wagner (with effectively seven parameters) has standard deviations of the order of E-5. Note, that although the latter requires determination of four parameters within the NL-LS routine, the exponents in (17.8) have been determined by optimizing with respect to the very same substances, so that in fact this equation is founded on seven estimates altogether. We should again be reminded that while the other approaches compared in Table 17.2 have some theory-based foundations, the RMM model is purely pragmatic and theory-free.

Table 17.2. Example 1- Comparison of residual standard deviations for vapor data [all residuals relate to log(P)]

Method (No. of parameters) (eq.)	Oxygen	Argon	Nitrogen
Antoine (3) (17.6)	4.915E-3	7.008E-3	8.729E-3
Extended Riedel (5) (17.7)	4.804E-4	5.190E-4	5.250E-4
Wagner (7) (17.8)	8.508E-5	2.096E-4	2.118E-4
RMM (3) (17.3)	4.689E-3	6.692E-3	9.032E-3

                                    RESPONSE MODELING METHODOLOGY

17.2.2. Example 2 - Temperature dependence of solid heat capacity

Heat capacity, C_p, is measured in J/mol-K. In this example, RMM is used to model temperature dependence of solid heat capacity. The available data relate to solid Acetonitrile (Puntam *et al.*, 1965, n=46), solid Ammonia (Overstreet and Giauque, 1937, n=42), solid Ethanoic acid (Martin and Andon, 1982, n=108) and solid Butanoic acid (Martin and Andon, 1982, n=104). Since solid Acetonitrile has a phase transition point at T=208.98K, which prevents the graph from being monotone increasing, only the data below this value were used for the fitting. This reduced the sample size to n=30. Similarly, the sample size for Butanoic acid was reduced to n=59. The analyses results are shown in Table 17.3.

Note that some observations were deleted due to outlying residuals in the linear regression part of the analysis. However, they were included in determining the parameters in the NL-LS analysis. Also, as with the vapor data (Example 1), the estimation process, as well as calculation of the residual standard deviations, is conducted in terms of the log-transformed response.

Table 17.3. Example 2- Results from the analysis of solid heat capacity data ("STD Res." relates to residual standard deviation from NL-LS; Upper entries relate to fitting of $\log(C_p)$, bottom entries to fitting of C_p).

Substance (n) (C_p range)	Linear Regr. Results			NL-LS Results			STD Res.
	F-Ratio	Excluded Obs.	$\sigma_{\varepsilon 1}$	α	λ	μ_2	
Acetonitrile (n=30) (0.41-19)	3.3E4 3.1E4	26-30 -- 29-30	1.927 2.645	31.00 11.16	-.8022 -.5591	-35.27 -16.06	0.03505 0.2888
Ammonia (n=42) (0.18-12)	1.8E5 5.4E4	40-42 -- 42	1.466 1.440	15.42 7.953	-.5835 -.4322	-22.77 -14.05	0.02558 0.1050
Butanoic acid (n=59) (5.8-135)	5.9E3 2.2E4	58- 59 -- 58-59	11.42 5.901	8.261 1.367	-.4994 -.0881	-10.62 -.9757	0.05450 3.968
Ethanoic acid (n=108) (2.7-94)	9.9E3 2.9E4	105-108 -- 105-108	17.26 9.453	20.68 8.289	-.7598 -.5583	-8.931 -9.823	0.03685 2.372

Since it is customary in modeling C_p to use the response in the original scale, Table 17.3 includes also results of fitting heat capacity data in the original scale (C_p). The model for C_p is the approximate mean of the response, $E(Y)$, given in (17.2), with $\sigma_{\varepsilon 1}=0$. Note, that the approximate assumption made earlier in fitting $\log(Y)$ (namely, that if $\sigma_{\varepsilon 1}$ is small no weighting is required in the NL-LS part of the procedure) may become questionable when C_p is used as the response. Therefore, as a general rule it is recommended that fitting will be applied to $\log(Y)$ rather than Y.

A comparative evaluation is now conducted. We adopt Daubert's assertion (1998) that for solid heat capacity "a simple polynomial in temperature is used" and "most data can be fitted with a linear equation with a quadratic necessary for a few systems". Since heat capacity is usually modeled in terms of C_p [rather than $\log(C_p)$], for comparison purposes we use C_p as the response, and our sample data are fitted with polynomials of increasing degree. This allows assessing the degree of the polynomial that is equivalent, in terms of the resultant accuracy, to (17.2) (the RMM model). Furthermore, the relative merits of the new approach in terms of stability can be evaluated.

As alluded to earlier, high order polynomials can become unstable (indicated by confidence intervals that are larger than the respective estimates' values). Quite often instability is not detected based on confidence intervals, however adding more polynomial terms may introduce unwarranted inflections that result in poor interpolation or extrapolation. Use of extreme care is recommended when fitting third and fourth order polynomials (Daubert, 1998).

Table 17.4 exhibits results for the comparative analysis of solid heat capacity data. The present approach is compared to modeling Cp by linear, quadratic and third degree polynomials.

We realize that the new approach generally provides accuracy comparable to second order polynomial (the maximum order recommended by Daubert, 1998), with better accuracy for most of the cases compared.

Figures 17.2 and 17.3 are plots of C_p for the best-fitting and the worst-fitting cases (Ammonia and Butanoic acid, respectively).

Table 17.4. Example 2- Comparative evaluation of the standard deviations of residuals

Substance (n)	Standard Deviation of residuals from modeling Cp as:			
	RMM (17.1)	Linear	2nd order polynomial	3rd order polynomial
Acetonitrile (n=30)	0.2888	0.9181	0.3512	0.1185
Ammonia (n=42)	0.1050	0.2192	0.09257	0.08747
Butanoic acid (n=59)	3.968	4.401	4.125	0.7987
Ethanoic acid (n=108)	2.372	6.449	3.818	0.9829

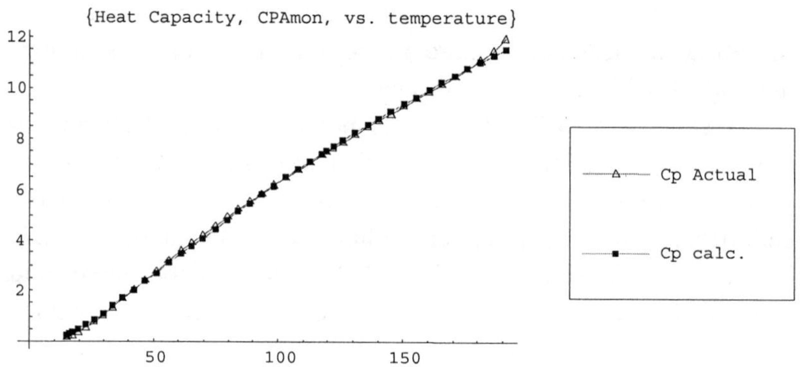

Figure 17.2. Plot of exact and calculated solid C_p for Ammonia

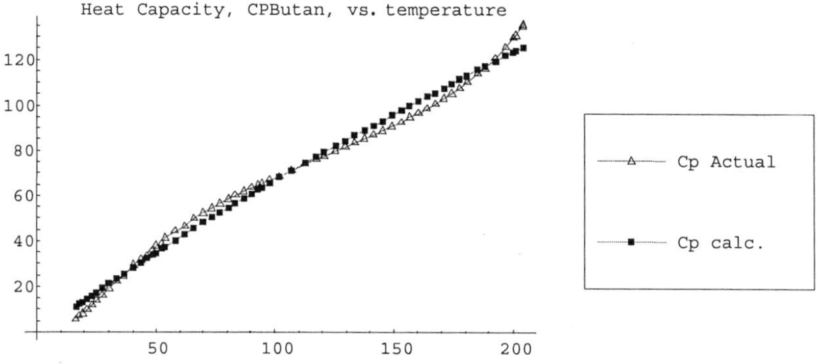

Figure 17.3. Plot of exact and calculated solid Cp for Butanoic acid

17.3. Applying RMM to a Chemo-Response - Second Variation

A variation of the RMM model, introduced in Chapter 9, is

$$Y = (M) \exp\{(\alpha/\lambda)[\eta^{\lambda}\exp(\lambda\varepsilon_1/\eta) - 1] + \varepsilon_2\}. \qquad (17.9)$$

For the data addressed in this section, we use (17.9) for modeling.

Assuming that the two normal errors, ε_1 and ε_2, have zero mean and standard deviations $\sigma_{\varepsilon 1}$ and $\sigma_{\varepsilon 2}$, respectively, and that they derive from a bi-variate normal distribution with correlation ρ, one can write (refer to Section 9.1):

$$\varepsilon_1 = \sigma_{\varepsilon 1}Z_1 \, , \, \varepsilon_2 = \sigma_{\varepsilon 2}[\rho Z_1 + (1-\rho^2)^{(1/2)} Z_2],$$

where Z_1 and Z_2 are two independent standard normal variables. Introducing for ε_2 back into (17.9), and assuming $\eta=1$, $\rho=\pm 1$ (thus, in fact, obtaining an inverse normalizing transformation, INT), we obtain

$$Y = (M) \exp\{(\alpha/\lambda)[\exp(\lambda\sigma_{\varepsilon 1}Z) - 1] + (\sigma_{\varepsilon 2}\rho)Z\}, \qquad (17.10)$$

where we replaced Z_1 by Z. Note, that the coefficients of Z can have different signs (when $\rho=-1$), a point that we have related to earlier.

Although (17.10) is a model of *random* variation (since $\eta=1$), we use it here for modeling *systematic* variation by treating Z as the independent variable, expressed in standard units defined by

$$Z = (X-M_x)/\sigma_x \, , \qquad (17.11)$$

where M_x and σ_x, respectively, are the median and the standard deviation of the independent variable, X. Equation (17.10) thus become

$$Y = \exp\{A + B[\exp(CZ) - 1] + DZ\} \qquad (17.12)$$

where $\{A,B,C,D\}$ are parameters that need to be determined.

The underlying *approximating* principle in this adaptation of the RMM model is that one does not have to distinguish between systematic and random variation in modeling the response. This may be achieved by modeling Y in terms of a single standardized r.v., however the latter represents the standardized independent variable, namely, the agent that transmits systematic variation to the response. In fact, this is the model adopted in Shore, Brauner and Shacham (2002), where an INT was used for modeling. At the time that the paper was submitted for publication, the INT was already known and well tested as a model for non-normal distributions (Shore, 2000). It was only later that we have learned that the INT can be derived as a special case of the RMM model (see details in Chapter 20). We reproduce in this section some of the numerical examples given in Shore, Brauner and Shacham (2002) with the allied analyses.

17.3.1. Example 1 - Temperature dependence of vapor pressure

Wagner (1973) and Wagner *et al.* (1976) had published high precision data of vapor pressure for Oxygen, Argon and Nitrogen. For the latter two, the ranges of the data were from the triple point to the critical point, and for Oxygen from the normal boiling point to the critical point. These data were introduced in Section 17.2 and used for modeling.

Further information for characterization of these data, together with the results of fitting (17.12) to these data, are given in Table 17.5.

It can be seen that the regressed model fits the data well. The variance [based on $\log(P)$] ranges from (1.76)E-7 (for Oxygen) up to (2.04)E-6 for Nitrogen. All the confidence intervals on the model parameters are smaller by at least two orders of magnitude than the parameter values.

The performance of model (17.12) for wide range vapor pressure data was compared with the performance of Antoine equation, the extended Riedel and Wagner equations [Eqs. (17.6)-(17.8), respectively].

Table 17.5. Example 1 - Data characterization and fit of (17.12), vapor pressure data

Parameter	Oxygen		Argon		Nitrogen	
	Value	Confidence Interval	Value	Confidence Interval	Value	Confidence Interval
A	0.3059587	9.79E-05	2.502048	2.73E-04	1.841746	5.08E-04
B	-0.855077	0.0049375	-0.85777	0.0092287	-0.96351	0.0128371
C	-0.603461	0.0015651	-0.71857	0.0033763	-0.70534	0.0040061
D	0.5336712	0.0015301	0.626196	0.0034159	0.626073	0.0046721
Variance	1.76E-07		4.33E-07		2.04E-06	
Sample size (n)	183		57		68	
T- units	K		K		K	
T- range	90.188	154.581	83.804	150.651	63.148	126.2
T- Median	125.1		120.1		97.04	
Pressure units	Mpa		bar		bar	
Pressure range	0.10128	5.04337	0.6895	48.578	0.1252	34.002
Median ln(P)	0.3008		2.5047		1.84	

The results of this comparison are shown in Table 17.6. It can be seen that the accuracy of INT (17.12) is orders of magnitude higher than that of the 3-parameter Antoine equation, and only slightly lower than the accuracy of the 5-parameter extended Riedel equation. Wagner's equation provides the most accurate correlations for Oxygen, Argon and Nitrogen. However, this equation was actually fitted (by stepwise regression) for the very same substances, as detailed earlier.

Table 17.6. Example 1 - Comparison of vapor pressure correlation equations

Correlation equation (No. of parameters)	Variance at optimal solution		
	Oxygen	Argon	Nitrogen
Antoine (3)	4.06E-05	5.09E-05	7.85E-05
Extended Riedel's (5)	1.38E-07	1.83E-07	2.73E-07
Wagner's (4)	6.32E-09	4.62E-08	4.73E-08
INT (4)	1.76E-07	4.33E-07	2.04E-06

It can be concluded that the "general-purpose" (17.12) favorably compete with the most accurate vapor-pressure equations for representing data covering a wide range of temperature.

17.3.2. Example 2 - Heat capacity of solids and liquids

Model (17.12) was fitted to heat capacity data of solid Ammonia (Overstreet and Giauque, 1937), solid Acetonitrile (Putnam *et al.*, 1965) and liquid Butanoic acid (Martin and Andon, 1982). For the latter, one outlying measurement was removed from the data. Solid Acetonitrile has a phase transition point at T=208.98 K, and only the data below that temperature was used in the regression. Some additional information for characterization of these data is shown in Table 17.7.

Since heat capacity data are usually modeled by C_p (not the logarithmic scale), heat capacity on the original scale has been used for modeling the data. The regression results are also shown in Table 17.7.

Table 17.7. Example 2 - Data characterization and fit of (17.12), heat capacity (H.C.) data

Parameter	Ammonia (solid)		Acetonitrile (solid)		Butanoic acid (liquid)	
	Value	Confidence Interval	Value	Confidence Interval	Value	Confidence Interval
exp(A)	6.0609261	1.72E-04	7.1819251	9.92E-02	187.57592	5.79E-02
B	-0.1107228	2.17E-05	-0.253187	0.029361	0.0310311	0.0153563
C	-2.1619705	2.12E-04	-2.9628732	0.186825	0.3461693	0.0857935
D	0.3230733	2.56E-05	0.2947711	0.012295	0.0526128	0.0027317
Variance	0.0038055		0.0192783		0.0127228	
Sample size (n)	42		30		31	
T- units	K		K		K	
T- range	15.04	191.09	15.63	208.98	272.75	373.06
T- Median	95.88		62.105		324.4	
H.C. units	cal/K-mol		cal/K-mol		J/K-mol	
H.C. range	0.176	11.94	0.409	18.7	169	209
Median Cp	6.0015		7.1025		187.2	

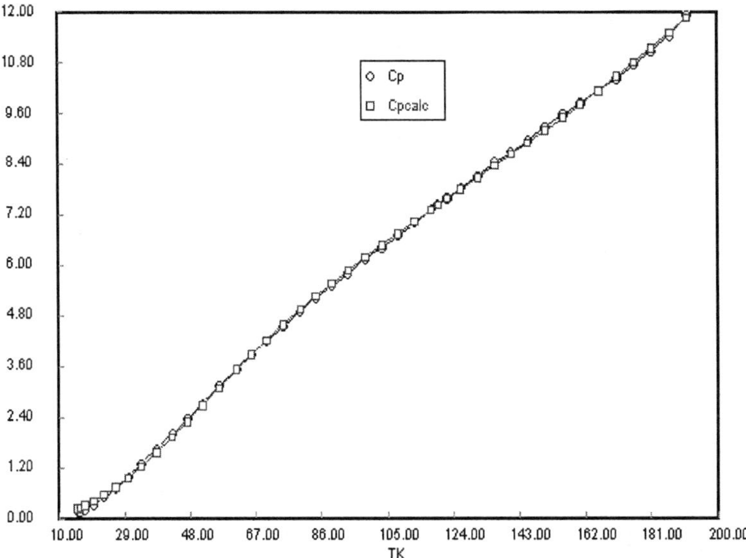

Figure 17.4. Experimental data and calculated curve [using (17.12)] for solid Ammonia's heat capacity

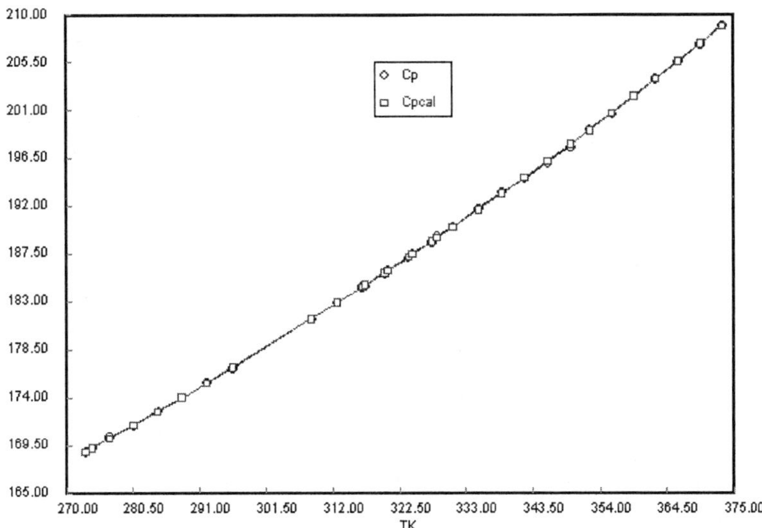

Figure 17.5. Experimental data and calculated curve [using (17.12)] for liquid Butanoic acid's heat capacity

It can be seen that the regressed model fits the data well. The variance ranges from 0.0038 (for solid Ammonia) up to 0.019 for solid Acetonitrile. All the confidence intervals for the model parameters are smaller than the respective parameters' estimates.

The experimental heat capacity data and calculated values are plotted versus temperature in Figure 17.4 (for solid Ammonia) and in Figure 17.5 (for liquid Butanoic acid). For the case of the Butanoic acid, all the experimental data points are virtually located on the calculated curve. For the solid Ammonia there are some small deviations relative to the experimental data. It should be emphasized that the shapes of these curves are completely different from the shape of the vapor pressure curve. For solid Ammonia, the curve is moderately concave and for liquid Butanoic acid the curve is slightly convex.

Heat capacity data for liquids and solids are usually modeled with polynomials (Daubert, 1998). High order polynomials are often needed to represent the data accurately, but high order polynomials can become unstable and involve unwarranted inflections, resulting in poor interpolation or extrapolation. Use of extreme care is recommended (Daubert, 1998) when fitting 3rd and 4th order polynomials.

The performance of the RMM model (17.12) for the heat capacity data was compared with the performance of polynomials of various orders. The results of this comparison are summarized in Table 17.8. In this table the variance and the relative variance [defined as 1 for (17.12)] of the various fittings are shown.

Table 17.8. Example 2 - Comparison of the heat capacity correlation equations

Correlation:	Variance at optimal solution (relative variance)					
	Ammonia		Acetonitrile		Butanoic acid	
5th order polynom.	0.0019	(0.4969)	0.0084	(0.4380)	unstable	
4th order polynom.	0.0026	(0.6772)	unstable		unstable	
3rd order polynom.	0.0083	(2.1693)	0.0157	(0.8130)	unstable	
Eq. (17.12)	0.0038	(1.0000)	0.0193	(1.0000)	0.0127	(1.0000)
2nd order polynom.	0.0090	(2.365)	0.1325	(6.8724)	0.0151	(1.1901)

The variance of (17.12) is smaller than that of the 2nd order polynomial in all cases. Attempting to fit a higher order polynomial to the data of liquid Butanoic acid yields unstable models where the confidence intervals on the parameters are larger than the parameter values. For the case of 3rd order polynomials, in one case (solid Ammonia) the variance is larger than that of the fitted (17.12), in the other case it is smaller. Increasing the order of the polynomials yields an unstable model in one case and further reduction of the variance in other cases. Thus, (17.12) in general yields better fits than low order polynomials (2nd or 3rd order). Higher order polynomials may become unstable or enable further reduction of the variance, however high order polynomials tend to introduce unwarranted inflections, as mentioned earlier.

17.3.3. Other temperature-dependent properties

To test the capability of the INT (7.12) to model the complete variety of curve-shapes that appear in modeling of physical and thermodynamic property data (see a detailed demonstration in Daubert, 1998) additional properties were also modeled with (17.12). These included viscosity of vapor and liquid and surface tension of liquid. The quality of the representation was similar to the quality of representation for vapor pressure and heat capacity data.

Characterization of the shapes of various curves that may be represented with high precision using (17.12) is given in Table 17.9. We realize that vapor pressure, heat capacity and vapor viscosity are increasing functions of temperature, while liquid viscosity and surface tension are decreasing functions. Vapor pressure and liquid heat capacity are convex functions (high convexity for vapor pressure, slight convexity for heat capacity), while viscosity, surface tension and solid heat capacity are concave functions (high concavity for liquid viscosity, moderate concavity for vapor viscosity and solid heat capacity, and slight concavity for surface tension).

Table 17.9. Shapes of curves representing temperature-dependent properties

No.	Property	Change with increasing temperature	Shape of the curve
1	Vapor pressure, liquid	increasing	highly convex
2	Heat capacity, solid	increasing	moderately concave
3	Heat capacity, liquid	increasing	slightly convex
4	Viscosity, gas	increasing	moderately concave
5	Viscosity, liquid	decreasing	highly concave
6	Surface tension, liquid	decreasing	slightly concave

Given the properties of the RMM model, discussed in earlier chapters, one can assert that RMM can accurately model the entire range of function-shapes that appear in physical and thermodynamic temperature-dependent data for pure substances.

References

[1] Daubert, T. E. (1998). Evaluated equation forms for correlating thermodynamic and transport properties with temperature. *Industrial Engineering and Chemistry Research*, 37, 3260-3276.

[2] Martin, J. F., Andon, R. J. L. (1982). Thermodynamic properties of organic oxygen compounds, Part LII. Molar heat capacity of Ethanoic, Propanoic and Butanoic acids. *J. Chem. Thermodynamics*, 14, 679-688.

[3] Myers, R. H. , Montgomery, D. C., Vining, G. G. (2002). *Generalized Linear Models, with Applications in Engineering and the Sciences*. John Wiley & Sons.

[4] Osborn, A. G., Douslin, D. R. (1966). Vapor pressure relations of 36 sulfur compounds present in petroleum. *Journal of Chemical Engineering Data*, 11(4), 502-509.

[5] Overstreet, R., Giauque, W. F. (1937). Ammonia. The heat capacity and vapor pressure of solid and liquid. Heat of vaporization. The Entropy values from thermal and spectroscopic data. *J. Amer. Chem. Soc.*, 59, 254-259.

[6] Putnam, W. E., McEachern, D.M. Jr, Kilpatrick, J. E. (1965). Thermodynamic properties of Acetonitrile. *J. Chem. Phys.*, 42(2) 749-755.

[7] Shore, H. (2000). General control charts for variables. *International Journal of Production Research*, 38(8), 1875-1897.

[8] Shore, H. (2003). Response Modeling Methodology (RMM)- A new approach to model a chemo-response for a monotone convex/concave relationship. *Computers and Chemical Engineering*, 27(5), 715-726.

[9] Shore, H., Brauner, N., Shacham, M. (2002). Modeling physical and thermodynamic properties via inverse normalizing transformations. *Industrial Engineering and Chemistry Research*, 41, 651-656.

[10] Wagner, W. (1973). New vapor pressure measurements for argon and nitrogen and a new method for establishing rational vapor pressure equations. *Cryogenics*, 13, 470.

[11] Wagner, W., Ewers, J., Pentermann, W. (1976). New vapor pressure measurements and a new rational vapor pressure equation. *J. Chem. Thermodynamics*, 8, 1049-1060.

Chapter 18

Forecasting S-Shaped Diffusion Processes

18.1. Introduction

Diffusion processes may be found in various areas of corporate activities, such as the time-dependent behavior of cumulative demand of a new product, or the adoption rate of a technological innovation. In most cases, the proportion of the population that has adopted the new product by time t behaves like an S-shaped curve, which resembles the sigmoid curve typical to many known statistical distribution functions. This analogy has motivated the common use of the latter for forecasting purposes.

In this chapter we use the inverse normalizing transformations (INTs), introduced in Chapter 20, to model S-Shaped diffusion processes related to market penetration data. Some of the 47 data sets, recently assembled from published sources by Meade and Islam (1998), and kindly provided to us by the authors, are used to demonstrate the accuracy obtained by forecasting via the INTs. The stability and accuracy of these forecasts, relative to current forecasting models, are discussed elsewhere (Benson and Shore, 2004). Since most of these models are given in Appendix A, the reader may wish to carry out these comparisons, based on data provided here and in the book's website (relate to the preface for details).

An explanation is required as to why we use here the INTs, which are placed in a later chapter. Forecasting via S-shaped market penetration models requires *relational* modeling, where the response is the proportion of the population, P, which has adopted the new product by a

297

specified time, T. Conversely, we may wish to forecast T, given P. In both cases, relational modeling is needed. By contrast, the INTs have been originally derived as models of *random* variation. This is why they are developed in Part IV of the book, dedicated to the modeling of random variation. In this chapter, the INTs are adapted to be used for relational modeling, and the natural place for this application is Part III of the book, to which this chapter belongs. A similar adaptation was done in the previous chapter (Section 17.3), where an INT was used to model chemical data.

It may be beneficial at this point to turn to Chapter 20, and get acquainted with the INTs used here for modeling S-shaped processes.

18.2. Theoretical Background for S-shaped Diffusion Processes

Diffusion processes are stochastic processes that describe the time-dependent evolution of the diffusion of a random phenomenon, such as the penetration into the market of a new product, or the adoption rate by the relevant population of a technological innovation. Most often, the response of interest is expressed as the time-dependent percentage of the population, where the diffusion has already taken place. For example, the proportion of the population that has already purchased the new product by time T. The forecasting problem is that of forecasting, at time T, the value of the response at time T+L, given the adoption history up to and including T. More formally, denote by p_t the cumulative diffusion rate by time $t \leq T$ ($0 \leq p_t \leq 1$). Then the required forecast is that of $E(P_{T+L} \mid p_1, p_2, \ldots, p_T)$.

Occasionally, inverse forecasts are also needed, namely, forecast the time, T_p, by which the cumulative adoption rate would surpass a given desirable threshold value, p, given the adoption history. Formally, find estimate for $E(T_p \mid t_1, t_2, \ldots, t_k)$, where T_p is the time by which the cumulative penetration rate has reached the value p, and adoption history is given by k time-points of the time-series, $\{(t_1, p_1), (t_2, p_2), \ldots, (t_k, p_k)\}$. For example, forecast $T_{0.5}$, the time by which half the target population has already adopted the technological innovation.

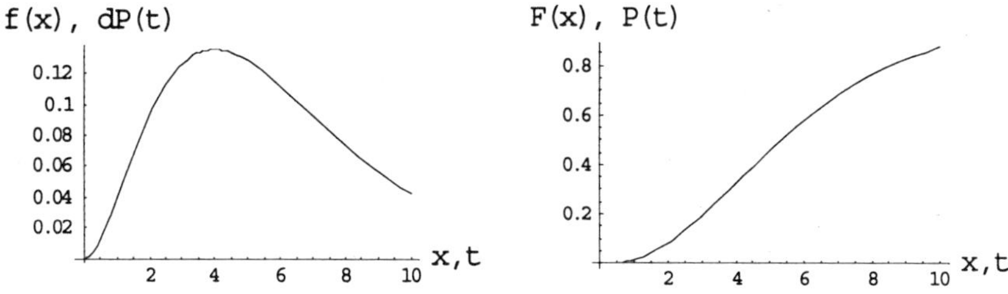

Figure 18.1. A typical S-Shaped curve (as applied to a distribution function and to a diffusion process)

A typical S-shaped diffusion process is displayed in Figure 18.1. The left-hand plot shows typical adoption *rates*, dP(t), as a function of time, and the right-hand plot shows *cumulative* adoption rate, P(t), as a function of time. One realizes that both plots resemble typical graphs that can be obtained, respectively, from plotting of the density function, f(x), and of the cumulative distribution function (CDF), F(x), of a random variable that is either symmetric or right-skewed (having positive skewness measure).

Various approaches have been pursued to model S-shaped diffusion processes so that the selected model, after proper estimation of its parameters, could be used to generate the required forecasts. In the following, we give a brief description of the main models, based on a more comprehensive survey given in Meade and Islam (1998, henceforth MI). Refer to this source for further details.

Models of diffusion processes used for forecasting may be divided into three major categories.

The first includes the familiar trend curve,

$$P_t = f_1(t) + \varepsilon_t, \tag{18.1}$$

where $f_1(t)$ is a function of time, which describes a sigmoid curve, and the error term, ε_t, is considered to be homoscedastic (homogenous, time independent, variance), namely, $E(\varepsilon_t)=0$, $Var(\varepsilon_t)= \sigma_\varepsilon^2$.

The second category comprises linearized trend curves, which assume that some non-linear transformation of P_t may be expressed as a linear function of t:

$$f_2(P_t) = a + bt + \varepsilon_t. \tag{18.2}$$

This approach may be likened to the response-normalizing approach, used for linear regression analysis, or to the link function, used in generalized linear models (GLM). Note, however, that for the latter it is a non-linear transformation of the response *mean* that is assumed to be expressible in terms of the linear predictor. In our case (18.2), as in the normalizing approach, the non-linear transformation is applied to the data, $\{P_t\}$ (not to the mean function). MI note (based on Young and Ord, 1989) that for some data sets the error structure in (18.2) may be more appropriate than that of (18.1).

A third category of models use non-linear auto-regressive (AR) relationships of first order, namely,

$$P_t = f_3(P_{t-1}) + \varepsilon_t .\tag{18.3}$$

Occasionally, models of the third group are obtained from those of the first by setting

$$dP_t = P_t - P_{t-1} = \partial f_1(t) / \partial t + \varepsilon_t ,\tag{18.4a}$$

or from those of the second category by setting

$$[df_2(P_t) / dP_t][dP_t / dt] = b,$$

from which, with $dt=1$, we obtain, approximately,

$$dP_t = P_t - P_{t-1} = b / [df_2(P_{t-1}) / dP_{t-1}] + \varepsilon_t .\tag{18.4b}$$

Based on a comprehensive literature review, MI have identified altogether twenty-nine models, which belong to either of these three categories. A close examination of these models reveals that about half (15) are either trend curve models or linearized trend models. A common feature shared by all these models (and AR models derived thereof) is the attempt to formulate a relationship between P_t and t either by expressing P_t explicitly in terms of t (trend curve models), or by finding a non-linear transformation of P_t that linearizes the relationship with t (linearized trend models).

Surveying commonly used models, which belong to these categories, shows that the majority of the models are based on an adaptation of some known statistical distribution (like the logistic, the log-normal or Gompertz). This is due to the fact that most trend models (or derivatives thereof) take advantage of the fact that the S-shaped curve, characteristic to a diffusion process, also typifies most known statistical distributions.

Given a distribution function F(x), where X is the random variable, a plot of F(x) *vs.* x would most likely reveal a typical sigmoid-shaped curve, similar to that displayed in Figure 18.1. If we substitute P_t for F(x) and t for x, relationships that have originally been derived to depict distribution functions may serve to describe the relationship $P_t=f(t)$. As a result, some well known and commonly applied distributions, like the normal, the logistic, the extended logistic, Gompertz, the log-normal and Weibull, are routinely used to model P_t in S-shaped diffusion processes.

MI classify the various models according to the position of the point of inflexion of the S-curve, since the timing of this point coincides with the maximum rate of diffusion (or penetration). They distinguish between symmetric models (inflexion point at $P_t=0.5$, like the normal and the logistic models), non-symmetric models ($P_t<0.5$ but constant, like Gompertz and linearized Gompertz), and flexible models (where the inflexion point can occur anywhere within the same model, including $P_t=0.5$; Typical examples are the log-normal, Weibull and the extended logistic). MI denote these classes by class II_1 (symmetric), class II_2 (asymmetric) and Class II_3 (flexible).

Based on a carefully selected subset of seven models that belong to the different classes (out of the original twenty-nine), MI introduce the following classification:

Class II_1 (symmetric): Simple Logistic, Mansfield (AR);
Class II_2 (asymmetric): Gompertz, Floyd (AR);
Class II_3 (flexible):
    - Class II_{3A}: Weibull, Extended Logistic (B);
    - Class II_{3B}: Cumulative Lognormal.

Class II_3 is subdivided based on results from stepwise discriminant analysis which has shown that the lognormal "behaves differently from the other flexible curves" (therein, p. 1122). The models, which may serve for comparison with the new approach, are detailed in Appendix A (where the models are numbered for easy reference).

A major implication of this classification is that different data sets (or different assumptions regarding these sets) would require the adoption of different models that would reflect the actual conduct (or assumptions) related to the inflexion point. Thus, for a given data set, different statistical distributions may show different degrees of consistency (or goodness-of-fit) with respect to the data.

The need to select different models for different data sets, indeed the opposite of what empirical modeling is all about, is reflected in MI method, as explained in their paper. Based on the above definitions of the various classes, they develop a classification scheme which would allow a forecaster to determine membership probabilities of a given data set to any of the above four classes of models. The need for such a scheme arises from the fact that "the evidence so far suggests that, for a given data set, it is difficult to identify the model underlying (or approximating to) the data generation process. However, it has been demonstrated that it is possible to assign the data set to a particular class of models with a probability of about 60%" (MI, p. 1122-3). This motivated MI to develop a scheme for generating a combined forecast based on carefully selected subset of seven forecasting models (as these were given earlier). A simple average of forecasts from all models belonging to a specific class is first calculated, and then the forecasts from the different classes are combined in a weighted average, using as weights membership probabilities derived via discriminant analysis.

As the authors clearly state, experience shows that in many applications simple averaging of the candidate forecasts was the best weighting scheme for a combined forecast. However, comparison with simple averaging shows that "the superiority of the weighting scheme based on model classification is very clear" (therein, p. 1124).

The need to select an appropriate *distributional* model for forecasting purposes is not unique to modeling of diffusion processes. The recent transition to real-time monitoring of industrial processes, in the framework of statistical process control (SPC), poses a similar difficulty. For a process that is non-normally distributed, determination of control limits for *real-time* SPC requires characterization of the underlying process distribution, namely, selecting a distributional model. However, in an industrial context it is unrealistic to expect that engineers (or operators) of SPC would engage themselves with the issue of which distribution best describes the process random behavior. A general methodology for constructing control charts in a non-normal environment is needed that may be applied in a uniform fashion, irrespective of the underlying process distribution. Efforts in this direction are described in Shore (2000a), where a new methodology is developed based on the INTs. This methodology is described in detail later in the book (Section 22.3).

In this chapter, a similar approach is pursued. We develop a general approach for forecasting S-shaped diffusion processes that would not require selecting, out of a given set of potential models, a certain model that best fits the data. Empirical modeling is again called for.

18.3. Modeling and Forecasting S-shaped Processes

An INT is a quantile relationship that expresses a given quantile, y_P, of a random variable (r.v.), Y, in terms of the corresponding standard normal quantile, z_P, namely: $y_p=g(z_p)$, where $P=F(y_P)=\Phi(z_P)$, and F(.) and $\Phi(.)$ are the distribution functions of Y and Z, respectively.

As explained in Chapter 20, the INT is a special case of the general RMM model. It is denoted the "origin" INT. Several off-spring INTs have been developed from the origin INT, which have a reduced number of parameters. It has been demonstrated (Shore, 2000ab, 2001) that the new INTs faithfully represent variously-shaped distributions, having skewness (asymmetry) measure that ranges from zero (symmetric) to over eleven (for the exponential this measure is only two!). The normal and the log-normal are exact special cases of the origin INT and of some of its derivatives. Furthermore, the Weibull and gamma distributions have been shown to be well represented.

Some of the INTs do not have explicit inverse function, namely, a function that describes z_p in terms of y_p. For others, the response (y_P) is given in terms of a single z_P, which allows derivation of an explicit expression for the inverse function. We denote the latter a normalizing transformation (NT), similarly with the well-known Box-Cox normalizing transformation.

The ability to derive an NT from an INT is important since it allows derivation of forecasts where the previous "Given" and "Forecasted" interchange roles. The two sets of transformations, the INTs and the

NTs, thus provide the required flexibility needed to construct both forecasts and their inverse, as explained earlier.

The success of the INTs (and the NTs) in representing distributions with different shapes (including, in the parlance of MI, symmetric, asymmetric and flexible distributions), suggests that the new INTs and NTs may successfully model S-shaped diffusion processes, and therefore be used to forecast future patterns of behavior for such processes. This approach would spare the forecaster the need to specify in advance an assumed distributional model, and eliminate the requirement of preparing a combined forecast based on a weighted scheme, as done in MI.

The INTs and the NTs, which will later be used for forecasting, are now introduced, and the analogy between modeling distribution functions and modeling S-shaped diffusion processes addressed.

Consider the basic RMM model (with $\eta=1$), re-introduced here for convenience:

$$Y = \exp\{(\alpha/\lambda)[(1+\varepsilon_1)^\lambda - 1]+\mu_2+\varepsilon_2\}$$

$$= \exp\{(\alpha/\lambda)[(1+\sigma_{\varepsilon1}Z_1)^\lambda - 1] + \mu_2 + \sigma_{\varepsilon2}[\rho Z_1 + (1-\rho^2)^{(1/2)} Z_2]\},$$

$$W = \log(Y) = (\alpha/\lambda)[(1+\sigma_{\varepsilon1}Z_1)^\lambda - 1] + \mu_2 + \sigma_{\varepsilon2}[\rho Z_1 + (1-\rho^2)^{(1/2)} Z_2]. \quad (18.5)$$

Introducing $\rho=\pm1$, we may obtain from (18.5) the "origin" INT (for simplicity, the subscript p, used earlier to denote the p-th quantile, is henceforth deleted). The origin INT is

$$y = \exp\{(\alpha/\lambda)[(1+\sigma_{\varepsilon1}z)^\lambda - 1] + \mu_2 + \sigma_{\varepsilon2}\rho z\},$$

$$w = (\alpha/\lambda)[(1+\sigma_{\varepsilon1}z)^\lambda - 1] + \mu_2 + \sigma_{\varepsilon}\rho z]. \qquad (18.6)$$

From these, "off-spring" INTs may be derived (refer to Chapter 20):

$$y = (M) \exp\{(B)[\exp(Cz) - 1]+ Dz\}, \qquad (18.7)$$

$$y = (M) \exp\{(B)[\exp(Cz) - 1]\}, \qquad (18.8)$$

$$y = (M) \exp\{(B/C)[(1+Az)^C - 1]\}, z > -1/A, \qquad (18.9)$$

$$y = (M) \exp[ABz / (1+Az)], \qquad (18.10)$$

$$y = M[1 + z / (BC)]^B, \qquad (18.11)$$

where M is the response median ($z=0$).

Note that (18.11) is derived from (18.9) by assuming in the latter C=0. The parameter A is then replaced by 1/(BC) so that the inverse of (18.11) results in a Box-Cox transformation (refer to Section 4.2, where this transformation is addressed). Also note that for (18.8)-(18.11), z can be expressed explicitly in terms of y. This implies that the inverse of these expressions may be used as normalizing transformations (NTs).

Both INTs and NTs will later be employed for forecasting (Section 18.4). The following NTs will be used:

Model NT1 [from (18.10)]:
$$z = (A) \log(y/M) / [B-\log(y/M)]. \qquad (18.12)$$
Model NT2 [from (18.11)]:
$$z = (A/B) [(y/M)^B -1]. \qquad (18.13)$$

Model NT2 is a Box-Cox normalizing transformation, with a re-scaled y so that the medians of Y and Z are preserved, namely, for y=M (the median), we have z=0. Other NTs may be similarly derived from the INTs given in (18.8)-(18.11).

It is now possible to define formally the procedures to fit a model and use it for forecasting. The procedures are based on the correspondence between $\{z, y\}$, the quantiles in the quantile-relationships given above by the INTs [(18.6)-(18.11)], and by the NTs [(8.12)-(18.13)], and $\{t, P_t\}$, where P_t is the cumulative relative demand at time t. To fit any of the INTs (or NTs) to a time series representing cumulative relative demand, we transform the given data by two different procedures, each representing a different forecasting objective.

Procedure I: For a Given Time Series $\{P_t\}$- Model and Forecast T_P in Terms of a specified P

This requires fitting an INT to the given data. The following procedure is pursued:

(A) *Fitting the model*

- Replace P (associated with the INT) with p_t and find the corresponding z_t. This implies finding for each p_t

$$\Phi^{-1}(p_t) = z_t , \qquad (18.14)$$

where $\Phi^{-1}(.)$ is the standard normal *inverse* CDF;
- Replace y, in the expression for the INT, by t;
- Fit an INT to the new data set, $\{z_t, t\}$, using non-linear least-squares (NL-LS).

(B) *Forecast T_P, given P*

Find $\Phi^{-1}(P)= z$; Introduce z into the fitted (estimated) INT to forecast T_P.

Sum up: Fitting an INT to a given time series allows us to forecast the time, T_P, when cumulative demand is expected to cross a desirable threshold value, P.

Procedure II: For a Given Time Series $\{P_t\}$- Model and Forecast P_T in Terms of a specified T

To forecast P_T, given T, a "normalizing transformation" (NT) needs to be estimated, namely, a quantile relationship that expresses z in terms of y. The analogy between the normalizing transformation $z=f(y)$, and the estimation and forecasting problem, is as follows:

(A) *Fitting the model*

- For each observation, t, replace y in the NT with t;
- For each observation, p_t, replace z in the NT with $z_t= \Phi^{-1}(p_t)$;
- Fit an NT to the new data set, $\{t, z_t\}$, using NL-LS.

(B) *Forecast P_T, given T*

- Replace y in the NT with T;
- Use the fitted NT to find $z_T = f(T)$;
- Find $P_t = \Phi(z_T)$.

The correspondence between the original variables of the INT and the NT and the time series variables, $\{t,\ p_t\}$, is given by the following fundamental transformations, which summarize the forecasting process:

Given P, forecast T (with a fitted INT): $P = \Phi(z_T) \rightarrow z_T \rightarrow T$

Given T, forecast P (with a fitted NT): $T \rightarrow z_T \rightarrow \Phi(z_T) = P$,

where the arrows imply transformations, either by the fitted INT or NT, or via available tables of the standard normal distribution.

18.4. Numerical Examples

The data sets used by MI (1998) were analyzed by the new approach in a thesis work by Benson (2002). A list of these data sets, as given in MI, is given in Appendix B. The results of Benson's work were summarized in Benson and Shore (2004). In this section, we take one relatively short data-set, used by MI, in order to demonstrate implementation of the above procedures.

The data-set is listed as item 3 in Appendix B. It relates to the purchase of B&W TV sets in the USA (source is Lindberg, 1982). The data are given in Table 18.1 and displayed in Figure 18.2. Penetration of B&W TV in the USA is easily recognized as an S-shaped diffusion process.

18.4.1. Forecasting T_P, given P

In accordance with Procedure I, one of the INTs given in Section 18.3 is fitted to the data. Using the four-parameter INT (18.7), we obtain after fitting via NL-LS:

Table 18.1. Source Data and Forecasts based on INTs (to forecast T_P) and NTs (to forecast P_T)

Period (t)	Cumulative demand (Propr.) (P_t)	Forecasting T_P (Given P)	Forecasting P_T (Given T)	z_t values (exact)	z_t values (from fitted NT)
1	0.004	1.347	0.0029	-2.652	-2.761
2	0.023	1.943	0.034	-1.995	-1.823
3	0.09	2.826	0.11	-1.341	-1.232
4	0.235	4.066	0.214	-0.7225	-0.7921
5	0.342	4.913	0.330	-0.4070	-0.4392
6	0.447	5.800	0.443	-0.1332	-0.1428
7	0.557	6.872	0.545	0.1434	0.1136
8	0.645	7.916	0.633	0.3719	0.3400
9	0.718	8.997	0.706	0.5769	0.5432
10	0.786	10.30	0.767	0.7926	0.7277
11	0.832	11.47	0.815	0.9621	0.8969
12	0.859	12.33	0.854	1.076	1.053
13	0.871	12.78	0.885	1.131	1.199
14	0.888	13.49	0.909	1.216	1.335

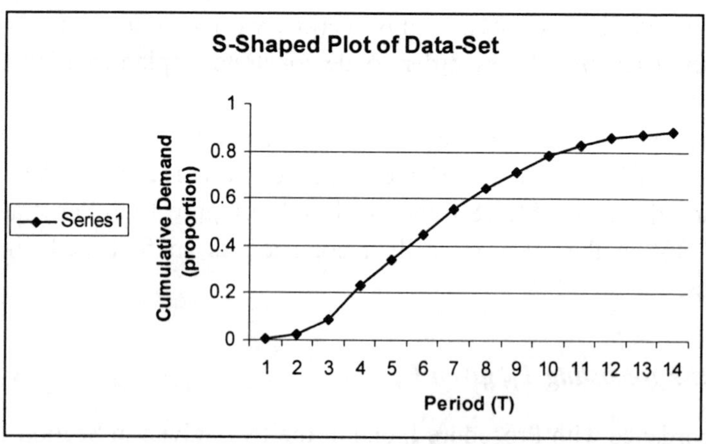

Figure 18.2. Purchase of B&W TV sets in the USA

$$t = (M) \exp\{(B)[\ \exp(Cz) - 1] + Dz\}, \qquad (18.15)$$

where

$$M = 6.293,\ B = 142.8,\ C = -0.01289,\ D = 2.454.$$

Results from the fitted INT are given in Table 18.1. Deviations of forecasts of T are restricted to about ± 0.5 [only one observation, (14), slightly exceeds this limit]. This implies that simple rounding would deliver virtually error-free forecasts.

In constructing these forecasts, all observations participated in the fitting procedure. A more adequate way to assess the effectiveness of the fitted INT would be to apply the fitting procedure to a small proportion of the data (say, the first third), and then evaluate the accuracy of the forecasts for the rest of the observations. Due to the small data-set used in this example relative to the number of parameters that need to be estimated (4) this is not implemented in this case. Instead, we use the first 8 observations to forecast the last six.

We obtain for the parameters' estimates:

$$M = 6.412,\ B = 2.728,\ C = 0,\ D = 0.6110.$$

One can easily find out that again good accuracy is obtained for the last six observations. However, since C is actually zero, this has implications for the correct structure of the model (C near zero was obtained also in the earlier fitting). As happened several times before in this book, we are "compelled" to modify the model according to the story told by the parameters' values. Taking account of the value of C (and the corresponding unstable value of B), the modified model is [replace B by B/C in (18.15), and then set C=0]

$$t = (M) \exp(Bz + Dz) = (M) \exp(\beta z), \qquad (18.16)$$

where M and β are two parameters that need to be estimated. We realize that analyzing the data via the INT points to the log-normal model as probably the best fitting model. The reader may wish to re-fit this model.

Results shown in Benson and Shore (2004) clearly demonstrate, via several effectiveness measures applied to variously-sized data-sets, that the INTs may serve as powerful forecasting tools for S-shaped diffusion processes.

18.4.2. Forecasting P_T, given T

Relating to the same B&W TV data, it is required to prepare forecasts for P_T, the cumulative demand proportion, expected at time period T. Fitting NT2 (18.13), we obtain

$$z = (A/B) \, [(t/M)^B - 1], \qquad\qquad (18.17)$$

with estimated parameters

$$M = 6.541, \ A = 1.665, \ B = 0.1355.$$

Values of the forecast z, with corresponding forecast P_T values, are given in Table 18.1. Good forecasts are again observed.

Deriving confidence intervals for the parameters' estimates, as well as analyzing the forecast errors, are essential ingredients in any analysis of forecasts. The reader is encouraged to conduct these analyses based on the given data and results in Table 18.1. NL-LS procedures embedded in statistical software packages routinely provide all the tools needed for such analyses.

References

[1] Benson, D. (2002). Forecasting product market penetration using inverse normalizing transformations. An M.Sc. Thesis Report. Supervisor: Haim Shore. Department of Industrial Engineering and Management. Ben-Gurion University.

[2] Benson, D., Shore, H. (2004). Forecasting an S-shaped diffusion process via normalizing transformations. A Technical Report. Department of Industrial Engineering and Management. Ben-Gurion University.

[3] Box, G. E. P., Cox, D. R. (1964). Analysis of transformations. *J. Royal Statistical Society* B, 26, 211-243.

[4] Lindberg, B. C. (1982). International comparison of growth in demand for a new durable consumer product. *J. Marketing Research,* 19, 364-371.

[5] Meade, N., Islam, T. (1998). Technological forecasting-model selection model stability and combining models. *Management Science,* 44(8), 1115-1130.

[6] Shore, H. (2000a). General control charts for variables. *International J. Production Research,* 38(8) 1875-1897.

[7] Shore, H. (2000b). Three approaches to analyze quality data originating in non-normal populations. *Quality Engineering,* 13(2), 277-291.

[8] Shore, H. (2001). Inverse normalizing transformations and a normalizing

transformation. In *Advances in Methodological and Applied Aspects of Probability and Statistics,* Book, N. Balakrishnan (Ed.), V. 2 , pp. 131-146. Gordon and Breach Science Publishers. Canada.

[9] Young, P., Ord, K. (1989). Model selection and estimation for technological growth curves. *International J. Forecasting*, 5, 501-513.

Appendix A. Current Forecasting Models

1. Simple Logistic: $P_t = \dfrac{1}{1 + c \cdot \exp(-bt)} + \varepsilon_t, b > 0, c > 0.$

2. Gompertz: $P_t = \exp(-c \cdot \exp(-bt)) + \varepsilon_t, b > 0, c > 0.$

3. Cumulative Log-normal:

$$P_t = \int_0^t \frac{1}{y\sqrt{2\Pi}\sigma} \exp\left(-\frac{(\ln(y)-\mu)^2}{2\sigma^2}\right) dy + \varepsilon_t, \sigma > 0.$$

4. Weibull: $P_t = 1 - \exp\left(-\left(\frac{t}{\alpha}\right)^\beta\right) + \varepsilon_t, \alpha > 0, \beta > 0.$

5. Extended Logistic (B):

$$P_t = \frac{1 - c\frac{p}{q}\exp(-(p+q)t)}{1 + c\exp(-(p+q)t)} + \varepsilon_t, p > 0, q > 0.$$

6. Mansfield (AR): $P_t - P_{t-1} = bP_{t-1}(1 - P_{t-1}) + \varepsilon_t, b > 0.$

7. Floyd (AR): $P_t - P_{t-1} = bP_{t-1}(1 - P_{t-1})^2 + \varepsilon_t, b > 0.$

Appendix B. Description of Data Sets (refer for relevant references to MI, 1998)

set #	Product or innovation	size	Original source
1	Colour TV USA	31	Lee et al. 1972
2	CATVUSA	29	Lee et al. 1972
3	TV B&W USA	14	Lindberg 1982
4	TV B&W Denmark	19	Lindberg 1982
5	TV B&W France	23	Lindberg 1982
6	TV B&W Finland	15	Lindberg 1982
7	TV B&W Norway	16	Lindberg 1982
8	TV B&W Sweden	15	Lindberg 1982
9	Product-1	40	Ralph and Linford 1972
10	Product-2	20	Ralph and Linford 1972
11	Product-3	25	Ralph and Linford 1972
12	Product-4	23	Ralph and Linford 1972
13	Product-5	27	Ralph and Linford 1972
14	Product-6	35	Ralph and Linford 1972
15	Central Heating UK	16	Monthly Digest of Statistics UK.
16	Ranges, USA	46	Merchandising Week, February 1972.
17	Coffee maker, USA	23	Merchandising Week, February 1972.
18	Heating pads, USA	32	Merchandising Week, February 1972.
19	Hot plates,USA	40	Merchandising Week, February 1972.
20	Water heaters, USA	30	Merchandising Week, February 1972.
21	Bed coverings,USA	23	Merchandising Week, February 1972.
22	Blender, USA	23	Merchandising Week, February 1972.
23	Disposer, USA	23	Merchandising Week, February 1972.
24	Freezer, USA	23	Merchandising Week, February 1972.
25	Mixer, USA	23	Merchandising Week, February 1972.
26	Ranges, built in, US	15	Merchandising Week, February 1972.

Appendix B (Cont'd)

set #	Product or innovation	size	Original source
27	Electronic switching system, UK	25	British Telecom
28	Electronic switching system, USA	18	Lee et al. 1972
29	Man made fibre, US	26	Martino 1983
30	Telex, Singapore	17	Quaddus 1986
31	Computer, Singapore	14	Quaddus 1986
32	Steam Vessel, USA	31	Ralph and Linford 1972
33	Cat. Cracking (oil)	27	Linstone and Sahal 1976
34	Steel, Oxy.process: for Spain	14	Poznanski 1983
35	Steel, Oxy.process: for World	19	Poznanski 1983
36	Steel, Oxy.process: for Belgium	18	Poznanski 1983
37	Steel, Oxy.process: for France	21	Poznanski 1983
38	Steel,Oxy.process:for Netherlands	22	Poznanski 1983
39	Steel, Oxy.process: for Sweden	21	Poznanski 1983
40	Plastic 1 gal milk	24	Young 1993
41	Plastic 0.5 gal milk	23	Young 1993
42	Man made fibre, UK	21	Harrison and Pearce 1972
43	US telephones	56	Young 1993
44	Domestic Elec, US	24	Young 1993
45	Dishwasher, US	23	Merchandising Week, February 1972.
46	Clothes dryer,US	23	Merchandising Week, February 1972.
47	Metal Vessels, US	17	Martino 1983

Part IV

MODELING RANDOM VARIATION - APPLICATIONS

Chapter 19

RMM Distributional Approximations

19.1. Introduction

This is the first of five chapters that comprise Part IV of the book. These chapters demonstrate RMM as a platform for empirical modeling of random variation. The claim of RMM for a "universal" character has been supported in earlier chapters by showing that a wide spectrum of existing scientific and engineering relational models, as well as current statistical models of random variation (namely, statistical distributions), are special cases of the RMM model (refer to Chapters 2 and 12). It is therefore appropriate that this part opens with a chapter that demonstrates the capability of the RMM error distribution to deliver satisfactory representation to variously shaped statistical distributions, which a practitioner may encounter in her/his daily routine.

In Chapter 10 fitting procedures for the RMM error distribution were developed. These procedures are used here to derive the RMM parameters that deliver best fit to some specified distributions. In Section 19.2 the RMM is fitted to some known distributions, assuming that the RMM errors are either normal or log-normal. The latter assumptions are shown to be essential to the effectiveness of the RMM model by assuming, in Section 19.3, that the errors are from another distribution, for example, the logistic distribution. The good accuracy obtained with normal or log-normal errors vanishes. What this might imply is discussed. In Section 19.4 new closed-form non-polynomial RMM-based approximations for the Poisson quantile function and for the CDF of the standard normal distribution are developed.

317

19.2. Fitting RMM with Normal or Log-normal Errors

In this section the RMM quantile function is fitted to a set of known statistical distributions, pursuing the original formulation of the RMM model, namely, the error terms, ε_1 and ε_2, are either normal or log-normal.

The RMM error quantile-function is fitted using the "Quantile-Matching" procedure (refer to Section 10.3 for details). We opt for this approach because when the distribution modeled by RMM is completely specified (no sample estimates are used in the fitting procedure), quantile-matching seems to be most convenient for identifying numerically the RMM parameters. The resulting goodness-of-fit is judged (in this chapter) mainly by how well the first four moments of the fitted RMM error distribution match the exact moments.

The moments of the fitted quantile function are numerically evaluated using the simple expression for the r-th non-central moment

$$\mu_r' = \int_{-\infty}^{\infty} y(z)^r \phi(z)dz , \qquad (19.1)$$

where y(z) is the fitted quantile function, and ϕ is the d.f. of the standard normal distribution.

From (19.1), the mean, the variance, the skewness measure, Sk, and the kurtosis measure, Ku, are (refer to Stuart and Ord, 1987, Section 3.14)

$$\mu = \mu_1',$$

$$V(Y) = \mu_2' - (\mu_1')^2 ,$$

$$Sk = \sqrt{\beta_1} = E[(Y-\mu)^3] / \sigma^3 = [\mu_3' - 3 \mu_2' \mu_1' + 2 (\mu_1')^3] / \sigma^3 ,$$

$$Ku = \beta_2 - 3 = E[(Y-\mu)^4] / \sigma^4 - 3$$

$$= [\mu_4' - 4 \mu_3' \mu_1' + 6 \mu_2' (\mu_1')^2 - 3 (\mu_1')^4] / \sigma^4 - 3 . \qquad (19.2)$$

For the normal distribution: Sk=0, Ku=0.

Two versions of the quantile-function are employed in the fitting procedure (refer to Chapter 9):

$$Y = (M) \exp\{(\alpha/\lambda)[(1+\varepsilon_1)^\lambda - 1] + \varepsilon_2 \}, \qquad (19.3)$$

$$Y = (M) \exp\{(\alpha/\lambda)[\exp(\lambda\varepsilon_1) - 1] + \varepsilon_2\}, \qquad (19.4)$$

where $\{M, \alpha, \lambda\}$ are parameters, the error terms, ε_1 and ε_2, are from a bi-variate normal distribution with standard deviations $\sigma_{\varepsilon 1}$ and $\sigma_{\varepsilon 2}$, respectively, and correlation ρ, that is,

$$\varepsilon_1 = \sigma_{\varepsilon 1} Z_1, \quad \varepsilon_2 = \sigma_{\varepsilon 2}\, [\rho Z_1 + (1-\rho^2)^{(1/2)} Z_2], \tag{19.5}$$

where Z_1 and Z_2 are independent standard normal variables.

The RMM quantile functions, (19.3) and (19.4), have two error terms. To fit a univariate distribution, introduce $\rho=\pm 1$ into (19.5). Both quantile functions are now expressed in terms of a single standard normal variable. This property is helpful in fitting the quantile functions via the quantile-matching procedure.

Re-scaling the parameter λ in terms of $\sigma_{\varepsilon 1}$, namely, $\lambda = \lambda_0/\sigma_{\varepsilon 1}$ (rescaling ensures reasonably small values for λ_0), we obtain from (19.3)-(19.5) for the quantile, y, in terms of the corresponding standard normal quantile, z:

$$y = (M)\, \exp\{(\sigma_{\varepsilon 1}\alpha/\lambda_0\,)[(1+\sigma_{\varepsilon 1} z)^{(\lambda 0/\sigma \varepsilon 1)} - 1] + (\sigma_{\varepsilon 2}\rho)z\}, \tag{19.6}$$

$$y = (M)\, \exp\{(\sigma_{\varepsilon 1}\alpha/\lambda_0)[\, \exp(\lambda_0 z) - 1] + (\sigma_{\varepsilon 2}\rho)z\}, \tag{19.7}$$

where M is the median (z=0), and ρ may be either 1 or -1.

To demonstrate the capability of the quantile-functions, (19.6) and (19.7), to deliver good representation to diversely-shaped distributions, they are fitted to a representative set of distributions, using quantile-matching.

Twelve distributions are included in the sample set. These include the normal, the F-Ratio distribution, beta, gamma (3 distributions), Weibull (4 distributions), Rayleigh and extreme value. The selected set has (positive) skewness values that range from 0 (normal) to over 11 (Weibull).

For each distribution in the testing set, seventy-one quantile values of z={-4.0(0.1)3.0}, with corresponding quantile values of the approximated distribution, are identified. These are then incorporated in a non-linear least-squares (NL-LS) procedure to determine the parameters of the fitted quantile function. Goodness-of-fit is assessed by calculating numerically, from the fitted quantile function, the first four moments, namely, the mean, the variance and the skewness and kurtosis measures [using (19.1) and (19.2)]. Numerical integration extends from $P = (2.866)10^{-7}$ ($z = -5$) to $P = 1 - (2.866)10^{-7}$ ($z = 5$).

We note here that there are other "Distance measures" in the literature that aim to judge how well one distribution, $f_a(y)$, may model (approximate) another given distribution, f. Often this distance is evaluated via *non-negative divergence*, or *pseudo-distance*, measures, designated by $D(f_a, f)$. For example, Karian and Dudewicz (2000), use the L_p-norm,

$$\|f_a(y) - f(y)\|_p = \left(\int |f_a(y) - f(y)|^p \, dy \right)^{1/p}, \, p \geq 1,$$

with $p=1$ or 2, to assess the quality of the fit of the generalized Lambda distribution. Other commonly used distance measures, and a good review of current literature, may be found in Warick (2004).

Applying NL-LS to the quantile data-sets, generated from the sample of distributions described earlier, parameters and calculated moments are obtained. These are displayed in Table 19.1. The parameters shown relate to the following re-parameterized quantile-functions

$$y = (M) \exp\{[B/(C/A)][(1+Az)^{(C/A)} - 1] + Dz\}, \qquad (19.6a)$$

$$y = (M) \exp\{B[Exp(Cz)-1] + Dz\}, \qquad (19.7a)$$

where M is the exact median. [On the reason for expressing the exponent by (C/A), relate to the comment preceding (19.6).]

In Table 19.1, exact moments are in the upper entries (in bold). Second and third entries from the top relate to fitting by (19.6a) and by (19.7a), respectively (with Z normal). Lower entries relate to fitting (19.7a) with a logistic error term (this will be discussed in Section 19.3).

Examining Table 19.1, one realizes that all first four moments are well preserved in the fitted quantile functions. Note, that only for the beta distribution the parameter D is negative, implying $\rho = -1$. For the F-ratio distribution, the value of "A" is of the order of 10^{-9}, implying that the fitted RMM in fact approximates the F-ratio distribution by a log-normal distribution. For the normal, (19.6a) provides an exact expression, wherefore the *parameters* in Table 19.1 are the exact ones ($A = 2/14 = 0.1429$).

Table 19.1. Parameters and moments (ex. and app.) for the fitted quantile-functions.
Exact moments are in bold.

Distribution	Parameters				Moments			
	A	B	C	D	Mean	Var.	Sk	Ku
Normal					**14.00**	**4.000**	**0**	**0**
μ=14, σ=2	0.1429	1	0	0	14.00	4.000	0	0
		-.1916	-.3298	.07907	14.00	3.987	-.0035	.01662
Beta					**0.1667**	**0.06313**	**1.658**	**1.714**
α=0.2, β=1	.003822	1217	-1.062	-.2498	0.1679	0.06284	1.631	1.654
M=.03125		-4.365	-1.062	-.2463	0.1679	0.06284	1.631	1.654
		-3.741	-1.369	-.08095	0.1784	0.06666	1.465	0.9980
F-Ratio					**1.182**	**0.6518**	**2.593**	**16.52**
n₁= 10,	.0000	--	--	--	--	--	--	--
n₂= 13		97.23	-0.00964	1.544	1.190	0.6504	2.441	12.19
M=.9842		-297.0	-0.01700	-4.389	1.163	0.5824	1.528	3.624
Gamma					**14.00**	**28.00**	**.7559**	**.8571**
α= 7, β=2	.01596	18.03	-.1543	.0970	14.00	28.00	.7559	.8555
M=13.34		-1.419	-.1860	.1216	14.00	28.00	.7558	.8557
		-487.2	-.01037	-4.675	13.91	25.42	.6334	.4215
Gamma					**6.000**	**12.00**	**1.155**	**2.000**
α= 3, β=2	.02490	18.56	-.2316	.1426	6.000	12.00	1.155	1.998
M=5.348		-1.532	-.2787	.1781	6.000	12.00	1.155	1.998
		-315.5	-.02193	-6.277	5.969	11.79	.8769	.5834
Gamma					**.5000**	**1.000**	**4.000**	**24.00**
α= 1/4, β=2	.07299	40.31	-.7512	.3799	.4996	1.001	4.011	24.23
M=.08735		-3.502	-.8128	.4499	.4995	1.002	4.026	24.56
		-3.464	-1.014	.3241	.5123	.8368	2.872	10.52
Weibull					**4.790**	**18.20**	**1.676**	**4.040**
α=1.125,	.03677	22.09	-.4020	.2091	4.790	18.19	1.677	4.044
β=5		-1.672	-.4615	.2503	4.790	18.20	1.678	4.052
M=3.610		-2.773	-.4354	-.06555	4.788	17.98	1.258	1.464
Weibull					**5.000**	**25.00**	**2.000**	**6.000**
α=1, β=5	.03954	23.16	-.3991	.2334	5.000	25.00	2.000	6.002
(exp. dis.)		-1.888	-.4603	.2805	5.002	25.00	2.002	6.019
M=3.466		-2.519	-.5082	.02868	5.006	24.26	1.467	2.272
Weibull					**5.665**	**50.99**	**2.815**	**12.74**
α=0.8, β=5	.04509	25.55	-.3916	.2846	5.665	50.99	2.815	12.72
M=3.162		-2.394	-.4556	.3450	5.667	50.99	2.818	12.78
		-2.526	-.6036	.1539	5.703	46.57	1.960	4.761
Weibull					**16.62**	**2723**	**11.35**	**287.6**
α=0.4, β=5	.06806	34.86	-.3581	.5014	16.61	2719	11.14	258.2
M=2.000		-4.979	-.4427	.6592	16.61	2723	11.20	262.8
		-4.512	-.6649	.4182	15.89	1405	5.297	44.73

Table 19.1 (cont'd)

Distribution	A	B	C	D	Mean	Var.	Sk	Ku
Rayleigh					**2.507**	**1.717**	**.6311**	**.2451**
σ=2	.01015	44.13	-.4312	.1263	2.508	1.714	.6340	.2505
M=2.355		-.9810	-.4494	.1336	2.508	1.714	.6336	.2491
		-10.41	-.1399	-.8503	2.515	1.729	.5177	-.2921
Extreme Value					**3.577**	**1.645**	**1.139**	**2.400**
α=3, β=1	.0006	.8994	-1.686	.3493				
M=3.366		-.0003	-1.691	.3494	3.577	1.670	1.117	2.316
		-276.7	-.005657	-1.232	3.542	1.439	.9547	1.605

An interesting observation regards the consistently small values obtained for the parameter "A" (near zero). Recall that as "A" goes to zero, we have lim $_{A\to0}(1+Az)^{1/A}$= exp(z). Given that in the numerical search for best fitting parameters, fitting (19.6a) may occasionally tumble over imaginary values in the expression $(1+Az)^{1/A}$ (due to negative values), the small values of "A" justify a transition from (19.6a) to (19.7a). Refer also to a discussion in Section 9.5.

To further demonstrate the accuracy achieved, we display in Figure 19.1 (at the end of the chapter), for nine distributions, the "Exact" and "Approximate" (fitted) quantile values, on the left, and an error-plot ("Approximate" minus "Exact"), on the right. All plots refer to fitting of (19.7a). The accuracy displayed in Figure 19.1 again justify using (19.7a) as a general platform for modeling random variation.

19.3. Fitting with a Logistic Error Term

When the RMM model was axiomatically derived in Chapter 7, it was assumed that the error terms were either normal or log-normal.

Will the RMM model still be effective, as a general platform for empirical modeling of random variation, had we discarded the assumption about the normality of the error terms?

To examine this hypothesis, we re-fitted the set of distributions given in Table 19.1, using for Z the symmetrically-distributed standardized logistic distribution, with inverse CDF

$$z= (\sqrt{3}/\pi)\log[P/(1-P)], \qquad (19.8)$$

where P=F(z), and F is the standard logistic CDF.

Lower entries in Table 19.1 display parameters and moments, obtained by fitting (19.7) with Z quantiles given by (19.8). Moments of the fitted quantile function are uniformly less accurate than those derived under the assumption of a normal error. This is unexpected since the logistic has longer tails (higher value for the kurtosis measure) than the normal, and one would therefore expect this r.v. to deliver better representation for skewed distributions with high kurtosis values, like some of the distributions in Table 19.1. This is not the case. Even though the structure of the non-linear model used in the fitting procedure remains intact, the skewness and kurtosis measures for the logistic-based fitted quantile functions are appreciably less accurate than those based on the normal error.

How can this be explained? Does this result suggest some associations between the normal distribution and other distributions, where the latter are currently known to be non-related to the normal (like Weibull)? It is hard to say. This and other issues regarding some possible implications of the RMM approach need some further study.

19.4. Approximations for the Normal and the Poisson Distributions

In this section RMM-based non-polynomial approximations are developed for the quantile of the Poisson distribution and for the CDF of the standard normal distribution (with errors of 10^{-5} to 10^{-7}, for the latter).

There are two objectives in developing these approximations.

First, the developed approximations are simple enough to be easily integrated into stochastic optimization models, like inventory models. Analytic investigation of the properties of these models becomes more feasible. Examples for the application of simple approximations to study inventory models are given in Chapter 23.

Secondly, we wish to demonstrate that RMM may be used as a general platform for modeling random variation not only for "regular" straightforward cases, like the *quantiles* of continuous random variables (as shown in the previous section), but also to other cases which are not intuitively evident. Approximating the CDF of the normal distribution and the quantiles of a *discrete* r.v. are two such cases.

In Section 19.4.1 the approximation for the Poisson quantile is developed, and in Section 19.4.2 the approximations for the CDF of the standard normal distribution.

19.4.1. Approximating the Poisson quantile

Fitting (19.7a) via NL-LS, with samples of Poisson quantiles and Poisson parameter $\lambda = \{1/2, 1, 2, 5, 10, 20\}$, parameters' values and approximate quantile values are obtained, as shown in Table 19.2.

To clarify how the *exact* quantile values were prepared for the fitting procedure, suppose that $\lambda = 1$. For the Poisson quantile value y=2, we have P= 0.9197. The corresponding standard normal quantile is z= 1.403.

The two quantiles, $\{1.403, 2\}$, comprise one point in the sample-set used in the NL-LS procedure.

Observing Table 19.2, we note the high accuracy for small λ, and the high accuracy that is preserved for extreme tail values (like y=1 for λ=20). Also of interest is the relatively stable error obtained for the (exact) quantile values of y=0. This suggests that in models, where a normal approximation is used for transformed Poisson data, values of zero should be replaced by 0.2945 (the average of the extreme values in Table 19.2).

To appreciate the accuracy of this simple approximation, one may compare it to the normal approximation, commonly implemented to approximate a Poisson variable.

Table 19.2. Parameters and approximate quantiles obtained from fitting (19.7a) to the Poisson distribution (with parameter λ)

	PARAMETERS				QUANTILES						
λ	M	B	C	D	0	1	2	4	6	8	10
1/2	.1794	-2.786	-.5660	.1925	.281	1.02	1.97	4.01	6.01	7.99	10.0
1	.4785	-1.934	-.6212	.2078	.284	1.02	1.97	4.01	6.01	7.99	10.0
2	1.315	-1.112	-.6903	.2265	.289	1.02	1.97	4.01	6.00	7.99	10.0
5	4.345	-.3190	-.8111	.2581	.296	1.02	1.97	4.01	6.01	7.98	10.0
10	9.325	-.0612	-.9325	.2892	.303	1.02	1.97	4.01	6.01	7.98	10.0
20	22.25	-.0039	-1.089	.3285	.308	1.02	1.97	4.01	6.01	7.98	10.0

19.4.2. Approximating the CDF of the standard normal distribution

Suppose that a random variable, Y, with distribution function P=F(y), has explicit inverse distribution function y= F^{-1}(P). Fitting the quantile function of the RMM error distribution, which is expressed in terms of a single Z [like (19.6a) or (19.7a)], would result in P being expressed explicitly in terms of a standard normal quantile. In other words, an approximation for the CDF of the standard normal distribution may be derived. This is the basis for the derivation of approximations to the CDF of the standard normal distribution in this sub-section.

Approximations for the normal CDF abound in the literature. A good review of these approximations appears in Johnson *et al.*(1994). Two relatively recent examples may help in evaluating the merits of the new ones, once developed.

Cao and Chen (1986) give an approximation which is a simplified version of the six-parameter approximation in Abramowitz and Stegun (1970). Their four-parameter approximation is

$$\Phi(z) = \begin{cases} \phi(z)(a_1 s + a_2 s^2 + a_3 s^3), & z \le 0 \\ 1 - \phi(z)(a_1 s + a_2 s^2 + a_3 s^3), & z > 0, \end{cases} \qquad (19.9)$$

where $\Phi(z)$ is the standard normal CDF, and

$$s = 1/(1+p|z|), \; p = 0.33267, \; a_1 = 0.4361836,$$

$$a_2 = -0.1201676, \; a_3 = 0.9372980.$$

This approximation has an absolute error less than 10^{-5}.

More recently, Bagby (1995) suggested the following approximation:

$$\Phi(z) = (1/2) \pm (1/2) \{1 - (1/30)[\, 7 \exp(-z^2/2) + 16 \exp(-z^2(2-\sqrt{2}))$$

$$+ (7 + (\pi/4)z^2) \exp(-z^2)\,] \}^{(1/2)}, \qquad (19.10)$$

where a minus sign refers to $z \le 0$. The absolute error is less than $(3)10^{-5}$.

RMM-based approximations are now developed. In Shore (2004a), equation (19.7a) was used to derive an approximation to the CDF of the standard normal distribution. The point of departure was an RMM modeling of the exponential quantile function. Introducing for the quantile of the exponential distribution (with parameter λ=1), in terms of the corresponding CDF, F(y)= P= 1-exp(-y), into (19.7a), we obtain

$$y = -\log(1-P) = (M) \exp\{B[\exp(Cz) - Dz\}. \qquad (19.11)$$

From this, an explicit expression for P in terms of the corresponding standard normal quantile, z, is

$$P_1(z) = 1 - \exp\{-\log(2)\exp[B[\exp(Cz)-1] + Dz]\}, \qquad (19.12)$$

where we introduced $\mu_2=\log(M)=\log[-\log(1/2)]$, to ensure that when z=0, P=1/2. [the reason for indexing P(z) will be made clear shortly]

Fitting (19.12) by a sample of standard normal quantiles, with the respective CDF values (for P>1/2 only), we obtain for the parameters in (19.12)

$$B = -1.81280, \ C = -0.472245, \ D = 0.294549. \qquad (19.13)$$

For z>0, this simple three-parameter approximation has maximum error of $+2(10)^{-5}$, around z=0.2, and it is smaller (in its absolute value) for larger z (though not monotonously so). This is shown in the error-plot of Figure 19.2 (reproduced from Shore, 2004a).

To enhance the accuracy of (19.12), a constraint is imposed that all approximations need to fulfill, that is, that the Taylor series expansion of the approximation will have the same structure as the Taylor series expansion of the exact CDF of the standard normal distribution.

Expanding the standard normal density function around z=0 and integrating term by term, we obtain for the exact CDF

$$\Phi(z) \cong 1/2$$

$$+\frac{z}{\sqrt{2\pi}} - \frac{z^3}{6\sqrt{2\pi}} + \frac{z^5}{40\sqrt{2\pi}} - \frac{z^7}{336\sqrt{2\pi}} + \frac{z^9}{3456\sqrt{2\pi}} - \frac{z^{11}}{42240\sqrt{2\pi}}$$

$$+\frac{z^{13}}{599040\sqrt{2\pi}} - \frac{z^{15}}{9676800\sqrt{2\pi}} + \frac{z^{17}}{175472640\sqrt{2\pi}} + O[z]^{18} \qquad (19.14)$$

By contrast, a Taylor expansion of (19.12), with parameters given in (19.13), results in the following expansion:

$$P_1(z) \cong (1/2) + 0.39894\,z + 0.0076223\,z^2 - 0.066490\,z^3 - 0.0041994\,z^4$$

$$+ 0.0099735\,z^5 + 0.0011527\,z^6 - 0.0011827\,z^7 - 0.00021105\,z^8$$

$$+ 0.00011292\,z^9 + 0.000029109\,z^{10} - 8.75067\,E(-6)\,z^{11} + O[z^{12}]$$

$$(19.15)$$

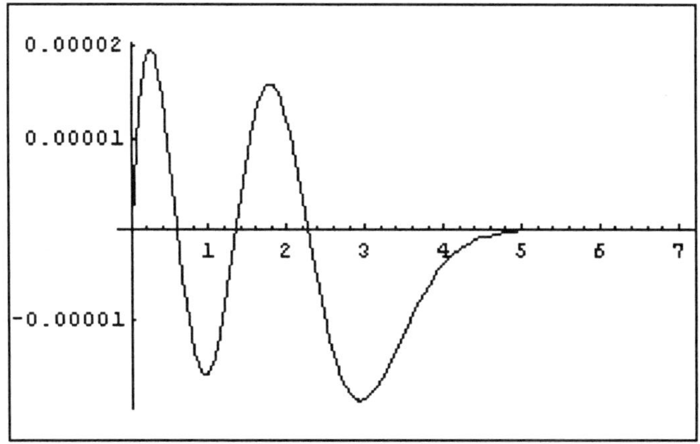

Figure 19.2. Error plot (app. minus ex.) of approximation (19.12) with parameters given by (19.13) (z>0)

To have similar structure for both expansions, the even-order power terms in (19.15) need to be eliminated. Re-write (19.12) as
$P_1(z)= 1- g(z)$. From the definition of the CDF it is obvious that $g(z)$ is the tail area at the point $Z=z$. Furthermore, due to the symmetry of the normal distribution, we can write, instead of (19.12):

$$P_2(z) = 1 - [g(z)+g(-z)]/2. \qquad (19.16)$$

A Taylor series expansion of (19.16) will contain only the even-order terms of (19.15). Therefore, to eliminate these terms in (19.15), define the approximation

$$P(z) = (1/2) + P_1(z) - P_2(z) = [1 + g(-z) - g(z)] / 2, \qquad (19.17)$$

which now contains only odd-order terms in the Taylor expansion, as required by (19.14). [the "1/2" in (19.17) is added to obtain the correct location for the value of P.]

$P(z)$ may now be fitted to a sample of values taken form the standard normal distribution. Re-defining $g(z)$ in (19.12) with the added parameter A (implying that we do not impose $\eta=1$ in deriving (19.4) from the original RMM model), we obtain

$$g(z) = \exp\{-\log(2)\exp[B[\exp(A+Cz)-1] + Dz]\}, \qquad (19.18)$$

with the fitted parameters' values

$$A = 0.072880383, B = -2.36699428,$$

$$C = -0.40639929, D = 0.19417768. \qquad (19.19)$$

Introducing from (19.19) into (19.18) and thence into (19.17), we obtain an approximation for the CDF of the standard normal distribution, the error plot of which is given in Figure 19.3. The maximum absolute error reduced from $2(10)^{-5}$ (refer to Figure 19.2) to $2(10)^{-6}$. Furthermore, a Taylor expansion of the approximation shows that the coefficients of (19.14) are well preserved in the expansion of (19.17).

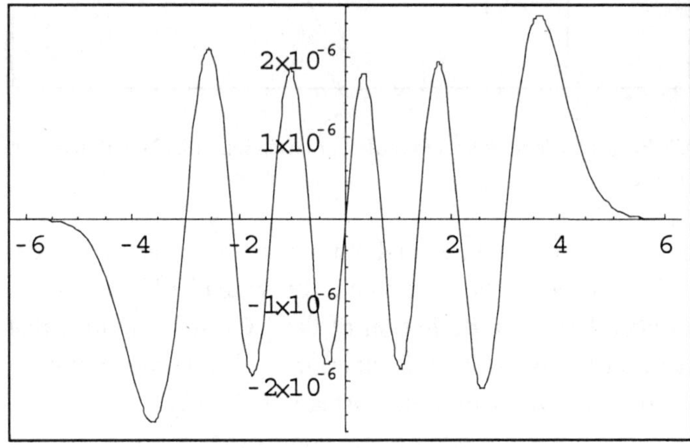

Figure 19.3. Error plot of approximation (19.18) with parameters given by (19.19)

In a similar fashion an RMM-based approximation for the standard normal CDF may be derived from (19.6a). Re-parameterized we have for $g(z)$, with parameters $\{\alpha, \lambda, S_1, S_2\}$:

$$g(z) = \exp\{-\log(2)\exp\{[\alpha / (\lambda/S_1)][(1+S_1 z)^{(\lambda/S1)} - 1] + S_2 z\}. \quad (19.20)$$

Expanding (19.20) into a Taylor series, we realize that to preserve the coefficient of z in the expansion of the *exact* normal CDF (19.14), we need to have

$$\alpha = [(2/\pi)^{(1/2)} / \log(2) - S_2] / S_1 . \qquad (19.21)$$

Introducing back into (19.20) and thence into (19.17), an approximation with three parameters, $\{\lambda, S_1, S_2\}$, is obtained. Fitting similarly to the earlier fitting, the parameters' values are

$$\lambda = -0.61228883, \; S_1 = -0.11105481,$$

$$S_2 = 0.44334159, \; \alpha = -6.37309208. \tag{19.22}$$

[the value of α was derived from (19.21) with the above values of S_1 and S_2; Also the coefficients of z in (19.20), S_1 and S_2, have different signs, a possibility that is anticipated from (19.7)].

Introducing from (19.22) into (19.20) and thence into (19.17), the error plot in Figure 19.4 is obtained. The maximum absolute error is reduced from $2(10)^{-6}$ (Figure 19.3) to $6(10)^{-7}$. Although (19.20) has only three parameters that need to be determined from the fitting procedure (α is not), it has accuracy better than the four- parameter approximation (19.18).

As a final comment, we note two interesting features of the new approximations for the normal CDF.

First, they are non-polynomial. This implies that their high accuracy emanates not from the large number of parameters that characterize polynomial approximations (refer to Johnson *et al.*, 1994), but possibly because the RMM original model is in some way related to the normal distribution. This shows in that the accuracy obtained from

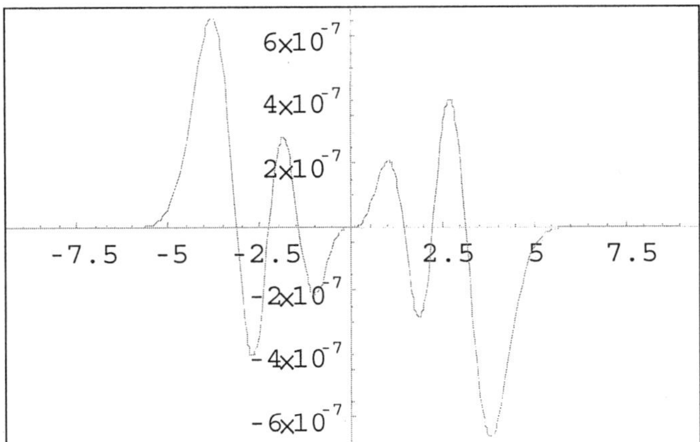

Figure 19.4. Error plot (app. minus ex.) of approximation (19.20) with parameters given by (19.22) (-9<z<9)

approximation (19.20) (derived from the original RMM model), is higher than obtained from approximation (19.18) (developed from *a variation* of the original RMM model). Furthermore, if the normal errors in the RMM model are replaced by other errors (for example, logistic errors), the high accuracy of the above approximations vanishes. This and other related issues are discussed with some numerical examples in Shore (2004b).

Secondly, the approximations to the standard normal CDF were developed from the exponential distribution, with the mediation of the RMM model. No further assumptions were made in deriving these approximations. This seems to indicate that the exponential variable, the RMM model and the normal variable, are somehow inter-related. This is another subject discussed in Shore (2004b). More research is needed.

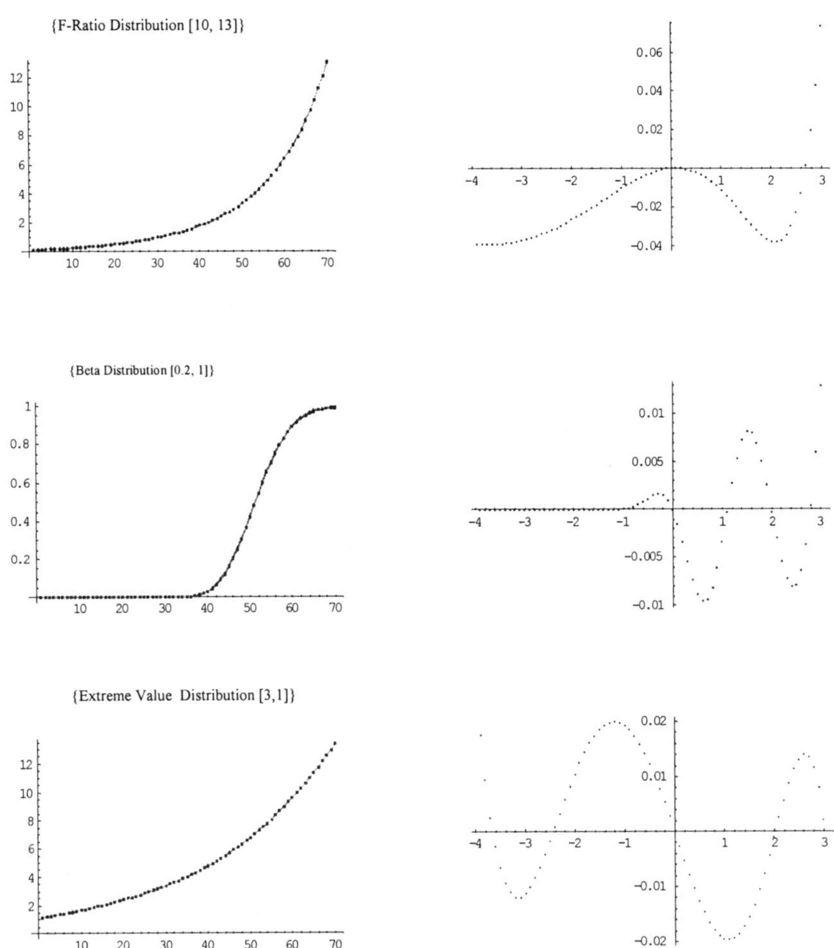

Figure 19.1. Exact and approximate (fitted) quantile functions vs. quantile sequential number (left), and the error plot (*vs.* z, right), for some of the distributions in Table 19.1 (Cont'd on the next two pages)

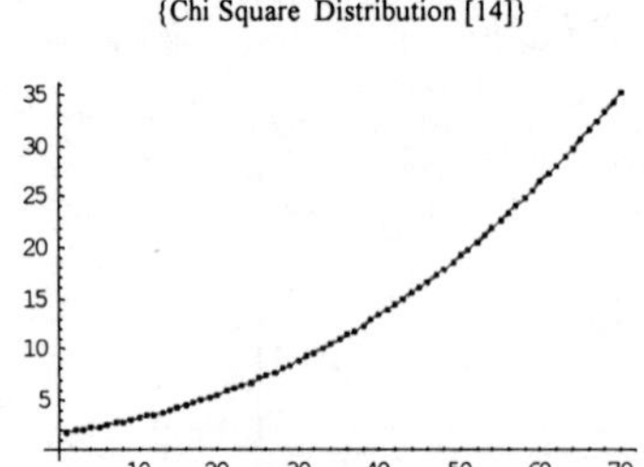
{Chi Square Distribution [14]}

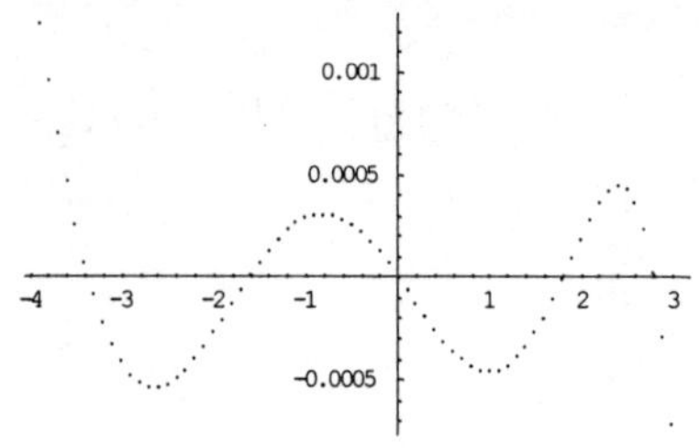

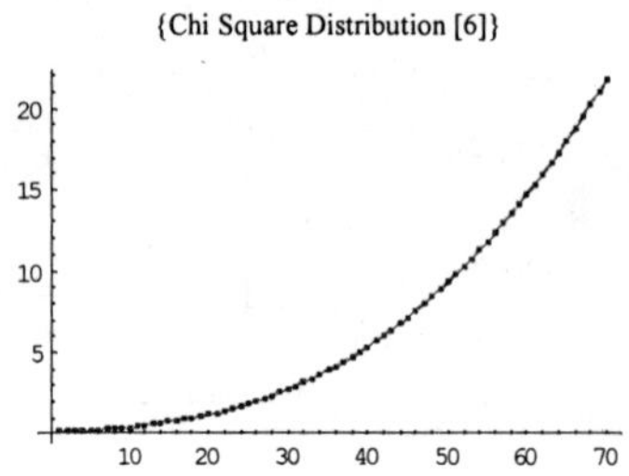
{Chi Square Distribution [6]}

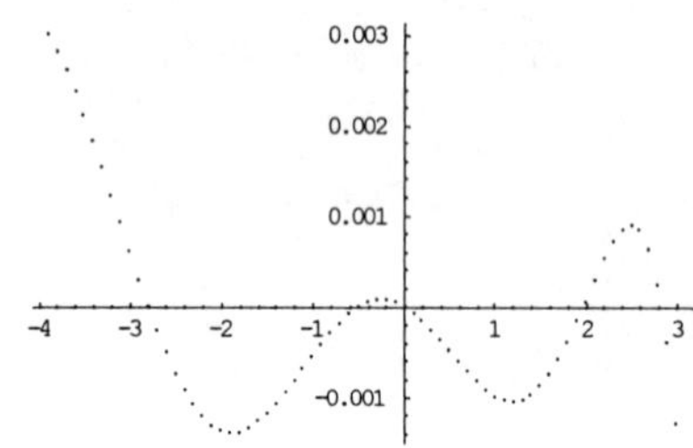

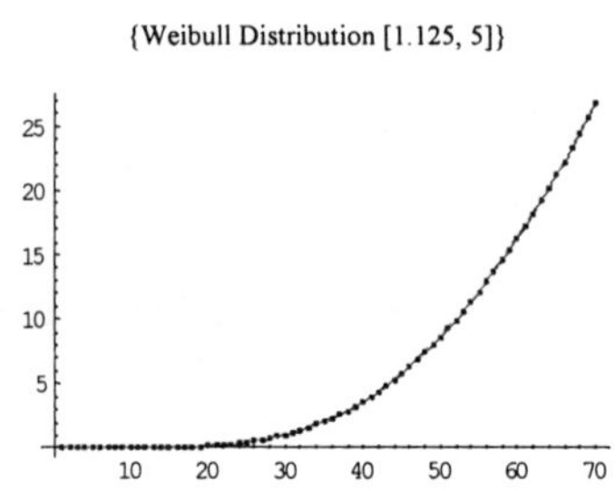
{Weibull Distribution [1.125, 5]}

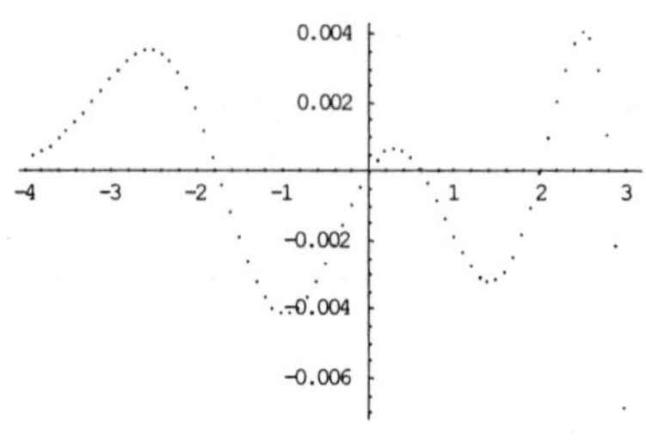

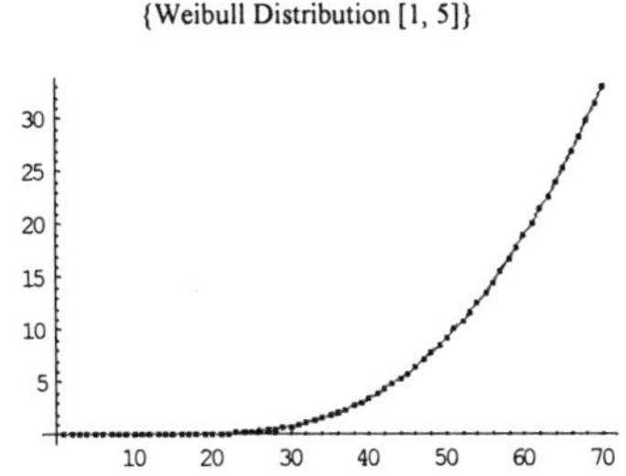
{Weibull Distribution [1, 5]}

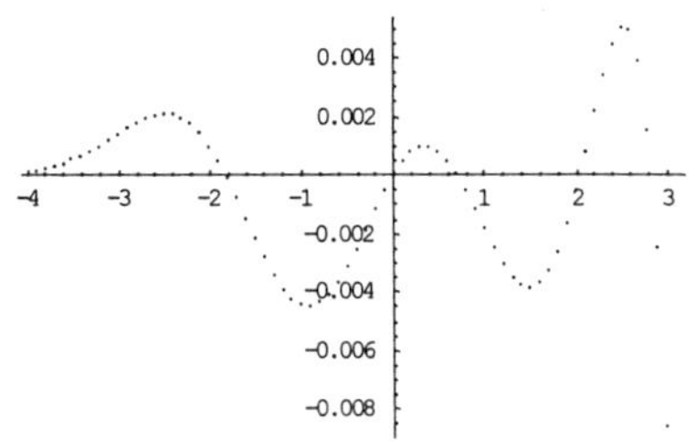

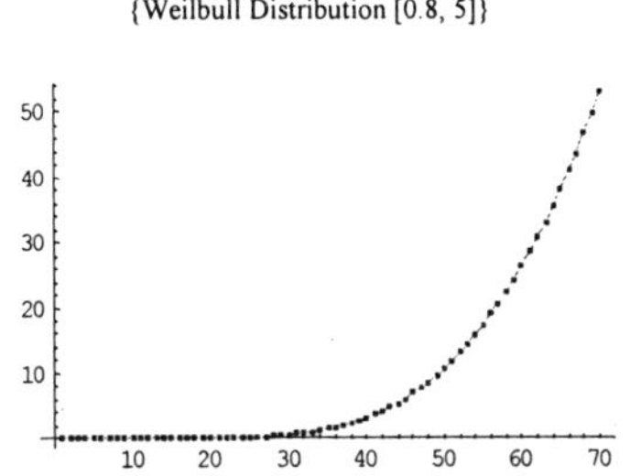
{Weilbull Distribution [0.8, 5]}

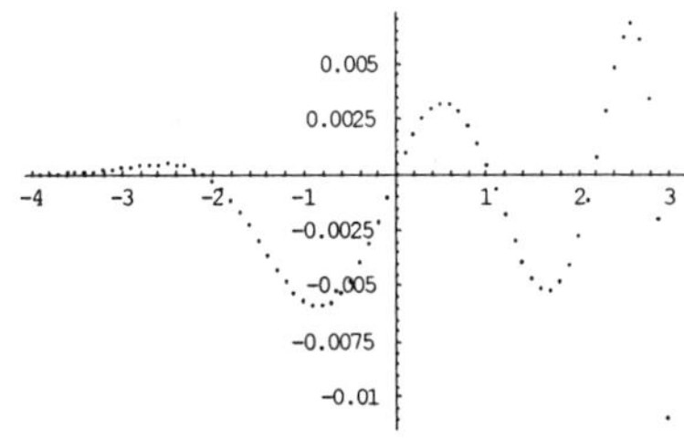

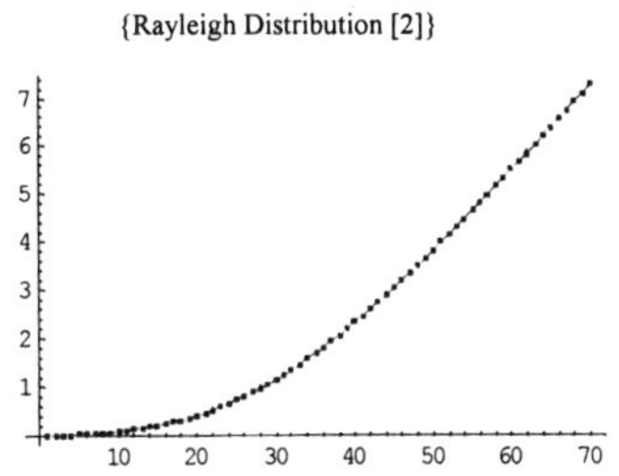
{Rayleigh Distribution [2]}

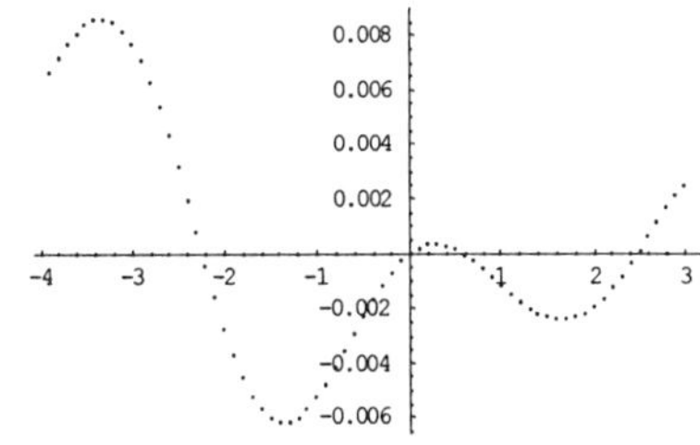

References

[1] Abramowitz , M., Stegun, I. A. (1970). *Handbook of Mathematical Functions.* National Bureau of Standards. New-York, Dover Publications.

[2] Bagby, R. J. (1995). Calculating normal probabilities. *American Mathematical Monthly,* 102, 46-49.

[3] Cao, J. H., Chen, K. (1986). *Introduction to Reliability Mathematics.* Beijing. Academic Publishers.

[4] Johnson, N. L., Kotz, S., Balakrishnan, N. (1994) *Continuous Univariate Distributions — Volume 1.* 2nd edition. John Wiley & Sons.

[5] Karian, Z. A. , Dudewicz, E. J. (2000). *Fitting Statistical Distributions: The Generalized Lambda Distribution and Generalized Bootstrap Methods.* CRC Press, Boca Raton, Florida, USA.

[6] Shore, H. (2004a). Response Modeling Methodology (RMM)- Current distributions, transformations and approximations as special cases of the RMM error distribution. *Communications in Statistics (Theory and Methods),* 33(7), 1491-1510.

[7] Shore, H. (2004b). Response Modeling Methodology (RMM)- on the relationship between the RMM model and the assumption of normal (or lognormal) errors. Under review.

[8] Stuart, A., Ord, J. K. (1987). *Kendall's Advanced Theory of Statistics. V. 1: Distribution Theory.* Charles Griffin & Company, Ltd., London.

[9] Warick, J. (2004). A data-based method for selecting tuning parameters in minimum distance estimators. *Computational Statistics and Data Analysis.* In press.

Chapter 20

Inverse Normalizing Transformations

20.1. Introduction

An inverse normalizing transformation (INT) is a quantile relationship that expresses the quantile of a non-normal r.v. in terms of the corresponding quantile of a standard normal variable. The word "inverse" implies that an INT is the inverse of a normalizing transformation. While the latter normalizes the distribution of a non-normal r.v. by applying to it some transformation, the INT intends to represent a non-normal variable by an explicit function of a normal variable. For example, the Box-Cox transformation, addressed in Section 4.2, is a "normalizing transformation". Its inverse, namely, an expression for the original response in terms of the normal variable, is an INT.

The concept of the INT was first introduced in Shore (2000a, 2001). The idea behind its introduction was to develop a widely applicable INT that could represent a wide spectrum of differently-shaped distributions, commonly assumed in the deployment of control charts for variables (in Statistical Process Control). The assumption was that the new INT would enjoy the same degree of "universality" as achieved by effective normalizing transformations (like the Box-Cox transformation). Later, the RMM model had been developed and found to include the new INT as a special case (refer to Chapter 7). In a way, the development of the INT was a precursor to the development of the RMM model. The unique properties of the INT, relative to the more general RMM model, are discussed in this chapter, and some parameter-reduced off-spring INTs developed and demonstrated.

335

In Section 20.2, derivation of the "origin" INT from the basic RMM model is briefly outlined. In Section 20.3 we discuss the problem of using four-moment matching in empirical modeling of random variation. This problem, already addressed in Chapter 5, has motivated the development of parameter-reduced off-spring INTs. These are introduced in Section 20.4. In Section 20.5 distribution fitting procedures are developed based on matching of moments (partial and complete) of second degree at most. The last Section 20.6 develops some normalizing transformations (NTs), derived from the INTs of Section 20.4.

The INTs of this chapter have been adapted for relational modeling in two earlier chapters that deal with applications to chemical engineering and to forecasting (Chapters 17 and 18, respectively). They are again applied in the context of modeling random variation in Chapters 22 and 23.

20.2. Derivation of the "Origin" INT

The basic RMM model is [refer, for example, to (19.3)]

$$Y = (M) \exp\{(\alpha/\lambda)[(1+\varepsilon_1)^\lambda - 1] + \varepsilon_2\}, \qquad (20.1)$$

where $\{M, \alpha, \lambda\}$ are real-valued parameters, and the error terms, ε_1 and ε_2, are from a bi-variate normal distribution with standard deviations $\sigma_{\varepsilon 1}$ and $\sigma_{\varepsilon 2}$, respectively, and correlation ρ, namely,

$$\varepsilon_1 = \sigma_{\varepsilon 1} Z_1, \ \varepsilon_2 = \sigma_{\varepsilon 2}[\rho Z_1 + (1-\rho^2)^{(1/2)} Z_2], \qquad (20.2)$$

where Z_1 and Z_2 are independent standard normal variables. Introducing from (20.2) into (20.1), and assuming $\rho = \pm 1$, (20.1) becomes, with Z_1 replaced by Z,

$$Y = (M) \exp\{(\alpha/\lambda)[(1+\sigma_{\varepsilon 1} Z)^\lambda - 1] + \sigma_{\varepsilon 2} \rho Z\}. \qquad (20.3)$$

Equation (20.3) is now a function of a single standard normal variable, and therefore can be perceived as an INT. We denote (20.3) the "Origin" INT to distinguish it from "Off-spring" parameter-reduced INTs, to be developed later. In Section 19.2, the capability of (20.3) to represent well a variety of differently-shaped distributions was demonstrated.

20.3. Four-Moment Matching - The Problem and a Solution

Four-moment fitting has been introduced in Section 5.3 as a widely applied approach to represent a distribution by a member of a known parameter-rich family of distributions (like Pearson or Johnson). The theoretical basis for using moment-matching to fit a known distribution to another, where the latter is either completely specified or only partially known (via sample data), has been spelled out. The current quality-engineering literature is in fact inundated by reports about various "distribution-free" techniques (more aptly denoted "Procedures for empirical modeling of random variation"), where the main instrument to achieve the "distribution-free" property is moment-matching by some parameter-rich family of distributions, like Pearson, Johnson or Burr.

When sample data are needed to estimate moments, the four-moment matching approach becomes problematic. This is due to the high sampling errors associated with estimating high-degree moments (refer to a brief discussion in Section 5.3). The author of this book has for a long time advocated avoiding the use of four-moment matching, suggesting, instead, the use of low-degree moments, partial and complete. This eliminates the use of sample estimates (of moments) with high mean-squared-errors (MSEs).

In this section, the "Sampling Problem" is re-addressed, and we elaborate on how the problem has been formerly solved, and describe briefly how it is solved for the new INTs (in Section 20.5).

As a departure point, consider Process Capability Analysis (PCA), as it is currently applied to non-normal populations.

PCA, essentially, is an attempt to relate the natural variability of a process to the technical (engineering) requirements. This entails knowing both the "end-points" of the process distribution (commonly defined by the 0.135-th and the 99.865-th percentiles of the process distribution), and the end-points of the engineering-related allowable variation in the measured variable. The assumption of normality is commonly adopted for the process variable, and this determines the coefficients of 3 (or 6) in the denominators of most basic process capability indices (PCIs).

For a process distribution having moderate to extreme skewness, the normality assumption is no longer valid, and the common PCIs cannot be used. One possible solution is to identify the true underlying distribution and then estimate its parameters with available data. There is a general consensus today that this is not a viable approach, and that a general methodology is required, which can produce good estimates of PCIs, without the need to identify the true underlying distribution. It is here where divergent attitudes have led to different solutions.

A widely used solution, which has also attracted much research interest, is to replace the unknown distribution with a known three- or four-parameter distribution, and estimate the unknown parameters of the latter by available sample data. Most commonly, this is done by moment matching, where the first four moments of the fitted distribution are made to coincide with the corresponding true moments, estimated from sample data. This is Clements' method (Clements, 1989) for non-normal PCA, which is now incorporated in most statistical packages that offer a statistical quality-control module. Clements procedure applies four-moment matching to a member of the Pearson four-parameter family of distributions, and then uses the fitted distribution in order to calculate the PCI. Kotz and Johnson (2002) note that "applications of this kind of method, with various distributional forms, have been quite numerous since 1992" (18 references are provided!).

From a theoretical point of view, this approach seems to be the most desirable. It is well known that, except for some well defined cases, two statistical distributions are identical if all their respective moments are equal. Naturally, the higher the number of moments two distributions share in common, the more similar their general pattern of behavior. In particular, if two distributions share the same first four moments, their general behavior is hardly distinguishable from one another. This was established and demonstrated in various papers that appeared over the years in the statistical literature, for example, Pearson, Johnson and Burr (1979; See details in Section 5.3.2). A four moment matching seems, therefore, like the best guarantee that the basic properties of an unknown process distribution be preserved in the approximating fitted distribution.

Unfortunately, this approach entails a serious problem, the "Sampling Problem". Although matching of the first four moments is a desirable property, one can rarely achieve it in practice because sample estimates of third- and fourth-degree moments are notoriously known for their large sampling errors. If used in PCA, sample estimates of skewness and kurtosis may lead to extremely biased estimates of the actual PCI.

Consider how Statistical Process Control (SPC) currently deals with non-normality.

For attributes data, it is commonly assumed that the underlying distribution is known. Once its parameters are estimated, all moments can be calculated (instead of being estimated). For instance, in constructing control limits for a p chart, the parameter p is first estimated, and then the mean and the variance are calculated from known expressions derived from the binomial distribution. While Shewhart control charts can be conveniently applied to sample statistics from attributes data that follow weakly skewed distributions (like the binomial or the Poisson), they cannot be applied to distributions with possibly moderate skewness, like the negative binomial, or to distributions with extreme skewness, like the geometric distribution.

Various solutions attempted to remedy this shortcoming of the Shewhart chart by suggesting alternative charts with non-symmetric control limits. A review of some of these solutions is given in Woodall (1997) and in Shore (2000b). As the latter comments, these solutions rarely incorporate the actual skewness of the monitoring statistic (measured by the third standardized moment) in the calculation of the control limits, neither is skewness explicitly preserved in the approximating distribution, used to determine the control limits.

For variables data, the problem with current solutions (including the use of Shewhart charts) is further exacerbated. First, if one wishes to monitor a process based on a Shewhart chart for individual observations, and the underlying process distribution is extremely skewed, the normalizing effect of the sample average is non-existent. This renders the use of a Shewhart chart inadequate. Secondly, unlike the case with attributes data, where skewness and kurtosis are often known (granted that estimates of the parameters have been obtained), with variables data the underlying process distribution itself is rarely known. This implies that no expressions are available to *calculate* the skewness and kurtosis measures (as can be done with attributes data). Sample estimates of these measures need to be extracted directly from available sample data. As a result, unreliable control limits may ensue.

In a more general context, the problem of distribution fitting by moment-matching, using sample estimates of high-degree moments, is relevant to all application areas, where the *unknown* non-normal distribution needs to be replaced by a *known* parameter-rich distribution.

For example, it now starts to be recognized that in inventory analysis, knowledge (or specification) of the first two moments of the underlying r.v.s does not guarantee attaining the optimal solutions and in many cases knowledge of the third moment (skewness) is indispensable [relate to a recent paper by Lau and Lau (2003), and references therein, and also to Shore, (2004a)].

To resolve the "sampling error" problem, associated with the use of sample estimates of high-degree moments in fitting procedures, an alternative approach has been developed by us some years ago. The new approach uses fitting by moment-matching *without* resorting to estimates of high-degree moments. It has been deployed in developing estimating (fitting) procedures for several new distributions, which have been introduced and reported about in the quality-engineering literature and in the statistics literature. A review of these distributions and the allied distribution-fitting procedures can be found in Shore (2000c, and references therein; Refer also to Chapter 5). It has been amply demonstrated (Shore, 1995, 1996, 1998ab) that the new approach consistently delivers better results in terms of reduced MSEs for functions, where quantiles derived from the fitted (estimated) distribution are incorporated.

The essence of the new approach is to fit a four parameter distribution, yet to employ in the fitting procedure moments of second degree at most. This may be implemented in two fashions. First, dividing the fitted distribution into two halves, one can fit each half by moment-matching, using *partial moments* of second degree at most. Alternatively, moment-matching, using moments of second degree at most, can be applied both to the response on the original scale *and* to the response in the log-transformed scale. In both alternatives, no moments of degree three or higher are employed in the fitting procedure. For some of the fitting procedures, the new approach utilizes moments of first degree only.

MSEs associated with moments of distributions fitted by the new approach are smaller than the respective MSEs, associated with distributions fitted by matching of direct sample estimates of the first four moments. In fact, in Shore (1996) it was demonstrated that the new approach provides a better estimate of skewness (derived from the fitted distribution by numerical integration) than that derived from a direct sample estimate. This implies that if skewness is important for modeling in inventory analysis (as related earlier), then a two-moment fitting procedure should be preferred to estimating skewness via a direct sample estimate of skewness. Another implication is that the two-moment fitted distribution would better represent the underlying unknown distribution than any sample-based four-moment fitting, irrespective of the actual distribution used in the fitting routine. The problem of the large MSEs associated with sample estimates of skewness and kurtosis is thus resolved.

Besides the empirical cumulative evidence for the validity of the new approach, it is also founded on a sound theoretical basis. Each half distribution is a distribution onto itself. Moment-matching for each half is justifiable on the same grounds that moment-based fitting is justified when applied to the complete distribution. The resulting fitted distribution would preserve the mean and the variance both for the complete distribution and for each half individually. It has been shown that when the approximated distributions are completely specified (no sample estimates are used), the new approach delivers good representation to widely used distributions, irrespective of which distribution is employed for the fitting (Shore, 2000c). In particular, fitting INTs by moment matching based on partial moments, or with moments in the log-transformed scale, has proved to result in good fit, where the latter is judged by how well the first *four* moments are preserved in the fitted INT.

This is demonstrated in the next two sections, where the new off-spring INTs are developed (Section 20.4) and fitting procedures are derived (Section 20.5).

Comment. The relative advantage of the two-moment procedure holds for small to medium sample sizes. As the sample size becomes large this advantage becomes less pronounced. See some demonstration and a thorough discussion of the "Sampling Problem" in Shore (2004b).

20.4. Parameter-Reduced INTs

The motivation for using parameter-reduced off-spring INTs is obvious. By using an INT with a smaller number of parameters, a smaller number of sample estimates of moments to identify the parameters are required. This implies using moments of lower degree and, consequently, attaining smaller MSEs for functions that incorporate quantiles derived from the fitted INT (like PCIs). A pre-requisite for a parameter-reduced INT to qualify as a general platform for modeling random variation is its demonstrable capability to preserve well the first four moments of the distribution to which it is fitted. This capability is demonstrated for every INT derived in this section.

20.4.1. Off-spring INT I

The origin INT (20.3) has five parameters $\{M,\alpha,\lambda,\sigma_{\varepsilon 1},\sigma_{\varepsilon 2}\rho\}$. One variation of the origin INT, already introduced in previous chapters, is (with $\rho=\pm 1$)

$$Y = (M)\exp\{(\alpha/\lambda)[\ \exp(\lambda\varepsilon_1) - 1] + \varepsilon_2\}$$

$$= (M)\exp\{(\alpha/\lambda)[\ \exp(\lambda\sigma_{\varepsilon 1}Z) - 1] + \sigma_{\varepsilon 2}\rho Z\}. \qquad (20.4)$$

This INT, together with its counterpart based on conditioning $Z_1|Z_2{=}z_2$ (refer to Section 17.2), may be re-written as the quantile function [refer to (11.14)]

$$y = (M)\ \exp\{A[\exp(Bz)-1] + Dz\}, \qquad (20.5)$$

where either B or D (but not both) can be negative. INT (20.5) has only four parameters, including the response median, M (the response quantile that corresponds to $z{=}0$). The unique properties of (20.5) were discussed in Section 19.2, and its capability to represent well diversely-shaped distributions was demonstrated in Table 19.1. Moment-matching procedures, based on low-degree moments, were developed for INT I in Sub-sections 11.3.2 and 11.3.3.

20.4.2. Off-spring INT II

From Table 19.1, it is clear that the parameter "D" in (20.5) (which represents $\sigma_{\varepsilon 2}\rho$) is consistently small, ranging from 0.1, for near-symmetric distributions, to around 0.5, for extremely skewed distributions. Assuming D=0 (equivalently, assuming that $\sigma_{\varepsilon 2}$ is negligible), we obtain INT II:

$$y = (M) \exp\{(A)[\ \exp(Bz) - 1]\}. \tag{20.6}$$

With a specified median (M), INT II has two-parameters that need to be determined. Fitting (20.6) to a sample-set of distributions, using the same NL-LS quantile-based procedure used to fit (19.6) [or (20.3)], parameters' values with the associated first four moments are given in Table 20.1. Good fit is achieved, though it is worse than that obtained for (20.3) (compare to results in Table 19.1).

20.4.3. Off-spring INT III

Assuming D=0, this time with respect to the origin INT (20.3), we obtain, with re-defined parameters,

$$y = (M) \exp\{(B/C)[(1+Az)^{C} - 1]\},\ z > -1/A. \tag{20.7}$$

Unlike INT I and II, this INT preserves the normal and log-normal as exact special cases. For the former case, insert in (20.7) B=1, C=0. For the latter case (log-normal), assume C=1. Alternatively, assume that Az<<1, to obtain

$$y = (M) \exp\{(B/C)[\exp(CAz) - 1]\}. \tag{20.7a}$$

Further assume C=0, to obtain a log-normal quantile

$$y = (M) \exp(ABz). \tag{20.7b}$$

Fitting INT III, similarly to INT II in the previous sub-section, the resulting parameters and moments are displayed in Table 20.1. Note that the parameter "A" is occasionally large. This implies that caution must be exercised not to use in the fitted INT III (20.7) values of z smaller than -1/A.

Table 20.1. Parameters' values and the first four moments associated with fitted INT (exact moments are upper entries):

INT II (upper entries): $y(z)= (M) \exp\{(A)[\exp(Bz) - 1]\}$;

INT III (middle entries): $y(z) = (M) \exp\{(B/C)[(1+Az)^C - 1]\}$;

INT IV (lower entries): $y(z)= (M) \exp[ABz/(1+Az)]$.

Distribution	A	B	C	Mean	Var.	Sk	Ku
Normal $\mu=14$, $\sigma=2$	-0.9473 -- .0712	-.1551 -- 2.029	-- -- --	14.00 14.00 -- 14.00	4.000 4.135 -- 4.075	0 -0.0185 -- .0105	0 -0.0869 -- -.0425
Gamma $\alpha= 7$, $\beta=2$	-3.077 .0545 .0631	-.1262 7.072 6.105	-- -1.320 --	14.00 14.00 14.00 14.00	28.00 28.20 27.99 27.99	.7559 .7508 .7564 .7572	.8571 .8053 .8546 .8593
Gamma $\alpha= 3$, $\beta=2$	-3.311 .0911 .0965	-.1830 6.627 6.246	-- -1.126 --	6.000 6.016 6.001 6.001	12.00 12.03 11.99 11.97	1.155 1.156 1.156 1.159	2.000 1.925 1.998 2.017
Gamma $\alpha= 1/4$, $\beta=2$	-5.750 .4531 .3448	-.5194 7.527 8.902	-- -.8739 --	.5000 .4840 .4994 .4585	1.000 .9789 .9974 .8805	4.000 4.007 3.984 4.373	24.00 22.46 23.58 29.50
Weibull $\alpha=1$, $\beta=5$	-3.949 .1774 .1747	-.2815 6.450 6.554	-- -.9388 --	5.000 5.011 5.001 4.996	25.00 24.67 24.98 24.97	2.000 2.030 1.997 2.001	6.000 6.012 5.962 5.969
Weibull $\alpha=0.8$, $\beta=5$	-4.980 .1864 .1745	-.2779 7.684 8.200	-- -.8659 --	5.665 5.656 5.665 5.659	50.99 50.40 50.96 50.90	2.815 2.832 2.807 2.812	12.74 12.47 12.60 12.59
Weibull $\alpha=0.4$, $\beta=5$	-10.14 .2495 .1681	-.2705 11.62 16.82	-- -.5214 --	16.62 16.35 16.62 16.40	2723 2652 2712 2692	11.35 10.45 11.04 11.19	287.6 209.3 250.4 255.8
Rayleigh $\sigma=2$	-1.814 .1268 .1702	-.3194 4.511 3.338	-- -1.643 --	2.507 2.521 2.511 2.506	1.717 1.729 1.712 1.703	.6311 .6443 .6355 .6343	.2451 .1840 .2431 .2827
Extreme-Value $\alpha=3$, $\beta=1$	-26.47 .2364 .0060	-.0134 1.466 59.07	-- .9278 --	3.577 3.577 3.568 3.576	1.645 1.699 1.652 1.685	1.139 1.089 1.132 1.090	2.400 2.144 2.474 2.152

20.4.4. Off-spring INT IV

Examining the parameters' values obtained for INT III (Table 20.1), one notes that C is near -1 · for many of the exhibited distributions, notwithstanding their widely varying skewness. For example, for the gamma with Sk=0.7559, we obtain C=-1.320. For Weibull with Sk=2.815, the value of C is -0.8659. This empirical observation suggests reducing the number of parameters in INT III (20.7) by setting C=-1. The resulting INT IV is

$$y = (M) \exp[ABz / (1+Az)]. \tag{20.8}$$

Fitting INT IV, similarly to the INTs in the previous sub-sections, the resulting parameters and moments are displayed in Table 20.1. Note that the normal distribution is well represented, although, unlike INT III (20.7), INT IV does not reduce naturally to the normal case. Also note that even though (20.8) has only three parameters, the kurtosis measure is approximately preserved for most of the distributions in Table 20.1.

For all the parameter-reduced INTs, all first four moments are preserved to an acceptable degree, notwithstanding the algebraic simplicity of the new INTs.

20.5. Distribution Fitting Procedures

Moment-based fitting procedures for the off-spring INTs are developed. It is assumed that in implementing the fitting procedure only sample estimates of moments are available (the distribution for which we wish to fit an INT is unknown or unspecified). Therefore, the underlying principle in deriving these procedures is to employ only low-degree moments (second degree at most), averting thereby the high sampling inaccuracies associated with sample estimates of high-degree moments. The new procedures will later serve to develop general SPC schemes for non-normal environments (Section 22.1).

20.5.1. Fitting procedures for INT I (Section 20.4.1)

(I) *Fitting with the median, the mean of W and the partial means of Y*

Refer to Sub-section 11.3.2.

(II) *Fitting with the mean and variance of W and partial means of Y*

Refer to Sub-section 11.3.3.

20.5.2. A fitting procedure for INT II (Section 20.4.2); Fitting with the mean and the variance of W and the mean of Y

An expression for the variance of the log-transformed INT I was developed in Sub-section 11.3.3. Since INT II (20.6) is derived from INT I (20.5) by setting in the latter D=0 we obtain for the variance of W, modeled by INT II [refer to (11.26)],

$$Var(W) = A^2 [\exp(2B^2)-\exp(B^2)]. \tag{20.9}$$

From this we obtain

$$A = \delta\{Var(W) / [\exp(2B^2)-\exp(B^2)]\}^{(1/2)}, \tag{20.10}$$

where δ=-1 for B<0, and δ=1 for B$\geq$0. This definition of δ is necessary to ensure that y in (20.6) is a monotone function of z. Introducing for A into (20.6), and also for log(M) [refer to (11.23)]

$$\log(M) = \mu(W) - A[\exp(B^2 / 2) - 1] \tag{20.11}$$

we obtain an INT in one parameter (B) that may be identified by matching the means.

Results from this fitting procedure are shown in Table 20.2. Note, that W derived from INT II is a linear transformation of a log-normal variable. Any current parameter-estimation procedure for a log-normal variable may be employed to find an estimate for B.

Table 20.2. Parameter values and the first four moments associated with moment-fitted INT II (Section 20.5.2): Matching by mean and variance of W and by mean of Y; M unspecified. Upper entries are the exact moments.

Dis.	A	B	Mean	Var.	Sk	Ku
Normal			14.00	4.00	0	0
$\mu=14$, $\sigma=2$	-.9675	-.1492	14.00	4.00	-.0095	-.0827
Gamma			14.00	28.00	.7559	.8571
$\alpha= 7$, $\beta=2$	-3.090	-.1253	14.00	27.98	.7510	.8076
Gamma			6.000	12.00	1.155	2.000
$\alpha= 3$, $\beta=2$	-3.084	-.1978	6.000	11.94	1.108	1.707
Gamma			.5000	1.000	4.000	24.00
$\alpha= 1/4$, $\beta=2$	-5.653	-.5722	.5000	1.050	3.828	19.83
Weibull			5.000	25.00	2.000	6.000
$\alpha=1$, $\beta=5$	-3.375	-.3470	5.000	24.31	1.809	4.361
Weibull			5.665	50.99	2.815	12.74
$\alpha=0.8$, $\beta=5$	-4.239	-.3455	5.665	48.81	2.487	8.898
Weibull			16.62	2723	11.35	287.6
$\alpha=0.4$, $\beta=5$	-8.693	-.3383	16.62	2278	8.030	111.8
Rayleigh			2.507	1.717	.6311	.2451
$\sigma=2$	-1.673	-.3495	2.507	1.706	.5781	.0139
ExtremeValue			3.577	1.645	1.139	2.400
$\alpha=3$, $\beta=1$	-.0003	-108	3.577	1.642	1.107	2.244

20.5.3. A fitting procedure for INT III (Section 20.4.3); Fitting by the median, the mean of W and the mean and variance of Y

Taking log of both sides of (20.7), and then applying the expected value operator, the mean of the log-transformed response is

$$\mu(W) = \log(M) + (B)E\{[(1+AZ)^C-1]/C\}. \qquad (20.12)$$

Developing the expression inside the expected value operator, E(.), into a Taylor series in terms of Z, around zero, and then taking expectation for the first six terms (note that all odd-order moments of Z are identically zero), we obtain

$$E\{[(1+AZ)^C-1]/C\} \cong (1/2)A^2(C-1)+(1/8)A^4(C-1)(C-2)(C-3). \qquad (20.13)$$

Introducing back into (20.12) and thence for B into (20.7), we obtain INT III expressed in terms of M, $\mu(W)$, and the parameters A and C.

These parameters may now be identified by matching the mean and the variance of INT III with those of Y. Table 20.3 exhibits the transformation's parameters and the resulting skewness and kurtosis values. In terms of preservation of all first four moments, acceptable fit is achieved. For all cases the value of C is practically -1, including distributions which are nearly symmetrical (like Rayleigh). This finding supports the assumption used to derive INT IV.

20.5.4. Fitting procedures for INT IV (Section 20.4.4)

(I) Fitting by the median (M), and the mean and variance of W

Taking log of both sides of (20.8) we obtain

$$W-\log(M) = ABz \,/\, (1+Az). \tag{20.14}$$

Developing the RHS into a Taylor series around zero and taking expectation [introduce C=-1 into (20.13)], we have for the mean of the log-transformed response

$$\mu(W) = \log(M) + B(-A^2 - 3A^4). \tag{20.15}$$

Likewise, squaring both sides of (20.14), developing the RHS into a Taylor series and taking expectation, give

$$\mu_2(W) - 2[\log(M)]\mu(W) + [\log(M)]^2 = B^2(A^2 + 9A^4), \tag{20.16}$$

where $\mu_2(W)$ is the second non-central moment of W (second-degree moment about zero). Assuming that the median (M) and the first two moments of W are specified, we can identify the parameters A and B from the last two equations. Values of these parameters for the test-set of distributions, and the resulting first four moments of the fitted INT IV, are given in Table 20.3.

(II) Fitting by the median (M), and the means of W and Y

This fitting procedure is similar to the previous one. However once the parameter B is identified from (20.15) it is inserted into INT IV (20.8), wherefrom A may be identified by matching of means. Only first-degree moments participate in this fitting procedure.

Table 20.3. Parameter values and the first four moments associated with moment-fitted INTs (top to bottom): INT III (Sec. 20.5.3): Matching with the mean of W and mean and variance of Y; M specified; INT IV (Sec. 20.5.4-I): Matching with the mean and variance of W; M specified; INT IV (Sec. 20.5.4-II): Matching with the means of W and Y; M specified.

Distribution	A	B	C	Mean	Var.	Sk	Ku
				14.00	4.00	0	0
Normal	.0722	1.989	-1.000	--	--	.0019	-.0411
$\mu=14$, $\sigma=2$	.0721	1.994	--	14.00	4.01	.0030	-.0412
	.0707	2.076	--	--	4.18	.0201	-.0420
				14.00	28.00	.7559	.8571
Gamma	.0635	6.072	-1.000	--	--	.7542	.8454
$\alpha=7$, $\beta=2$	.0635	6.069	--	13.99	28.01	.7665	.8629
	.0636	6.066	--	--	28.01	.7553	.8532
				6.000	12.00	1.155	2.000
Gamma	.0978	6.181	-1.000	--	--	1.154	1.992
$\alpha=3$, $\beta=2$	.0977	6.177	--	5.998	11.97	1.158	2.002
	.0978	6.180	--	--	12.00	1.154	1.991
				.5000	1.000	4.000	24.00
Gamma	n.a	n.a.	n.a.	--	--	--	--
$\alpha=1/4$, $\beta=2$	n.a	n.a.	n.a.	--	--	--	--
	n.a.	n.a.	--	--	--	--	--
				5.000	25.00	2.000	6.000
Weibull	.1670	6.700	-1.002	--	--	2.022	6.106
$\alpha=1$, $\beta=5$	.1696	6.666	--	4.973	24.44	2.011	6.078
	.1701	6.697	--	--	25.04	2.028	6.187
				5.665	50.99	2.815	12.74
Weibull	.1701	8.364	-1.001	--	--	2.836	12.29
$\alpha=0.8$, $\beta=5$	.1696	8.332	--	5.618	49.73	2.827	12.80
	.1701	8.373	--	--	51.24	2.854	13.06
				16.62	2723	11.35	287.6
Weibull	.1702	16.71	-1.000	--	--	11.12	252.2
$\alpha=0.4$, $\beta=5$	.1696	16.66	--	16.62	2605	11.02	247.1
	.1702	16.73	--	--	2705	11.15	253.6
				2.507	1.717	.6311	.2451
Rayleigh	.1701	3.348	-1.000	--	--	.6378	.2869
$\sigma=2$	.1696	3.333	--	2.504	1.686	.6366	.2849
	.1703	3.342	--	--	1.709	.6364	.2862
				3.577	1.645	1.139	2.400
ExtremeValue	.0007	-497	-1.000	--	--	1.124	2.302
$\alpha=3$, $\beta=1$	n.a.	n.a.	--	--	--	--	--
	.0007	-492	--	--	1.628	1.118	2.278

Results from the latter for the test-set of distributions are given in Table 20.3. We realize that although the moments are less well preserved relative to procedures based on matching with higher-degree moments, we expect the low mean-squared-error associated with sample estimates of first degree to deliver accurate representation to the unknown approximated distribution.

The INTs developed in this chapter are used to develop general SPC schemes for non-normal populations in Section 22.1.

20.6. Normalizing Transformations

The INTs developed in earlier sections allow derivation of normalizing transformations (NTs), namely, expressions which depict the normal quantile in terms of the corresponding quantile of the original response.

From INT II (20.6):

$$z = (1/C)\log\{(1/B)\log(y/M)+1\}. \qquad (20.17)$$

From INT III (20.7):

$$z = (1/A)\{[(C/B)\log(y/M)+1]^{(1/C)} - 1\}. \qquad (20.18)$$

From INT IV (20.8):

$$z = (1/A) \log(y/M) / [B-\log(y/M)]. \qquad (20.19)$$

We leave it as an exercise to fit these NTs to the sample-set of distributions, used to fit the INTs, and to evaluate the relative effectiveness of these NTs. Note, that fitting (20.17)-(20.19) by a quantile-based NL-LS procedure is expected to deliver an expression for the random variable, Z, in terms of Y. Calculating numerically the moments of the fitted Z, using for these calculation the d.f. of the original response, Y, should result in moments close to those of the standard normal distribution.

References

[1] Clements, J. A. (1989). Process capability calculations for non-normal distributions. *Quality Progres,* 22(2), 49-55.

[2] Kotz, S., Johnson, N. L. (2002). Process Capability Indices- A Review, 1992-2000. *Journal of Quality Technology,* 34(1), 2-19.

[3] Lau, H. S., Lau, A. H. L. (2003). Nonrobustness of the normal approximation of lead-time demand in a (Q,R) system. *Naval Research Logistics,* 50(2), 149-166.

[4] Pearson, E. S., Johnson, N. L., Burr, I. W. (1979). Comparisons of the percentage points of distributions with the same first four moments, chosen from eight different systems of frequency curves. *Communications in Statistics (Simulation and Computation),* B8(3), 191-229.

[5] Shore, H. (1995). Fitting a distribution by the first two sample moments (partial and complete). *Computational Statistics and Data Analysis,* 19, 563-577.

[6] Shore, H. (1996). A new estimate of skewness with MSE smaller than that of the sample skewness. *Communications in Statistics (Simulation and Computation),* 25(2), 403-414.

[7] Shore, H. (1998a). A new approach to analyzing non-normal quality data with application to process capability analysis. *International Journal of Production Research (IJPR),* 36(7), 1917-1933.

[8] Shore, H. (1998b). Approximating an unknown distribution when distribution information is extremely limited. *Communications in Statistics (Simulation and Computation),* 27(2), 501-523.

[9] Shore, H. (2000a). General control charts for variables. *International Journal of Production Research (IJPR),* 38(8), 1875-1897.

[10] Shore, H. (2000b). General control charts for attributes. *IIE Transactions,* 32(12), 1149-1160.

[11] Shore, H. (2000c). Three approaches to analyze quality data originating in non-normal populations. *Quality Engineering,* 13 (2), 277-291.

[12] Shore, H. (2001). Modeling a non-normal response for quality improvement. *International Journal of Production Research,* 39 (17), 4049-4063.

[13] Shore, H. (2004a). A general solution for the newsboy model with random order size and possibly a cutoff transaction size. *Journal of the Operational Research Society,* 55(11), 1218-1228.

[14] Shore, H. (2004b). Non-normal populations in quality applications- A revisited perspective. *Quality and Reliability Engineering International,* 20(4), 375-382.

[15] Woodall, W. H. (1997). Control charting based on attribute data: bibliography and review. *Journal of Quality Technology,* 29, 172-183.

Chapter 21

Piece-Wise Linear Approximations

21.1. Introduction

The normal approximation is widely used to represent distributions that are approximately or asymptotically symmetrical. For example, in Statistical Process Control (SPC), all of Shewhart control charts are based on normal approximations for the distributions of the statistics that monitor the process. A p chart and a c chart are based on normal approximations to the binomial and the Poisson distributions, respectively.

Normal approximations are simple to apply: Just equate the mean and the variance of the fitted normal distribution with the respective moments of the "true distribution", and then use the former in the application of interest. The simplicity of the normal approximation, however, is often compromised by the realities that a practitioner encounters and sometimes ignores. For example, s/he may use a normal approximation for extremely skewed distribution because it is simple, and then obtain results that deviate appreciably from the correct values. A good example for that is the wide use of *traditional* process capability indices (PCIs), which are based on the normal assumption. If the process distribution is far from being symmetric, the PCI may deliver very deceiving information about the true capability of the process. If used in trade negotiations between suppliers and their customers, this may have adverse effect on the future of their relationships.

To circumvent the shortcomings of the normal approximation, various normal-based approximations have been suggested in the literature. One example is the Cornish-Fisher expansion (Stuart and Ord, 1987, Section 6.26). This expansion provides the quantile of a r.v., with known cumulants, in terms of these cumulants and a power series of the corresponding normal quantile:

$$x_p\text{-}z_p = l_1 + (1/6)l_3(z_p^{\,2}\text{-}1) + (1/2)l_2 z_p + (1/24)l_4\,(z_p^{\,3}\text{-}3z_p)$$

$$- (1/36)l_3^{\,2}\,(2z_p^{\,3}\text{-}5z_p) +... \tag{21.1}$$

In this equation, $\{z_p, x_p\}$ are the respective quantile values of the standard normal variable, Z, and of the standardized r.v., which we wish to represent by the Cornish-Fisher expansion: $X=(Y\text{-}\mu)/\sigma$. Also, l_r is the standardized r-th cumulant of Y:

$$l_k = \kappa_r / \sigma^r, \tag{21.2}$$

where κ_r is the r-th un-standardized cumulant, and σ is the standard deviation. Cumulants are functions of moments of Y about an arbitrary point, $\{\mu_r'\}$. A detailed explanation of their derivation and common applications may be found in Stuart and Ord (1987, Section 3.12). In particular, assuming that $\kappa_1=\mu_1'=0$ (namely, moments are calculated about the mean), it can be shown that the expressions for the first four (un-standardized) cumulants in terms of central moments (moments about the mean) are (therein, p. 88)

$$\kappa_1= \mu_1,\ \kappa_2= \mu_2,$$

$$\kappa_3= \mu_3,\ \kappa_4= \mu_4 - 3\mu_2^{\,2}, \tag{21.3}$$

where μ_r is the r-th central moment (for example, μ_2 is the variance). It is easily recognized that the third and fourth *standardized* cumulants, l_3 and l_4, are the traditional skewness meaure, Sk, and kurtosis measure, Ku, respectively.

While the Cornish-Fisher expansion and other normal-based approximations (refer to some references in Stuart and Ord, 1987, Section 6.26) may deliver satisfactory accuracy, they suffer from some undesirable features. First, they are often cumbersome to apply. For example, it is often difficult to determine, in advance, how many terms to include in a Cornish-Fisher approximation to obtain satisfactory accuracy. Secondly, they often require knowledge of high moments, which are difficult to estimate from sample data. Considering the Cornish-Fisher approximation, it is apparent from (21.1) that even if a very small number of terms is included in the expansion, knowledge of the skewness and kurtosis measures is required. However, sample estimates of these moments are associated with large sampling errors, as discussed in Section 20.3.

The linear piece-wise normal approximations developed in this chapter are intended to avoid the drawbacks of current normal or normal-based approximations. The piece-wise approximations were initially developed in the mid-eighties of the last century (Shore, 1986a). Several variations of the original presentation have since been developed (these will be alluded to, with references, later on). With the recent introduction of RMM, these approximations now seem like a natural derivative of the RMM error distribution, and thus can be perceived as part and parcel of the RMM model.

A unique feature of the new approximations is their capability to preserve all first three moments of the distribution to which they are fitted (namely, the mean, the variance and the skewness measure). As will be expounded later, this is a direct result of the use of a *piece-wise* linear approximation. Furthermore, quantiles other than the standard normal can be used in the approximations. In fact, quantiles of any symmetrically distributed random variable may replace the standard normal quantiles. This explains why, in the title of this chapter, we refer to the new approximations as "Piece-wise Linear Approximations", omitting any mention of "normal" as an essential element in these approximations.

In the next Section 21.2, the rational for the piece-wise linear approximations is introduced. When a normal variable is used in these approximations, they will be denoted, for brevity, the *modified normal approximations*. In subsequent sections we will develop in detail variations of the basic approximation and the allied fitting procedures. For reasons similar to those given in developing fitting procedures for the inverse normalizing transformations (Chapter 20), moment-matching procedures will be confined to moments (or their estimates) of second-degree at most.

21.2. The Basic Modified (Normal) Approximation

Consider a plot of the quantile function of any skewed random variable as a function of the corresponding standard normal quantile. Several such plots were given in Chapter 19, where it was demonstrated how well the RMM error quantile-function represents quantiles of commonly used skewed distributions. We re-introduce in Figure 21.1 one such plot. The plot relates to the gamma distribution, with parameters $\alpha=3$, $\beta=2$ (Sk= 1.1547). Each dot in the plot represents a pair of corresponding quantiles of Z, the standard normal variable (horizontal axis), and of U, the standardized gamma variable.

Examining the plot, two properties seem to stand out. First, the plot represents a monotone convex relationship, and, secondly, convexity is not very intense. The first property will be used later on, when the assumption of monotone convexity will lead to a certain structure for the coefficients of the modified normal approximation. In fact, since this structure can be shown to allow preservation of skewness (in addition to preservation of the mean and the variance), we use this outcome to "prove" the convexity property (no rigorous mathematical proof will be given). In this section we use the second property (mild convexity) to derive the modified normal approximation.

The mild convexity displayed in Figure 21.1, and other similar plots, suggests that perhaps this unknown relationship can be approximated by a piece-wise linear transformation of the standard normal quantile.

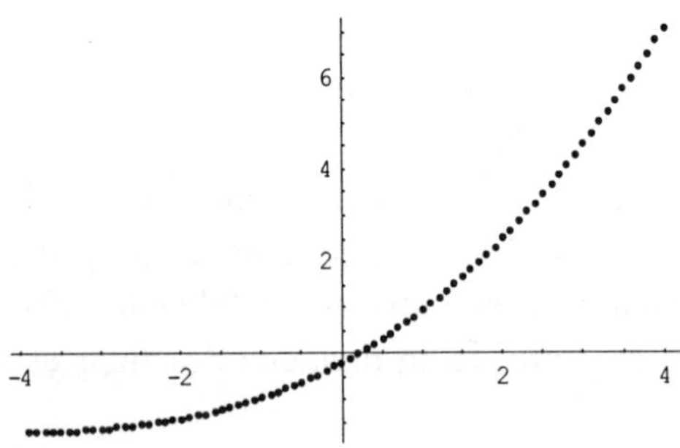

Figure 21.1. Quantile values of (z,u) (standard normal, standardized gamma)

A good division point is to have a separate linear transformation for $z<0$ and for $z \geq 0$. Later on the special benefits of this division will be discussed.

Accordingly, the basic modified normal approximation is

$$u = (y\text{-}\mu)/\sigma = \begin{cases} A_1 z + B_1, z < 0, \\ \\ A_2 z + B_2, z \geq 0, \end{cases} \qquad (21.4)$$

where u is the standardized quantile of Y, and {z,y} represent the P-th quantiles of Z, the standard normal variable, and of Y, namely, $F(y) = \Phi(z) = P$, and F and Φ are the CDFs of Y and Z, respectively.

Fitting procedures for this approximation, or variations thereof, are developed in Sections 21.3-21.5.

21.3. A Variation of the Basic Model with a Fitting Procedure

Consider re-writing the basic model as

$$y = \begin{cases} \mu + \sigma [(A - C)z - hC], z < 0 \\ \\ \mu + \sigma [(A + C)z - hC], z \geq 0, \end{cases} \qquad (21.5)$$

where {h,A,C} are parameters that need to be determined, and {μ,σ} are the mean and the variance of Y.

This approximation was first introduced in Section 5.2, where current families of distributions used in empirical modeling had been reviewed. The structure of (21.5) implies that we have assumed uniform convexity (the first property) since the slope for $z \geq 0$ is always larger than that for $z < 0$ (assuming C>0). If Y is a discrete r.v., the term $-1/(2\sigma)$ should be added to both parts on the right-hand-side to account for the fact that a quantile of a discrete variable (Y) is approximated by a continuous variable (Z).

Values for the parameters {h,A,C} are sought which preserve in approximation (21.5) the first three moments of Y, or, equivalently, the first three moments of the standardized variable $U=(Y\text{-}\mu)/\sigma$.

A modified approximation for the quatile, u, of U, is [from (21.5)]

$$u = \begin{cases} [(A - C)z - hC], & z < 0 \\[2ex] [(A + C)z - hC], & z \geq 0, \end{cases} \tag{21.5a}$$

Let us develop the first three moments of (21.5a). This requires knowledge of the partial moments of Z. Discarding the assumption that Z is normal, however assuming that it is a standardized symmetrically distributed variable, denote by $M_U(Z^r)$ the r-th upper partial moment of Z, namely,

$$M_U(Z^r) = \int_0^\infty z^r \phi(z) dz . \tag{21.6}$$

The upper partial mean is $M_U(Z)$. The lower partial mean is $M_L(Z) = \mu - M_U(Z)$. Also, since Z is symmetrical, $M_U(Z^2) = M_L(Z^2) = 1/2$.

For simplicity we will henceforth denote $M_U(Z^r)$ by M_r.

Due to the symmetry of the distribution of Z, for all odd-order moments, the sum of the upper and the lower r-th partial moments is zero (the lower partial moment is the negative of the upper partial moment). Similarly, even-order upper and lower partial moments (of the same order) are equal. Values of partial moments for various Z (like the standard normal or the standard logistic) will be provided later on.

It can easily be shown that for $h = 2M_1$, (21.5a) has zero mean, irrespective of the values of the other parameters.

To obtain a unit variance, we need to have for (21.5a)

$$\text{Var}(U) = 1 = A^2 + C^2(1 - 4M_1^2). \tag{21.7}$$

To preserve the skewness measure, we equate the given Sk (the skewness of U) with the skewness calculated from (21.5a) (refer for details to Shore, 1986a):

$$Sk = 6C(M_3 - M_1) + 2C^3(12M_1^2 M_3 - 4M_1^3 - 2M_3). \tag{21.8}$$

A solution, $\{A, C\}$, for the set of equations, (21.7) and (21.8), plus the value assigned to h earlier, ensures that (21.5a) preserves all first three moments of U. Since (21.8) depends on C only, and since this is a third-degree polynomial, we may find one explicit solution for C (the other two solutions include imaginary values). Re-writing (21.8) as

$$k_1 C^3 + k_2 C - Sk = 0, \tag{21.8a}$$

we obtain for the real-valued solution

$$C = \{-(2)(6)^{1/3} k_2 + [18k_1^{1/2}Sk + (2)(3)^{1/2}(4k_2^3 + 27k_1Sk^2)^{1/2}]^{2/3}\} /$$

$$\{(2)(3)^{2/3}k_1^{1/2}[9k_1^{1/2}Sk+(3)^{1/2}(4k_2^3 + 27k_1Sk^2)^{1/2}]^{1/3}\} \tag{21.9}$$

Introducing back into (21.7), we can solve for A, provided that

$$1- C^2(1-4M_1^2) \geq 0.$$

The expression in (21.8) indicates that preservation of skewness is possible by the algebraic structure assumed in (21.5a), even though mere linear transformations of Z are used. Of-course, regular conditions that ensure existence of a real-valued root for a third-degree polynomial have to be satisfied.

Consider some "candidates" for Z (for all $M_2=1/2$).

Case 1. Z is Standard Normal

For a standard normal Z, we have

$$M_1 = 1/ (2\pi)^{1/2} = 0.3989, \; M_3 = (2/\pi)^{1/2} = 0.7979. \tag{21.10}$$

We obtain for (21.7) and (21.8), respectively:

$$A^2 = 1-0.3635C^2,$$

$$Sk = 2.3940C - 0.6523C^3. \tag{21.11}$$

Introducing $k_1 = -0.6523$ and $k_2 = 2.3940$ into (21.9), C can be expressed in terms of Sk. The parameter "A" may thence be easily calculated from (21.7).

Case 2. Z is Standard Logistic

The standardized logistic variable is defined in Section 19.3. For this case we have

$$M_1 = 0.3821, \; M_3 = 0.90656. \tag{21.12}$$

Introducing into (21.7) and (21.8), results in:

$$A^2 = 1-0.4160C^2,$$

$$Sk = 3.1468C - 0.8959C^3. \tag{21.13}$$

These equations may be solved similarly to Case 1. Note, that for the logistic, the quantile, z_p, is expressed explicitly in terms of the CDF $F(z_p)$=P. This implies that the 100P-th percentile of Y, y_p, may be expressed via (21.5) explicitly in terms of P. This is a desirable property for certain applications, where the quantile of Y needs to be expressed explicitly in terms of the corresponding CDF. For instance, in stochastic inventory models, we may wish to express inventory level in terms of the required (or optimal) safety level (the probability of no shortages). This can easily be accommodated with (21.5). A demonstration will be given in Section 23.2.

Case 3. Shore's Approximations to the Standard Normal Inverse Distribution Function

A group of highly accurate approximations for the quantile function of the standard normal variable has been developed (Shore, 1982). All approximations are expressed in terms of the CDF of the standard normal distribution, and they share with the latter the same first five (complete) moments. Letting P= $\Phi(z)$, where Φ is the standard normal CDF, three of the approximations are

$$z = -5.530 \{ [P/(1-P)]^{0.1193} - 1\}, z\geq0, \qquad (21.14)$$

$$z = -0.4095\{(1-P)/P + \log[(1-P)/P] - 1\}, z\geq0, \qquad (21.15)$$

$$z = -0.4507 \log[(1-P)/P] + 0.2250, z\geq0. \qquad (21.16)$$

(coefficients in these equations are somewhat different from those in Shore (1982), due to the improved accuracy of today's procedures.)

Table 21.1 displays the first four upper partial moments [exact, and those obtained from (21.14)-(21.16)]. In addition to preserving the first five complete moments, the partial moments are also well preserved. These approximations may be easily incorporated in the piece-wise linear approximations.

An examination of the accuracy of the above approximate quantile functions, and comparison with alternative approximations, are given in Shore (1995a, 1996a, 2000). These references may be related to for some numerical comparisons.

Table 21.1. First four upper partial moments of approximations to the inverse distribution
function of the standard normal distribution

Source:	M_1	M_2	M_3	M_4
Standard Normal	$1/(2\pi)^{1/2}=$	1/2	$(2/\pi)^{1/2}=$	3/2
(exact)	0.3989		0.7979	
Eq. 21.14	0.4001	0.5000	0.7970	1.500
Eq. 21.15	0.4095	0.5031	0.7897	1.499
Eq. 21.16	0.4249	0.5000	0.7738	1.500

In Section 21.5 a fitting procedure is developed for the basic model
(21.4), which utilizes only moments (partial and complete) of second-
degree at most.

21.4. A Simplified Fitting Procedure

Model (21.5) has three parameters, which allow equality of the first three
moments with the approximated distribution (whether the latter is known
or only partially known via sample estimates of moments). Examination
of the values of the parameters obtained on using different Zs reveals
some regularities that may be exploited in order to achieve simpler
expressions, which require less complicated numerical calculations.

First, note that k_1C^3 in (21.8a) is usually very small (relate to values
of k_1C^3 obtained for various Zs and different values of Sk). This suggests
that one may neglect the second term in (21.8) to obtain explicitly for C

$$C = Sk / [6(M_3 - M_1)], \tag{21.17}$$

and thence for A [from (21.7)]

$$A = \{1 - [(Sk/6) / (M_3 - M_1)]^2 (1 - 4M_1^2)\}^{1/2}. \tag{21.18}$$

The parameters $\{A, C\}$, determined by (21.17)-(21.18), preserve the
mean and the variance of the approximated distribution. However, the
actual skewness of (21.5) (denote it by $\hat{Sk}$) is, from the RHS of (21.8)
after introducing for C from (21.17):

$$\hat{Sk} = Sk + 2 [(Sk/6) / (M_3 - M_1)]^3 (12M_1^2M_3 - 4M_1^3 - 2M_3). \tag{21.19}$$

Consider the cases examined earlier.

Case 1. Z is Standard Normal

From (21.17)-(21.19), respectively:

$$C = Sk/2.394 = 0.4177\ Sk,$$

$$A = (1 - 0.06342 Sk^2)^{1/2},$$

$$\hat{Sk} = Sk - 0.04754 Sk^3. \qquad (21.20)$$

Case 2. Z is Standard Logistic

$$C = Sk/3.147 = 0.3178\ Sk,$$

$$A = (1 - 0.04201 Sk^2)^{1/2},$$

$$\hat{Sk} = Sk - 0.02876 Sk^3. \qquad (21.21)$$

Case 3. Shore's Approximations to the Standard Normal

Expressions obtained are similar to those obtained in (21.20). The reader may wish to derive these expressions, based on the approximations given in (21.14)-(21.16).

We realize from the expressions for $\hat{Sk}$ that, provided the actual skewness of the monitoring statistic is not extremely high, Sk is well preserved by the approximate quantile function given in (21.5), with parameters given by (21.17) and (21.18). Also, if Sk is not very high, for all Zs we have A≈1. Introducing A=1 may further simplify the use of approximation (21.5).

We conclude this section with an observation regarding the algebraic structure of (21.5). Observing the parameters C and A, given in (21.17) and (21.18), respectively, one realizes that there is a close association between the sign of Sk and the convexity of the quantile function of Y [as approximated by (21.5)]. For positive Sk, the slope of (21.5a) for $z>0$

is larger than that for z<0, implying that the true relationship is uniformly convex (second derivative of the *exact* unknown expression for the quantile function is positive). Conversely, when Sk is negative, this implies that the quantile function is concave (second derivative is negative).

This observation is probably true also for the quantile function of the RMM error distribution, developed in Chapter 9, and for the inverse normalizing transformations, developed in Chapter 20. Note, however, that this observation is not a rigorous mathematical proof for the relationship between Sk and the convexity (concavity) of the actual quantile relationship between the response, Y, and Z.

A mathematical proof for the "inverse" observation, namely, that a convex (concave) transformation of a symmetrically distributed random variable results in positive (negative) skewness may be found in Van Zwet (1964).

21.5. A Fitting Procedure Using First- and Second-Degree Moments

To describe the two-moment fitting procedure, re-write (21.4) in terms of the quantile of the un-standardized response, Y:

$$y = \begin{cases} A_1 z + B_1, z < 0, \\[2mm] A_2 z + B_2, z \ge 0. \end{cases} \qquad (21.22)$$

Similarly to the definition of the r-th upper partial moment of Z, M_r, given in (21.6), the r-th upper partial moment of Y, $M_r(Y)$, is defined by

$$M_r(Y) = \int_M^\infty y^r f(y)\,dy, \qquad (21.23)$$

where f(y) is the density function of Y, and M is its median. Thus, to estimate, say, $M_2(Y)$ from sample data, we need to calculate the squared values of all data larger then the median, and then calculate the average of *all* data points, where values lower than the median are replaced by zero. For example, for the data $y=\{7,2,15,3\}$, an estimate of $M_2(Y)$ would be $(7^2+15^2)/4= 68.5$.

In this section, the parameters of (21.22) are determined (or estimated) by a moment-matching procedure based on matching first and second degree moments, partial and complete.

From (21.22), the partial lower and upper first- and second-degree moments may be easily calculated in terms of the partial first- and second-degree moments of Z (M_1 and M_2, respectively) and in terms of the parameters of (21.22). Equating these to the corresponding moments of Y, we obtain for the parameters of (21.22):

$$A_1^2 = \{(\sigma^2 + \mu^2) - M_2(Y) - 2[\mu - M_1(Y)]^2\} / [(1/2) - 2M_1^2],$$

$$B_1 = 2[\mu - M_1(Y) + A_1M_1],$$

$$A_2^2 = \{M_2(Y) - 2[M_1(Y)]^2\} / [(1/2) - 2M_1^2],$$

$$B_2 = 2[M_1(Y) - A_2M_1], \tag{21.24}$$

where for the standard normal $M_1 = 1/\sqrt{2\pi} = 0.3989$. (Relate to Shore, 1995a, for a detailed derivation of these expressions.)

A guiding principle for this fitting procedure is that both the *complete* moments (the mean and the variance) and the *partial* moments are preserved. No higher-degree moments take part in the fitting procedure. When variables data are used to estimate the response moments, cumulative evidence demonstrates that this procedure is associated with appreciably reduced mean-squared-errors (MSEs). For example, in Shore (1996a) it was demonstrated that if the skewness measure is calculated numerically from the fitted (21.22), with parameters determined by (21.24), the associated MSE is appreciably lower than that of a direct sample estimate of skewness. This indeed implies that irrespective of the distribution one wishes to fit, any three-moment fitting procedure will result in a worse fit than fitting by the two-moment procedure, outlined in this section (refer also to a discussion in Section 20.3).

21.6. Review of Related Published References

Endeavors to approximate a non-normal distribution by a piece-wise normal approximation have been reported with respect to statistical process control applied to non-normal distributions (Choobineh and

Ballard, 1987, and Bai and Choi, 1995). However, therein determination of the parameters is solely based on a sample estimate of the percentage of the (unknown) distribution that lies below the mean. No attempt at preserving the first three moments of the actual distribution is reported in these references.

The piece-wise linear approximations reported here, whether normal-based or otherwise, were first introduced in Shore (1986a), where they were applied to inventory analysis and for the determination of an optimal sample size in hypothesis testing. Later on, the approximations were used to derive simple explicit solutions for the newsboy problem (Shore, 1986b), to derive highly accurate formulae for sample-size determination in estimating binomial parameters (Shore, 1986c) and for constructing Shewhart-like control charts (Shore, 1991). A modified and much improved version of the latter, which pertains only to attributes data, has been developed more recently (Shore, 2000). The two-moment procedure has been developed some time ago (Shore, 1995a), and applied to solve problems related to inventory analysis and reliability engineering (Shore, 1995b, 1996b, respectively).

In Chapters 22 and 23, these approximations are employed to derive general control charts for attributes and for inventory analysis, respectively.

21.7. A Numerical Example

Assume that a certain process variable has a gamma distribution with parameters $\alpha=3$ and $\beta=2$. The median, M, and the related first three moments are

$$M = 5.348,\ \mu = 6.000,\ \sigma^2 = 12.00,\ Sk = 1.1547. \qquad (21.25)$$

It is easily verified by numerical integration that the first two partial moments of the gamma distribution, with the specified parameters, are

$$M_1(Y) = 4.319,\ M_2(Y) = 41.60. \qquad (21.26)$$

Using for Z the standard normal variable, we obtain from (21.20) for the parameters of (21.5):

$$h = 2M_1 = 0.7978,\ A = 0.957,\ C = 0.4823. \qquad (21.27)$$

Similarly, from (21.24), the parameters of (21.22) are

$$A_1 = 2.023, \ B_1 = 4.976,$$

$$A_2 = 4.864, \ B_2 = 4.756. \tag{21.28}$$

Introducing the parameters in (21.27) into (21.5), a piece-wise linear approximation is obtained for the quantile of Y in terms of the standard normal quantile. From this, the r-th non-central moment of Y is

$$\mu_r' = E(Y^r) = \int_{-4}^{0} [\sigma(A-C)z - 0.7978C\sigma + \mu]^r \, \phi(z)dz$$

$$+ \int_{0}^{4} [\sigma(A+C)z - 0.7978C\sigma + \mu]^r \, \phi(z)dz, \tag{21.29}$$

where $\phi(z)$ is the density function of the standard normal variable, and the limits of integration were set arbitrarily as z=-4 and z=4.

Similarly, introducing the parameters in (21.28) into (21.22), we obtain for the r-th non-central moment of Y:

$$\mu_r' = E(Y^r) = \int_{-4}^{0} [A_1 z + B_1]^r \, \phi(z)dz$$

$$+ \int_{0}^{4} [A_2 z + B_2]^r \, \phi(z)dz \tag{21.30}$$

From (21.29) and (21.30), the first four moments are

$$\mu = \mu_1', \ \sigma^2 = \mu_2' - (\mu_1')^2,$$

$$Sk = [\mu_3' - 3\mu_2' \, \mu_1' + 2(\mu_1')^3] / \sigma^3,$$

$$Ku = [\mu_4' - 4\mu_3' \, \mu_1' + 6\mu_2' \, (\mu_1')^2 - 3(\mu_1')^4] / \sigma^4 - 3 \tag{21.31}$$

Table 21.2 displays the resulting first four moments. In this table, (21.22) [the un-standardized (21.4)] is fitted by the two-moment fitting procedure, with parameters given by (21.24). Also, (21.5) is fitted by the three-moment procedure, with parameters calculated from (21.20) (namely, Sk is *approximately* preserved).

For both models, the first three moments are well preserved. Note, in particular, that for (21.22) only moments of second degree at most took part in the fitting procedure. However, the skewness measure is preserved.

Table 21.2. First four moments of Y (using a standard normal z)

Approximation used:	Mean	Variance	Sk	Ku
Eq. (21.22) (two moment fitting)	6.000	12.00	0.9539	0.7826
Eq. (21.5) (three moment fitting)	6.000	12.00	1.078	0.9093
Exact moments	6.000	12.00	1.155	2.000

Due to the semi-linear nature of these approximations, values of the kurtosis deviate considerably from the exact values. This implies that far-tail values will render low accuracy and therefore should not be calculated from these approximations.

References

[1] Bai, D. S., Choi, I. S. (1995). X and R control charts for skewed populations. *Journal of Quality Technology.* 27(2), 121-131.

[2] Choobineh, F., Ballard, J. L. (1987). Control limits of QC charts for skewed distributions using weighted variance. *IEEE Transactions on Reliability*, 36, 473-477.

[3] Shore, H. (1982). Simple approximations for the inverse cumulative function, the density function and the loss integral of the normal distribution. *Applied Statistics*, 31, 108-114.

[4] Shore, H. (1986a). Simple general approximations for a random variable and its inverse distribution function based on linear transformations of a nonskewed variate. *SIAM Journal on Scientific and Statistical Computing*, 7, 1-23.

[5] Shore, H. (1986b). General approximate solutions for some common inventory models. *Journal of the Operational Research Society,* 37(6), 619-629.

[6] Shore, H. (1986c). Approximate closed form solutions to some tests related to the binomial distribution. *The Statistician* (Journal of the Institute of Statisticians, England), 35(4), 471-478.

[7] Shore, H. (1991). New control limits that preserve the skewness of the monitoring statistic. *Transactions of the 45th Annual Quality Congress.* 45, 77-83.

[8] Shore, H. (1995a). Fitting a distribution by the first two sample moments (partial and complete). *Computational Statistics and Data Analysis*, 19, 563-577.

[9] Shore, H. (1995b). Setting safety lead-times for purchased components in assembly systems- a general solution procedure. *IIE Transactions*, 27, 634-637.

[10] Shore, H. (1996a). A new estimate of skewness with MSE smaller than that of the sample skewness. *Communications in Statistics (Simulation and Computation)*, 25(2), 403-414.

[11] Shore, H. (1996b). Optimum schedule for preventive maintenance- a general solution for a partially specified time-to-failure distribution. *POMS (An International Journal of the Production and Operations Management Society)*, 5(2), 148-162.

[12] Shore, H. (2000). General control charts for attributes. *IIE Transactions*, 32, 1149-1160.

[13] Stuart, A., Ord, J. K. (1987). *Kendall's Advanced Theory of Statistics, V. 1: Distribution Theory*. Charles Griffin & Co. Ltd. London.

[14] Van Zwet, W. R. (1964). *Convex Transformations of Random Variables*. Mathematisch Centrum, Amsterdam.

Chapter 22

General Control Charts

22.1. Introduction

The RMM original model and some of its derivatives (like the piece-wise linear approximations) may serve as good platforms to construct control schemes for statistical process control (SPC). In Section 22.2 general process control schemes for attributes are developed, and likewise for variables in Section 22.3.

The term "General" in the title of this chapter needs qualification. Traditional Shewhart charts assume approximate normality for the monitoring statistic, hence the ± 3 coefficients in the expressions for the control limits. The new control charts extend the spectrum of scenarios, where SPC may be implemented, to non-normal environments (or when approximate normality cannot be assumed). We use piece-wise linear approximations (Chapter 21) to derive general control charts for attributes, and we use inverse normalizing transformations (Chapter 20) to develop general control charts for variables.

Throughout this chapter *control limits* refer to the traditional Shewhart limits (the $\pm 3\sigma$ limits). *Probability limits* relate to other limits determined by some specified (required) false alarm probabilities (the probabilities of exceeding the limits when the process is in a state-of-control).

Good sources for SPC are Montgomery (2001) and Grant and Leavenworth (1996).

22.2. General Control Schemes for Attributes

22.2.1. Introduction

The term "Attribute", in statistical quality control, refers to a characteristic that is measured by counting. For example, the percentage of non-conforming items in a production line and the rate of arrival to a service facility are both attributes. Two widely used Shewhart control charts for attributes are the p chart, which monitors the *percentage* of occurrence of a certain attribute, or the c chart, which is the same for the *rate* of occurrence for a certain event of interest. These two charts are based on the binomial and the Poisson models, respectively. A basic principle underlying all of Shewhart-type control charts is that the distribution of the plotted monitoring statistic may be approximated by a normal distribution, with parameters that allow the approximating normal distribution to preserve the true mean and standard deviation (or their sample estimates). Since the binomial and the Poisson distributions may be approximated by the normal distribution, the p and the c charts may be "fashioned" after the traditional Shewhart model.

As evidenced by the widespread implementation of Shewhart control schemes, these have proved to be useful and effective. Indeed, numerous studies have repeatedly demonstrated the general robustness of the Shewhart control chart to deviation from normality (Burr, 1967; Schilling and Nelson, 1976; Balakrishnan and Kocherlakota, 1986; Wheeler, 1991; Yourstone and Zimmer, 1992; Shore, 1998, 2000a, and references therein).

There are, however, situations where ignoring skewness may adversely affect the performance of the traditional Shewhart control chart. Circumstances where non-normality may be critical to the proper functioning of Shewhart charts have been extensively studied and reported in the literature (refer to the previous references). This has motivated the development of various control schemes that take account of the non-normality of the monitoring statistic. Some recent examples for *variables* control charts are Bai and Choi (1995) and Grimshaw and Alt (1997).

Likewise, various solutions have been suggested for scenarios when the traditional Shewhart charts for *attributes* can not be implemented due to extreme skewness (of the distribution of the statistic used to monitor the process). Examples are the arc-sin transformation, applied to binomial data, the use of the "Q chart" for monitoring binomial and Poisson parameters (Quesenberry, 1991ab), or the use of the "g-chart" and the "h-chart" based on the geometric distribution (Kaminsky *et al.*, 1992; Benneyan, 2001ab). A good summary review of these efforts may be found in Woodall (1997) and in Ryan and Schwertman (1997).

While these methods take account of the departure of the monitoring statistic from normality, the allied control procedures often depart considerably from the practices exercised within the framework of Shewhart-type charts. Furthermore, the actual skewness of the monitoring statistic (as measured by the third standardized moment) rarely enters in the calculation of the control limits, neither is it explicitly preserved in the approximating distribution used to determine the control limits. For a monitoring statistic that is highly skewed, absence of the latter property may adversely affect the performance of the control chart.

These reasons have motivated the development of the new methodology to construct general control charts for attributes data. As the reader may already realize, the "general" here refers to the fact that normality, or approximate normality, is no longer needed. The skewness of the monitoring statistic, ignored by the Shewhart charts, is accounted for. As with all of Shewhart attributes control charts, it is assumed that the moments of the underlying distribution are known, once the parameters are estimated. For example, for the classical p chart, once we have a sample-estimate for p, the mean and the variance may be calculated, using expressions from the binomial model. These are then used to construct the control limits. Analogously, it is assumed here that once estimates of the parameters are available, one can calculate the mean and the variance, but also *the skewness measure*. This implies that no direct sample estimate of skewness is needed.

Extension of the methodology introduced here to variables data has been attempted (Shore, 1991). However, for variables data a common assumption is that the underlying distribution is unknown. Therefore all moments need to be *estimated* directly from sample data (instead of being *calculated*). Since estimating skewness is problematic (we have referred to this point in Chapter 11 and again in Section 20.3), better approaches are needed that avoid the large mean-squared-errors associated with sample estimates of skewness. These are developed, using inverse normalizing transformations, in the next Section 22.3.

Relative to current approaches, and in particular relative to Shewhart attributes control charts, the new methodology has three desirable features:

A. Far-tail quantile values used to determine probability limits, control limits, or other performance measures (like the average run length, ARL), are derived from a fitted distribution that preserves all first three moments of the plotted statistic. This property ensures that quantile values derived from the fitted distribution are highly accurate. Since for all of Shewhart attributes control charts a known probability model is assumed in developing the control limits, the new approach extends Shewhart charts by requiring that not only the mean and the standard deviation be specified but also the skewness measure. This poses no difficulty, given the known underlying probability model.

B. The control limits are expressed *explicitly* in terms of the mean, the standard deviation and the skewness measure. For example, given a binomial model with a known parameter p (or its estimate), the control limits are calculated in terms of the related mean, standard deviation and the skewness measure. Simplicity in calculating the control limits is achieved.

C. Quantile values needed for calculating probability limits are standard normal quantiles. This implies that although we take account of the skewness of the underlying distribution in the construction of probability limits, we still remain in the standard normal domain. Calculation of performance measures is thus very much along the guidelines used for regular Shewhart charts. Note, however, that quantiles of the standardized logistic distribution may be used, instead of standard normal quantiles. This may occasionally deliver better performance, relative to control charts based on the latter.

In the next Sub-section 22.2.2 the new (modified) control charts for attributes are developed, based on the piece-wise linear approximations of Chapter 21. In Sub-section 22.2.3 we simplify the expressions for the new control limits, and in Sub-section 22.2.4 "Skewness inflation factor" is introduced into the probability (control) limits in order to enhance their accuracy (relative to exact limits derivable from the exact distribution of the monitoring statistic). Sub-section 22.2.5 introduces explicit expressions for the new probability limits for some commonly encountered distributions (including those for which there currently exist Shewhart charts). In Sub-section 22.2.6 we evaluate the accuracy of the new control limits relative to the exact limits (calculated from the actual distribution) and relative to Shewhart charts.

22.2.2. Modified control limits for attributes

Let Y be a measured attribute with known mean (μ), standard deviation (σ) and skewness (Sk) (the latter is the traditional measure of skewness, namely, Sk$= \mu_3/\sigma^3$, and μ_3 is the third central moment). Denote by q_1 and q_2 (q_1, $q_2 < 1/2$) the required False Alarm Rates (FARs) with respect to the lower probability limit (LPL) and to the upper probability limit (UPL), respectively. This implies that for a stable process we require that

$$\text{LTA} = \Pr(Y<\text{LPL}) \leq q_1, \ \text{UTA} = \Pr(Y>\text{LPL}) \leq q_2, \qquad (22.1)$$

where LTA and UTA are the lower and the upper tail areas, respectively.

In defining these probabilities, a convention pursued by both Ryan and Schwertman (1997) and Quesenberry (1991b) is followed, according to which out-of-control probabilities represent the events of a point falling outside the control limits, not on or inside these limits. Accordingly, for any approximating quantile function used to construct probability limits, a unit should be added in calculating the *lower* probability limit.

Denote the P-th quantile of the standard normal variable by z_p.

To construct probability limits that achieve the required FARs, q_1 and q_2, we use the quantile function in (21.5), with parameters provided by (21.7) and (21.8). Given the discrete nature of Y and the definition of q_1, a "-1/2" should be added to both probability limits *and* a unit (1) should be added to the LPL (as alluded to earlier).

Accordingly, and given the specified μ, σ and Sk, the expressions for the probability limits of a general attribute control chart are

$$\text{UPL: } \mu + z_{1\text{-}q2}\,(A+C)\sigma - 0.7978C\sigma - 1/2,$$

$$\text{CL: } \mu, \tag{22.2}$$

$$\text{LPL: } \mu + z_{q1}\,(A\text{-}C)\sigma - 0.7978C\sigma + 1/2,$$

where A and C are the solution to (Case 1 in Section 21.3)

$$A^2 = 1\text{-}0.3635C^2, \quad Sk = 2.3940C - 0.6523C^3. \tag{22.3}$$

For the second expression in (22.3), extreme values are obtained at $C=\pm1.1061$, where Sk= ±1.76. This implies that a solution for {A, C} may be obtained in the range Sk= {-1.76, 1.76}. For most current Shewhart attributes charts this range is sufficient. If we wish to extend the applicability of the new control limits to cases with higher skewness (like the geometric distribution, with a skewness value near 2), we may use the standardized logistic quantile (Case 2 in Section 21.3)

$$z_p = (\sqrt{3} / \pi)\,\log[P/(1\text{-}P)], \tag{22.4}$$

where P is the value of the CDF. Relative to the normal variable, the logistic has longer tails. Therefore, its quantiles may be more appropriate in constructing control limits for highly skewed distributions.

Probability limits corresponding to (22.2) and (22.3) may easily be developed. The resulting expressions are

$$\text{UPL: } \mu + z_{1\text{-}q2}\,(A+C)\sigma - 0.7642C\sigma - 1/2,$$

$$\text{CL: } \mu, \tag{22.5}$$

$$\text{LPL: } \mu + z_{q1}\,(A\text{-}C)\sigma - 0.7642C\sigma + 1/2,$$

where A and C are the solution to [Eq. (21.13)]

$$A^2 = 1\text{-}0.4160C^2, \quad Sk = 3.1468C - 0.8959C^3. \tag{22.6}$$

For the second expression in (22.6), extreme values are obtained at $C=\pm1.0820$, where Sk= ±2.27. This implies that a solution for {A, C} may be obtained in the range Sk= {-2.27, 2.27}. This is a wider range than that associated with the normal.

Another benefit of using the logistic is that, relative to the normal, the approximate simplified expressions for A and C (developed in Section 21.4 and used in Sub-section 22.2.3) better preserve the skewness of the underlying distribution [compare expressions for Sk in (21.20), normal, with (21.21), logistic].

Employing traditional FARs associated with Shewhart charts, namely, $q_1 = q_2 = 0.00135$, we obtain from (22.2) for the *control* limits, based on standard normal quantiles (introduce $z = \pm 3$ for z):

$$\text{UCL: } \mu + 3\,(A{+}C)\sigma - 0.7978C\sigma - 1/2,$$

$$\text{CL: } \mu \tag{22.7}$$

$$\text{LCL: } \mu - 3\,(A{-}C)\sigma - 0.7978C\sigma + 1/2,$$

where A and C are determined as before (22.3).

For the logistic, the corresponding equations are (introduce $z = \pm 3.642$ for z):

$$\text{UCL: } \mu + 3.642\,(A{+}C)\sigma - 0.7642C\sigma - 1/2,$$

$$\text{CL: } \mu \tag{22.8}$$

$$\text{LCL: } \mu - 3.642\,(A{-}C)\sigma - 0.7642C\sigma + 1/2,$$

where A and C are determined as before (22.6).

Simplified limits will be introduced in the next Sub-section 22.2.3. We will confine ourselves to the normal case, which renders the new approach a natural extension of the existing Shewhart-charts. Also, the normal case will be more suitable for attributes that are approximately normally distributed, like the binomial and the Poisson. For very highly skewed distributions, or distributions that do not converge to the normal (like the geometric), using control limits based on logistic quantiles may prove to be more effective.

22.2.3. Simplified limits

The need to solve numerically for A and C [refer to (22.3)] may be cumbersome. Simpler limits can be derived (for either probability limits or control limits) which circumvent this difficulty. Also the simplified expressions allow expressing the limits explicitly in terms of the mean, the standard deviation and the skewness measure.

The simplified probability limits are based on the piece-wise linear approximations, developed in Section 21.4. Therein an approximation was developed for the quantile relationship between Z and the response, Y, expressed explicitly in terms of the mean, the variance, and the skewness measure. However, the latter is only approximately preserved. The reader may wish to consult Section 21.4 for the underlying theoretical basis of the expressions that appear in the formulae for the probability limits.

The resulting *probability* limits are

UPL: $\mu + z_{1-q2}\sigma\,[(1- 0.06342Sk^2)^{1/2} + 0.4177Sk] - (1/3)\sigma Sk - 1/2,$

CL: $\mu,$ (22.9)

LPL: $\mu + z_{q1}\sigma\,[(1- 0.06342Sk^2)^{1/2} - 0.4177Sk] - (1/3)\sigma Sk + 1/2.$

For $-1\leq Sk\leq1$, the skewness of the approximate quantile function, used to construct the probability limits in (22.9), does not deviate more than ±0.0475 from the exact value [refer to (21.20)]. Given that the underlying monitoring statistic is an attribute (and therefore assumes only integer values), this deviation is of little consequence.

Introducing $z=\pm3$ into (22.9), the *control* limits are

UCL: $\mu + 3\sigma(1- 0.06342Sk^2)^{1/2} + 0.9198\sigma Sk - 1/2,$

CL: $\mu,$ (22.10)

LCL: $\mu - 3\sigma(1- 0.06342Sk^2)^{1/2} + 0.9198\sigma Sk + 1/2.$

A further simplification may be introduced by observing that the square-root in (22.9) and (22.10) [the parameter A in (22.2)], is very close to 1 (for non-extreme values of Sk). Substituting this value for the square-root in (22.9), we obtain simplified probability limits:

UPL: $\mu + z_{1-q2}\sigma + \sigma Sk\,(0.4177z_{1-q2} - 1/3) - 1/2,$

CL: $\mu,$ (22.11)

LPL: $\mu + z_{q1}\sigma - \sigma Sk\,(0.4177z_{q1} + 1/3) + 1/2.$

Similarly, for the control limits:

$$\text{UCL: } \mu + 3\sigma + 0.9198\sigma Sk - 1/2,$$

$$\text{CL: } \mu, \tag{22.12}$$

$$\text{LCL: } \mu - 3\sigma + 0.9198\sigma Sk + 1/2.$$

Unlike equations (22.9) and (22.10), derived from a quantile function that preserves exactly the mean and the standard deviation of the monitoring statistic, the last two equations are based on a quantile function that preserves only the mean, while the standard deviation and the skewness are only approximately preserved. For $Sk \leq 1$, the standard deviation of the approximate quantile function, used to compute the probability limits and the control limits in (22.11) and (22.12), respectively, does not deviate more than 3.1% from the exact value (for explanation of this result, refer to Chapter 21, where the piece-wise linear approximations, on which these control limits are based, are developed in detail).

Since attributes control limits relate to an integer monitoring statistic, this is again of no meaningful consequence to the effectiveness of the control chart.

22.2.4. *Probability limits with "inflated" skewness*

Having derived the algebraic structure of the simplified probability limits, an attempt is now made to enhance the performance of these limits by adjusting the values of the coefficients. This adjustment is intended to improve the accuracy of probability limits associated with far-tail quantile values, like those of control limits.

As a departure point to develop the adjusting strategy, note that the general expressions for the probability limits, developed earlier, are based on a quantile function that is a linear transformation of the standard normal quantile (different transformations are used for $z_p < 0$ and for $z_p > 0$). It is to be expected that while this quantile function may provide good approximation for non-extreme quantile values, it will tend to lag behind the true values for far-tail probabilities, like those associated with control limits. Since higher skewness will increase the values derived from the quantile function [relate to the equations for the probability limits, developed earlier, like (22.12)], the above "mal-behavior" for extreme quantile values may be corrected if instead of the actual skewness an inflated value be introduced in the expressions for the probability limits. In other words, Sk in (22.11) or (22.12) will be replaced by (L_i)Sk (i=1 for the LPL and i=2 for the UPL), where L_i will be determined by a least-squares fitting routine, using far-tail quantile values taken from some common attributes distributions (specifically, the binomial, the Poisson, the geometric and the negative-binomial distributions). Using exact quantile values associated with tail-areas near the nominal 0.00135, the values of $\{L_i\}$ obtained exhibit a high degree of consistency, irrespective of the underlying attribute distribution. Furthermore, nearly equal values are obtained for L_1 and for L_2. Some values are shown in Table 22.1 (for definitions of parameters refer to Sub-section 22.2.5).

Given the close proximity between the values of L_1 and L_2, it is decided to set equal values for both coefficients at $L_1= L_2= 1.44$.

Accordingly, the final probability limits, after skewness inflation, are

$$UPL: \mu + z_{1-q2}\sigma + (1.44)\sigma Sk\,(0.4177z_{1-q2} - 1/3) - 1/2,$$

$$CL: \mu, \tag{22.13}$$

$$LPL: \mu + z_{q1}\sigma - (1.44)\sigma Sk\,(0.4177z_{q1} + 1/3) + 1/2.$$

Similarly, for the control limits, after skewness inflation, we obtain

$$UCL: \mu + 3\sigma + 1.324\sigma Sk\,-1/2$$

$$CL: \mu \tag{22.14}$$

$$LCL: \mu - 3\sigma + 1.324\sigma Sk\,+1/2$$

To assess the performance of these control limits, the associated ARLs need to be calculated. (ARL is a commonly used measure for the sensitivity of the control chart to the existence of out-of-control states.) This will be done in Sub-section 22.2.6, where the relative effectiveness of the new control limits is assessed.

Table 22.1. Some values of L_1 (for the LPL) and of L_2 (for the UPL)

Distribution	L_1	L_2
Poisson [λ=5(1)50]	1.4043	1.3438
Binomial [n=100(100)3000]	1.5109	1.3662
p=0.01		
p=0.1	1.5000	1.4039
p=0.3	1.5259	1.3690
Geometric	---	1.4206
[p=0.005,0.01,0.05,0.10,0.20]		
Negative-Binomial	1.3949	1.4568
[(k,p)={(10,.01),(50,.01), (100,.01),(3,.10),(50,.10)}]		

22.2.5. *Probability limits for some attribute distributions*

Applying the general expressions developed in the previous sub-section, Shewhart-like control limits may be constructed for some specified distributions. Modified control limits will be developed for the traditional np and c control charts, and new limits will be developed for attributes with geometric and negative-binomial distributions. For the latter, the probability functions, used in deriving the moments and the probability limits, are, respectively,

$$f(y) = p(1-p)^y , \ y = 0,1,2,..,$$

$$f(y) = \binom{y+k-1}{y} p^k (1-p)^y , \ y = 0,1,..., \qquad (22.15)$$

Note, that the negative-binomial random variable describes the numb er of "Failures" (Y) before the occurrence of the k-th "Success" in repeatedly performed independent Bernoulli trials (trials that result in "0" or "1"), with success probability p.

The results are summarized in Table 22.2. Since the probability function of the geometric distribution is a monotone decreasing function of y, there is no meaning to defining an LPL for this distribution. However, a UPL obviously makes sense since it will enable detecting improbable high values of Y (the monitoring statistic). Note that although probability (control) limits, based on *normal* quantiles, deliver

Table 22.2. Probability limits for the binomial, Poisson, geometric and negative-binomial.

Distribution	Moments			Probability Limits minus center-line	
	μ	σ^2	Sk	LPL - μ	UPL - μ
Binomial	np	$np(1-p)$	$(1-2p)/\sigma$	$z_{q1}[np(1-p)]^{1/2} - 1.44(1-2p)$ $\times (0.4177z_{q1} + 1/3) + 1/2$	$z_{1-q2}[np(1-p)]^{1/2} +1.44(1-2p)$ $\times (0.4177z_{1-q2} - 1/3) - 1/2$
Poisson	λ	λ	$1/\lambda^{1/2}$	$z_{q1}(\lambda)^{1/2} - 1.44$ $\times (0.4177z_{q1} + 1/3) + 1/2$	$z_{1-q2}(\lambda)^{1/2} + 1.44$ $\times (0.4177z_{1-q2} - 1/3) - 1/2$
Geometric	$(1-p)/p$	$(1-p)/p^2$	$(2-p)/(1-p)^{1/2}$	--	$z_{1-q2}[(1-p)^{1/2}/p]$ $+ 1.44 [(2-p)/p]$ $\times (0.4177z_{1-q2} - 1/3) - 1/2$
Negative Binomial	$k(1-p)/p$	$k(1-p)/p^2$	$(2-p)/[k(1-p)]^{1/2}$	$z_{q1}[k(1-p)/p^2]^{1/2} -1.44[(2-p)/p]$ $\times (0.4177z_{q1}+1/3) + 1/2$	$z_{1-q2}[k(1-p)/p^2]^{1/2} +1.44[(2-p)/p]$ $\times (0.4177z_{1-q2} - 1/3) - 1/2$

good performance for the upper limit of the geometric distribution, due to the high skewness of the latter (around 2), a preferable choice would probably be the use of control charts based on *logistic* quantiles.

Examining Table 22.2, the modified control limits seem to retain algebraic simplicity comparable to that of Shewhart charts. In the next sub-section the relative effectiveness of the new control charts is assessed.

22.2.6. Numerical assessment

In evaluating the effectiveness of the modified control limits, we use an optimality criterion, first introduced by Ryan and Schwertman (1997). The latter employed this criterion in order to formulate guide-lines for rounding fractional control limits obtained for attributes control charts. They suggest minimization of the following loss function (LF):

$$LF = |1/LTA - 1/0.00135| + |1/UTA - 1/0.00135|. \qquad (22.16)$$

This function expresses the sum of the absolute values of the deviations from the nominal value of the ARLs, associated with the control limits in a state-of-control (namely, 740.7). The ARL is the average number of samples taken from the time the process switches to an out-of-control state until a signal is indicated by the control chart (there is a point outside the control limits). For a process in control, the ARL should be equal to the reciprocal of the nominal FAR, namely, ARL=1/FAR=1/0.00135=740.7. The actual ARL will be the reciprocal of LTA, for the lower limit, and the reciprocal of UTA, for the upper limit. Due to the integer nature of the monitored attribute, these ARLs would not necessarily be equal to the nominal values, so the question of how to round fractional control limits remains open. The LF criterion is one solution to this question.

Given the exact underlying distribution, the exact (optimal) integer control limits may be identified, which minimize (22.16). In other words, integer control limits are selected that provide ARLs closest to the nominal value of 740.7.

Regarding the new control charts, their relative effectiveness may be assessed by comparing their control limits to the exact (optimal) control limits , derived from (22.16), and to Shewhart traditional control limits (for the latter, a continuity correction will be added, as done with the modified control limits).

— To obtain optimal control limits for the new control charts, the following procedure is suggested (the same procedure should be followed for the UCL and for the LCL):

- Round the fractional control limit, derived from (22.14), up and down to the nearest integer.

- Introduce the two consecutive integer values into (22.13) to find the corresponding approximate z_p values. Note that to derive z from the expression for the LPL, the +1/2 in (22.13) should be replaced by -1/2 (refer to the definition of the LPL in (22.1) and the comment that follows this equation).

- From a table of the standard normal CDF, identify P values (CDF values), associated with the calculated z_p. P represents the approximate LTA, for the LCL, and (1-P) represents the approximate UTA, for the UCL. These are inserted into the optimality criterion (22.16) to determine the integer optimal control limit (minimization of LF is required).

Comparison may now be conducted with the exact optimal control limits [derived from (22.16), based on the exact LTA and UTA]. Some values, obtained from various distributions, are displayed in Table 22.3.

Table 22.3. Comparison of exact with approximate optimal CLs for the binomial, the Poisson, the geometric and the negative binomial distributions. The normal approximation is also given (with continuity correction) under "Shewhart"

Distr.:	Exact CL		Z (exact)		New CL		Shewhart CL	
	LCL	UCL	Z_L	Z_U	LCL	UCL	LCL	UCL
Binom. (n,p)=								
(100,.005)	0	4	0.2683	3.601	-0.4457	4.209	0.1893	2.540
(500,.005)	0	8	-1.394	3.062	0.1557	8.178	-0.1995	6.830
(100,.01)	0	5	-0.3424	3.272	-0.1092	5.213	0.1594	3.755
(500,.01)	0	13	-2.480	3.218	-0.0261	13.08	-1.017	11.66
(100,.05)	0	13	-2.517	3.312	-0.0549	13.08	-0.9854	11.72
(500,.05)	12	41	-2.793	3.134	11.97	41.04	10.89	39.77
(100,.10)	2	20	-2.887	3.153	1.844	20.09	0.8391	18.96
(500,.10)	31	71	-2.918	3.050	30.94	71.04	29.93	69.96
(100,.40)	26	55	-2.821	3.127	25.92	55.10	25.68	54.82
(500,.40)	167	233	-2.995	3.037	166.95	233.0	166.7	232.8
Pois. λ =								
5	1	12	-1.746	2.875	1.166	12.18	0.5964	10.93
6	1	14	-2.112	2.989	1.117	14.14	0.3271	12.82
7	1	16	-2.442	3.103	1.027	16.09	0.0380	14.71
10	3	20	-2.314	2.950	3.094	20.11	2.183	18.83
15	6	27	-2.426	2.926	6.083	27.11	5.104	25.83
25	12	41	-2.732	3.041	12.00	41.06	10.84	39.71
35	19	53	-2.830	2.925	18.98	53.08	17.76	51.81
45	27	66	-2.786	3.013	27.00	66.05	25.81	64.72
50	31	72	-2.784	3.001	31.01	72.05	29.82	70.72
Geom. p=								
0.005	-	1317	-	3.000	-	1325	-	796.9
0.01	-	656	-	2.998	-	660.2	-	396.9
0.05	-	128	-	3.003	-	128.8	-	77.03
0.10	-	62	-	3.009	-	62.32	-	37.05
0.20	-	29	-	3.027	-	29.10	-	17.03
Neg. Bin (k, p)=								
(10, .01)	302	2201	-2.996	3.000	309.9	2197	47.95	1933
(50, .01)	3102	7327	-2.998	2.999	3103	7324	2841	7060
(100, .01)	7179	13151	-2.998	3.000	7179	13148	6917	12884
(3, .10)	1	101	-2.678	2.995	3.979	100.8	-17.51	75.72
(50, .10)	274	676	-2.986	2.997	274.2	675.7	249.2	650.6

The new approach provides control limits appreciably better than those provided by the normal approximation (on which the Shewhart control charts rest). In particular, note the accurate values obtained for the geometric distribution (skewness value near 2!), notwithstanding the use of normal quantiles in calculating these values.

To further assess the new general control charts for attributes, we have examined a larger sample of parameter values for the binomial distribution (the p control chart). For the binomial with parameters p={0.005, 0.01, 0.05, 0.1, 0.2, 0.3, 0.4, 0.5} and n=100(100)3000, exact optimal control limits, according to the LF criterion (22.16), have been calculated. These are compared to values obtained from (22.14) (using simple rounding) and with the classical Shewhart control limits (with the added continuity correction, again using simple rounding).

The distribution of the deviations from the optimal limits is given in Table 22.4. The new control limits for the binomial outperform the traditional limits, especially for small p values.

Table 22.4. Distribution (in %) of deviations from exact optimal control limits. For each given p, 30 cases were examined [n=100(100)3000]. Lower entries refer to Shewhart charts.

p	Exact=Approx. (zero dev.)		Exact≠Approx. (dev. of ±1)		Exact≠Approx. (|dev.| >1)	
	LCL	UCL	LCL	UCL	LCL	UCL
0.005	53	87	47	13	0	0
	3	0	87	77	10	23
0.01	77	83	23	17	0	0
	10	0	87	83	3	17
0.05	90	100	10	0	0	0
	10	0	90	90	0	10
0.10	97	90	3	10	0	0
	10	0	90	100	0	0
0.20	93	100	7	0	0	0
	33	3	67	97	0	0
0.30	93	100	7	0	0	0
	40	50	60	50	0	0
0.40	100	97	0	3	0	0
	90	63	10	37	0	0
0.50	100	100	0	0	0	0
	100	100	0	0	0	0

22.3. General Control Schemes for Variables

22.3.1. Introduction

Traditional Shewhart control charts are based on the approximate normality of the process monitoring statistic. There is a vast literature that deals with the sensitivity of Shewhart control charts (and others) to deviations from the normality assumption (see related discussion in Section 22.2.1). However, with the introduction of real-time process monitoring, the averaging effect associated with periodic sampling of the process no longer exists. This implies that control limits are required to reflect the true process distribution, and the control schemes in use can no longer rely on the asymptotic normality of the plotted statistics.

For a process distribution that is extremely skewed, the performance of the Shewhart-based SPC scheme may be adversely affected by the violation of the normality assumption. Modification of the traditional (Shewhart) methods for process control is therefore essential to guarantee acceptable levels of performance.

In Section 22.2 we offered some solutions for attributes control charts, and assessed their effectiveness. In this section, the search for solutions is extended to control charts for variables, which need to operate in a non-normal environment.

The reader may wonder, under these circumstances, why not use distribution-identifying methods to "reveal" the true process distribution. Once this had been accomplished and proper hypothesis testing validated that indeed the true distribution was identified, control limits may be plotted that coincide with the end-points of the actual process distribution (however these "end-points" are defined). While this approach looks tempting, it is not practical. Quality practitioners require an effective methodology that would deliver good performance characteristics irrespective of whether we know the true underlying process distribution. In other words, they require empirical modeling of the process random variation.

RMM has been claimed throughout to provide an effective platform for modeling random variation. This claim was demonstrated extensively in the first three chapters of Part IV of the book (Chapters 19-21). In particular, inverse normalizing transformations (INTs), derivatives of the RMM error distribution that had been introduced in Chapter 20, were shown to provide effective and easy-to-implement general representation for diversely-shaped distributions. In Section 22.3 the INTs are used to develop general control schemes for variables.

In constructing control charts for normal or non-normal environments, there are two options to pursue. While these options are always available, the SPC literature, it seems, often fails to distinguish between them.

(I) Control limits for the end-points of the process distribution

With this approach, we wish to model the random variation (the statistical distribution) of the process variable of interest. The control limits, based on the model, would then reflect the end-points of this distribution. Actual process monitoring would attempt to establish that the process distribution has not changed by establishing that the probability of exceeding the end-points remains small.

Let us demonstrate by an example from attributes charts. In monitoring high-quality processes, where the occurrence of a non-conforming item is an extremely rare event, it is customary nowadays to monitor the number of "good" (conforming) items between two consecutive non-conforming items. The monitoring statistic is commonly assumed to have a geometric distribution. Control limits based on the geometric distribution reflect the end-points of this distribution, as these are calculated from far-tail quantile values (see examples for the application of the g-chart, based on the geometric distribution, in Benneyan, 2001ab).

However monitoring a process in real-time by establishing the end-points of its distribution is not always effective (as may be implied by the previous geometric example). In particular, monitoring the end-points may not be sufficient if the process distribution has more than one parameter (as the geometric does). Consider a process distribution which is normal. For this process distribution, one may define end-points and construct control limits based on these, for example, $\mu+3\sigma$, for the upper control limit (UCL), and $\mu-3\sigma$, for the lower control limit (LCL). The control chart would then display individual observations taken from the process, and a signal be indicated if a point fell outside the end-points of the process distribution.

It is easily recognized that this approach is not adequate for a normal process distribution. Concurrent changes in both the mean and the variance may fail to be detected even when the process is obviously non-stable. For example, an upward shift in the mean accompanied by a reduction in the variance may fail to trigger an alarm (an out-of-limits observation) because the probability of exceeding the UCL may still be extremely small (as prior to the change). In other words, the actual upper end-point of the process distribution may remain the same even though the process distribution has changed (that is, the mean and the standard deviation have changed). Consequently, for a normal process distribution, monitoring the end-points of the process distribution may not be sufficient. The variance also needs to be monitored, as indeed we do in implementing a Shewhart control chart for individual observations.

To sum up, if a process distribution depends on more than one parameter, it would be more desirable to monitor the process via monitoring the stability of all the *parameters* of the process distribution, rather than via monitoring the end-points, as done with the geometric g-chart.

(II) Control limits for the parameters of the process distribution

This is the most common approach in SPC. It is assumed that the process distribution remains stable if its *parameters* are. Shewhart control schemes are of this sort. When the process mean and variance (or range) are being monitored via Shewhart charts, it is the process normal distribution which is monitored, since the latter depends exclusively on the mean and the variance. In fact, for any two-parameter distribution (either symmetric or otherwise), monitoring the mean and the standard deviation is sufficient to warrant a stable distribution.

To demonstrate this, consider a gamma process distribution, having parameters $\{\alpha, \beta\}$. The first four moments are

$$\mu = \alpha\beta, \ \sigma^2 = \alpha\beta^2, \ Sk = 2/\alpha^{1/2}, \ Ku = 6/\alpha.$$

For this distribution, stable mean and variance imply a stable distribution because the first two moments uniquely determine the parameters of the distribution. The problem in implementation is that using sample estimates of the mean and the variance to monitor the stability of the respective parameters do not necessarily allow use of traditional Shewhart charts. For example, for the above gamma distribution, it is known (Stuart and Ord, 1987, p. 362) that the distribution of the sample average, based on n observations, is also gamma with the first four moments

$$\mu = \alpha\beta, \ \sigma^2 = \alpha\beta^2/n, \ Sk = 2/(\alpha n)^{1/2}, \ Ku = 6/(\alpha n).$$

We realize that for small n, the skewness of the distribution of the sample average may still be too high to permit using a Shewhart chart for averages. A method for accommodating distributions of non-normal averages is needed. It is here that the INTs can be helpful.

22.3.2. INT-based control schemes for variables

The INTs, introduced in Chapter 20, may be used to construct general control charts for variables, irrespective of which of the two approaches, outlined in the previous subsection, we select to pursue.

If a control chart based on monitoring the "end-points" of the process distribution is desired, one can fit, based on the available data, an INT, and then calculate control limits based on the end-points of the estimated INT. Actual process monitoring would then be conducted by plotting individual observations, and detecting instability conveyed by observations beyond the control limits.

Using the second approach, we would decide what parameters uniquely define the INT, and then construct control limits for monitoring these parameters. An underlying principle here is that parameters that appear in any of the distributions of the statistics, used to monitor the INT parameters, are monitored too. This principle is implemented also when the traditional Shewhart charts are used. For example, in monitoring the mean of the process distribution via the sample average, we *also* monitor the variance because this is one of the parameters that determine the distribution of the sample average (the monitoring statistic).

If a certain INT is used to represent a process distribution, monitoring the parameters of this INT implies monitoring the parameters of the process distribution that took part in fitting the INT. To realize that, suppose that a certain fitting procedure for the selected INT requires specification of the response median, the response mean and the response variance (relate to the fitting procedures delineated in Section 20.5). This implies that in order to ensure the stability of the INT (and, by implication, the stability of the process distribution, represented by the INT), only these response parameters need to be monitored.

Two SPC schemes are suggested here. The schemes differ according to the INT used, and according to the allied fitting procedure. The latter determines which moments (or other estimable parameters) of the response distribution need to be monitored. Data requirements are also determined by the relevant fitting procedure.

The schemes displayed are not the only ones that one can conceive of, based on the INTs. Other schemes, perhaps with better performance properties, may be constructed. The following schemes should therefore be regarded just as demonstration of how the INTs may be used in the implementation of SPC to non-normal process distributions.

Scheme I. Data requirements- Estimates of M, μ and $\mu(W)$

Let Y be a process variable to be monitored, and let W be the log-transformed response, namely, $W=\log(Y)$. In Chapter 20 we have introduced a three parameter INT, re-introduced here for convenience:

$$W = \log(Y) = \log(M) + ABZ / (1+AZ), \qquad (22.17)$$

where M is the median and Z is a standard normal variable. Developing the right-hand-side in terms of Z into a Taylor series and taking expectation of the first five terms we obtain

$$\mu(W) = \log(M) + B(-A^2-3A^4). \qquad (22.18)$$

Introducing for B back into (22.17), W is expressed in terms of the median, the mean, $\mu(W)$, and the parameter "A":

$$W = \log(M) + [\mu(W)- \log(M)]AZ / [(-A^2-3A^4)(1+AZ)]. \quad (22.19)$$

The parameter "A" may now be determined by matching the response mean (μ) with $E[\exp(W)]$, as the latter is implied from (22.19).

From this matching procedure it is apparent that INT (22.17) is uniquely determined by M, $\mu(W)$ and μ. Estimates of these parameters would be used as the monitoring statistics to ensure the stability of the process distribution. The respective control scheme is referred to as Scheme I.

Alternatively, develop W into a Taylor expansion around μ and take expectation of the first three terms to obtain (the second term vanishes)

$$\mu(W) \cong \log(\mu) + (1/2)(\sigma/\mu)^2. \qquad (22.20)$$

Monitoring $\mu(W)$ may therefore be replaced by monitoring the mean and the variance of the process variable, Y. In other words, if (22.17) is a good model for the log-transformed process variable, $W=\log(Y)$, then monitoring the median, the mean and the variance of Y (the response on the original scale) would ensure that the process distribution is stable.

The implication of this result is that for skewed distributions, where the median is different from the mean, the traditional Shewhart scheme for variables, which comprises monitoring the mean and the variance, should be complemented by monitoring the response median. However, if the mean and the median coincide (the process distribution is symmetrical), the traditional Shewhart scheme is revoked (monitoring of the mean coincides with monitoring of the median).

A control scheme based on monitoring M, μ and σ (with an INT different from (22.17)], is given later in this section (refer to Scheme II).

The following procedure, related to Scheme I, is based on the assumption that INT (22.17) is fitted via M, μ and $\mu(W)$. Therefore these are the parameters that need to be monitored. It is also assumed that estimates of these parameters have distributions that can be adequately described by INT (22.17). The procedure below is divided into two parts:

- Phase 1: Fitting (estimating) the INTs and calculating control limits
- Phase 2: Implementing the control scheme.

1. Estimating the INTs and calculating the respective control limits

- Draw K independent samples of n observations each. Calculate for the j-th sample (j=1,2,..,K) the median, M_j, the mean, μ_j, and the mean of the log of the original observations, $\mu_j(W)$.
- Given the K values $\{M_j\}$, estimate their median, their mean and the mean of the log-transformed sample-medians. Repeat the same calculations for the sample-mean [based on $\{\mu_j\}$] and for the sample-mean of the log transformed observations [based on $\{\mu_j(W)\}$].
- Fit INT (22.17) for the distributions of the three statistics used to monitor the process (namely, the sample median, the sample mean and the sample mean of W). In the fitting procedures, use sample estimates derived from the preliminary set of K samples (as explained in the previous paragraph).
- From the three estimated quantile functions, find control limits to monitor M, μ and $\mu(W)$. (Introduce in the fitted INT, z=3 to obtain the UCL, and z=-3 to obtain the LCL.)

2. Applying the control charts

- Draw a sample of n observations and calculate the median, the average and the average of the log-transformed observations. Plot the results in the respective control charts.
- If any point exceeds the control limits- a change has occurred in the underlying process distribution. The nature and the magnitude of the change need to be investigated.

Comment. Because the underlying process distribution is unknown, there is no way of telling, in advance, to what degree the statistics used are correlated. It is a good idea to calculate these correlations. If some high correlations are observed between two statistics, one control chart or the other may be superfluous and should be discarded. Alternatively, the use of a multivariate statistic, like Hotelling's T^2, may be considered.

Note, that the problem of correlations, which can raise appreciably the actual false alarm rates, does not exist with Shewhart-based schemes. The reason for that is that under normality estimates of the mean and the variance are independent (uncorrelated).

The issue of possible correlations between statistics used for SPC in a non-normal environment needs further study.

A Numerical Example

Suppose that the process response follows a gamma distribution. As demonstrated earlier, the sample average also has a gamma distribution. Scheme I calls for monitoring the median and the two means of the response (on the original scale and on the log-transformed scale). In this example, we investigate the properties of the control chart for the sample average, assuming that the underlying true distribution is gamma. We use (22.17) as a model for the distribution of the sample average.

Since the latter is distributed as gamma, the median, the mean and the mean of the log-transformed sample average are all known. We may therefore fit INT (22.17), based on the exact parameters, and calculate thereof the control limits for the sample average (the 0.135 and the 99.865 percentiles of the fitted INT). As noted earlier, we do not need to monitor σ since, using (22.17), the standard deviation does not participate in determining the parameters' values for the fitted INT.

Table 22.5 presents four sets of control limits for the average-control-chart:

Set A: The exact limits, derived from the exact distribution of the sample-average (a gamma distribution, as expounded earlier).
Set B: Approximate limits derived from fitting the INT, given by (22.17), using in the fitting routine the exact median, mean and the mean of the log of the sample-average.
Set C: Shewhart limits for the X control chart, using the exact mean and standard deviation.
Set D: Shewhart limits for the X control chart, using the standard deviation calculated numerically from the fitted INT (22.17)

We realize that the first two sets of control limits are nearly identical, and so are the last two. The proximity in the values of the control limits

                                     RESPONSE MODELING METHODOLOGY

Table 22.5. Four sets of control limits for an average-control-chart (Sample size n=5).

Dis.: Gamma	Set A (exact)		Set B (Eq. 22.17)		Set C (Shewhart, ex. σ)		Set D (Shewhart, ap. σ)	
	LCL	UCL	LCL	UCL	LCL	UCL	LCL	UCL
$\alpha= 7, \beta=2$	7.95	22.2	7.94	22.2	6.90	21.1	6.88	21.1
$\alpha= 3, \beta=2$	2.39	11.7	2.39	11.7	1.35	10.6	1.35	10.6
$\alpha= 2, \beta=2$	1.23	8.87	1.23	8.87	.205	7.79	.204	7.80
$\alpha= 1, \beta=2$	.317	5.76	.314	5.76	-.683	4.68	-.683	4.68
$\alpha= 0.8, \beta=2$	.186	5.07	.184	5.07	-.800	4.00	-.799	4.00
$\alpha= 0.4, \beta=2$	.0212	3.56	.0192	3.56	-.897	2.50	-.896	2.50
$\alpha= 0.2, \beta=2$	.0005	2.64	.0003	2.66	-.800	1.60	-.801	1.60

corroborates our assertion that using Scheme I provides all the required information about the process distribution, and no monitoring of variability is needed.

To assess the performance of the X chart, where the limits are determined based on the specified median, mean and mean of the log of the sample-average (corresponding to Set B in Table 22.5), the parameter "α" of the gamma process distribution is changed to $(1+\delta)\alpha$, where

$$\delta = \{-50\%, -20\%, -10\%, 0\%, 10\%, 20\%, 50\%\}.$$

Each change, δ, incurs a change in the process mean from μ to $(1+\delta)\mu$ (refer to the expressions for the moments of the gamma distribution given earlier). Accordingly, *exact* ARLs associated with deviations in the mean are calculated for the selected gamma distributions.

Additionally, *approximate* ARLs are derived by re-fitting (22.17) (using the changed process-parameter), and thence identifying values of z associated with current (pre-change) control limits. From the z values the respective approximate ARLs may be computed. Finally, exact ARLs associated with traditional Shewhart X control chart are computed for comparison purposes.

The results are given in Table 22.6. Note, that for $\alpha \leq 1$, Shewhart lower control limit is negative. Therefore occasionally no LCL values are given for these values of α.

Examining Table 22.6, a few observations are noteworthy.

Table 22.6. ARLs associated with different deviations (δ) in the process mean, when control limits used are Set B (refer to Table 22.5). "Ex."- refers to the exact ARLs; "Ap."- refers to the approximate ARLs (calculated from the fitted INT, using post-change parameters' values); "Shew."- refers to Shewhart X chart, where control limits are defined by Set C (refer to Table 22.5). The ARLs are the exact values

Distribution Gamma ($\beta=2$)	Control limits:	-50%	-20%	-10%	0%		10%	20%	50%
			LCL		LCL	UCL		UCL	
$\alpha=7$	Ex.:	1.37	20.6	106	764	756	146	38.4	3.06
	Ap.:	1.37	20.4	104	741	741	144	37.9	3.06
	Shew.:	1.97	94.8	764	8765	255	58.0	17.7	2.14
$\alpha=3$	Ex.:	3.16	51.6	183	752	745	256	100	12.0
	Ap.:	3.16	51.4	182	741	741	254	99.8	11.9
	Shew.:	27.7	4707	3.7E4	3.5E5	179	70.0	31.2	5.35
$\alpha=2$	Ex.:	5.05	73.0	223	752	742	313	144	22.6
	Ap.:	5.05	72.8	222	741	741	312	144	22.5
	Shew.:	5157	1.3E7	2.3E8	4.6E9	149	70.4	36.4	7.92
$\alpha=1$	Ex.:	10.5	117	292	766	739	407	235	58.2
	Ap.:	10.5	116	286	741	741	407	235	58.2
	Shew.:	--	--	--	--	107	64.8	41.0	13.2
$\alpha=8$	Ex.:	12.8	132	315	778	739	435	267	75.6
	Ap.:	12.9	130	305	741	741	436	267	75.5
	Shew.:	--	--	--	--	96.8	62.0	41.3	14.9
$\alpha=4$	Ex.:	21.3	189	407	892	741	516	367	148
	Ap.:	21.3	171	353	741	741	516	367	148
	Shew.:	--	--	--	--	71.0	52.5	39.7	19.2
$\alpha=2$	Ex.:	34.9	332	715	1549	783	608	479	251
	Ap.:	25.6	209	391	741	741	585	466	248
	Shew.:	--	--	--	--	54.6	44.3	36.4	21.7

First, the ARLs computed from the fitted INT are nearly identical to those calculated from the *exact* post-change average-distribution. This implies that if the underlying distribution is not known, the ARLs may still be calculated if good estimates for the distributional parameters, needed to fit (22.17), are available.

A second observation refers to the mal-performance of the traditional Shewhart chart *vis-a-vis* false alarm rates. The ARLs in a state-of-control deviate considerably from the standard (nominal) value of 740.7 (comparable results are reported in Borror *et al.*, 1999). On the other hand, SET B control limits, calculated with no knowledge of the process variance, deliver *exact* ARLs that are close to the desired standard value.

A third observation refers to the inability of the Shewhart chart to detect downward shifts in the mean for $\alpha \leq 1$. Conversely, Set B control limits provide non-negative control limits that may be very instrumental in detecting large downward shifts (observe the small ARLs for large negative δ).

Finally, given the numerous publications which advocate the general applicability of the X chart, due to its robustness to underlying non-normality (refer for some references to Shore, 2000b), Table 22.6 clearly shows that with the new approach the performance of a control chart for the mean can be appreciably improved.

As emphasized earlier, Scheme I is introduced only for demonstrative purposes. No claim is extended that this is the best scheme that the new approach may produce. Rather, we wished to show that control charts, based on INTs, may deliver performance that is superior to traditional Shewhart charts. In particular, the skewness of the process distribution is well taken care of.

Scheme II. *Data requirements- Estimates of M, μ and σ*

In Chapter 20, the following four-parameter INT was developed, and it is re-introduced here for convenience:

$$W = \log(Y) = \log(M) + B[\exp(Cz)-1] + Dz, \qquad (22.21)$$

where M is the median. This INT has been shown to be particularly effective in representing quantile functions of both continuous and discrete random variables (Shore, 2000b, 2001).

Taking expectation of (22.21) we obtain (find details in Chapter 20)

$$\mu(W) = \log(M) + B[\exp(C^2/2)-1]. \qquad (22.22)$$

Introducing for B from (22.22) into (22.21), and assuming that M is specified, we obtain a quantile function in two parameters, C and D. The latter may now be determined by matching the means and the variances with those of the response variable, Y.

Recalling that monitoring $\mu(W)$ is approximately equivalent to monitoring μ and σ^2 (22.20), one learns that monitoring the median, the mean and the variance would guarantee the stability of the process distribution, as the latter is modeled by INT (22.21).

We leave it as an open exercise to construct a control scheme, based on (22.21), and study its properties for an assumed gamma distribution, similarly to the analysis conducted for Scheme I above.

References

[1] Bai, D. S., Choi, I. S. (1995). X and R control charts for skewed populations. *Journal of Quality Technology*, 27, 121-131.

[2] Balakrishnan, N., Kocherlakota, S. (1986). Effects of non-normality on X charts: Single assignable cause model. *Sankya*, B 48, 439-444.

[3] Benneyan, J. C. (2001a). Number-between g-type statistical quality control charts for monitoring adverse events. *Health Care Management Science*, 4, 305-318.

[4] Benneyan, J. C. (2001b). Performance of Number between g-type statistical quality control charts for monitoring adverse events. *Health Care Management Science*, 4, 319-336.

[5] Borror, C. M., Montgomery, D. C., Runger, G. C. (1999) Robustness of the EWMA control chart to normality. *Journal of Quality Technology*, 31(3), 309-316.

[6] Burr, I. W. (1967). The effect of non-normality on constants for X and R charts. *Industrial Quality Control*, 34, 563-569.

[7] Grant, E. L., Leavenworth, R. S. (1996). *Statistical Quality Control*. 7th Ed. McGraw-Hill, NY.

[8] Grimshaw, S. D., And Alt, F. B. (1997). Control charts for quantile function values. *Journal of Quality Technology*, 29 (1), 1-7.

[9] Kaminsky, F. C., Benneyan, J. C., Davis, R. D., Burke, R. J. (1992). Statistical control charts based on a geometric distribution. *Journal of Quality Technology*, 24, 63-69.

[10] Montgomery, D.C. (2001). *Introduction to Statistical Quality Control*, 4th Ed., John Wiley & Sons.

[11] Qesenberry, C. P. (1991a). SPC Q charts for a binomial parameter: Short or long runs. *Journal of Quality Technology*, 23, 239-246.

[12] Quesenberry, C. P. (1991b). SPC Q charts for a Poisson parameter λ: Short or long runs. *Journal of Quality Technology*, 23, 296-303.

[13] Ryan, T. P., Schwertman, N. C. (1997). Optimal limits for attributes control charts. *Journal of Quality Technology*, 29, 86-98.

[14] Schilling, E. G., Nelson, P. R. (1976). The effect of non-normality on the control limits of X charts. *Journal of Quality Technology*, 8, 183-188.

[15] Shore, H. (1991). New control limits that preserve the skewness of the monitoring statistic. *Transactions of the 45th Annual Quality Congress*, 45, 77-83.

[16] Shore, H. (1998). A new approach to analysing non-normal quality data with application to process capability analysis. *International Journal of Production Research*, 36, 1917-1933.

[17] Shore, H. (2000a). General control charts for attributes. *IIE Transactions,* 32, 1149-1160.

[18] Shore, H. (2000b). General control charts for variables. *International Journal of Production Research*, 38(8), 1875-1897.

[19] Shore, H. (2001). Modeling a non-normal response for quality improvement. *International Journal of Production Research,* 39 (17), 4049-4063.

[20] Stuart, A., Ord, J. K. (1987). *Kendall's Advanced Theory of Statistics, V. 1: Distribution Theory.* Charles Griffin & Co. Ltd. London.

[21] Wheeler, D. J. (1991). Shewhart's charts: Myths, facts and competitors. *Transactions of the ASQC 45th Annual Quality Congress,* 533-538.

[22] Woodall, W.H. (1997). Control charting based on attribute data: Bibliography and Review. *Journal of Quality Technology,* 29, 172-183.

[23] Yourstone, S.A., And Zimmer, W.J. (1992). Non-normality and the design of control charts for averages. *Decision Sciences,* 23, 1099-1113.

Chapter 23

Inventory Analysis

23.1. Introduction

The piece-wise linear approximations, introduced in Chapter 21, were used in the past for a wide array of inventory-analysis applications. It was generally assumed that the distributions of related random variables, like lead-time or lead-time demand, were unknown. However their first three moments were estimable so that the piece-wise linear approximations could replace the unknown exact distributions. Some references, where the piece-wise linear approximations were implemented to obtain optimal solutions for stochastic inventory-related problems include Shore (1986ab, 1995a, 1999, 2004a) and Moon and Silver (2000).

In this chapter, relationships and formulae, used in some of these references, are developed in Section 23.2. Only expressions, which seem to have potential applicability to other inventory-related stochastic optimization problems, are included.

An alternative approach to inventory analysis, based on a new four-parameter distribution, recently introduced, is developed in Section 23.3. Though this distribution is unrelated to the RMM model, it still demonstrates application of empirical modeling to solve inventory problems.

The last Sections 23.4 and 23.5 demonstrate application of the two approaches for some inventory-analysis optimization models.

23.2. First Approach - The Quantile Function and the Loss Function

In Chapter 21 the piece-wise linear approximation was derived

$$y = \begin{cases} A_1 z + B_1, z < 0, \\ \\ A_2 z + B_2, z \geq 0, \end{cases} \tag{23.1}$$

where y is the quantile of a random variable (r.v.), Y, having distribution function, F(y), and z is the corresponding quantile of Z, a standardized r.v. with a symmetric distribution (Sk=0).

Re-expressed in a different form, one may write for (23.1)

$$y = \begin{cases} \mu + \sigma [(A - C)z - hC], z < 0, \\ \\ \mu + \sigma [(A + C)z - hC], z \geq 0, \end{cases} \tag{23.2}$$

where $\{h, A, C\}$ are the solution of

$$h = 2M_1,$$

$$A^2 = 1 - C^2(1 - 4M_1^2),$$

$$Sk = 6C(M_3 - M_1) + 2C^3(12M_1^2 M_3 - 4M_1^3 - 2M_3), \tag{23.3}$$

M_i is the i-th upper partial moment of Z (the i-th degree moment of the upper half of the distribution of Z), and $\{\mu, \sigma, Sk)$ are, respectively, the mean, the standard deviation and the skewness measure of Y (Sk=0 for the normal). Refer for details to Chapter 21.

Equations (23.2) and (23.3) provide an approximation that preserves the first three moments of the distribution of Y (mean, variance and skewness). If Y is a discrete variable, "-1/2" should be added to the right hand side of both (23.1) and (23.2) to correct for continuity.

Simpler expressions may be obtained to calculate the parameters, as expounded in great detail in Chapter 21. Some of these expressions may now be used to derive approximate expressions for the loss function.

The loss function (LF) plays an important role in many stochastic optimization models. In particular, it is useful in the area of inventory analysis, where penalties for shortages constitute a regular component of the objective function in the optimization problem. The LF is defined by

$$L_Y(u) = \int_u^\infty (y-u)f(y)dy = \int_u^\infty [1-F(y)]dy \,, \; y \geq 0, \qquad (23.4)$$

where $L_Y(u)$ is the loss function at Y=u.

From the structure of the expression for the LF, it is readily recognized why it plays such a central role in inventory analysis. For example, if at a particular point in time, when a replenishment order is issued, available inventory is u, and lead-time demand (namely, demand for the stored items in the time interval until the order arrives) is a random variable, Y, with CDF, F(y), then obviously the LF expresses the expected shortage at the time that the order arrives. (Note that in this example the lead time is assumed to be constant.)

The unique properties of the piece-wise linear approximations allow derivation of explicit simple expressions for the LF.

Suppose first that Z is standardized logistic [refer to (19.8) or (22.4)]. From (23.2), with parameters given in (21.21) but A=1, we have

$$dy = \begin{cases} \sigma(1 - 0.3178Sk)dz, \; P < 1/2, \\ \sigma(1 + 0.3178Sk)dz, \; P \geq 1/2, \end{cases} \qquad (23.5)$$

For the standardized logistic

$$dz = (\sqrt{3}/\pi)[P(1-P)]^{-1}dP, \qquad (23.6)$$

where $P=\Phi(z)$, and Φ is the CDF of Z.

Introducing into (23.5) and thence into (23.4), the loss function associated with (23.2), with A=1, is

$$L_Y(P) = \begin{cases} (\sqrt{3}/\pi)(\sigma)\{(1\text{-}0.3178Sk)[\log(1/2)\text{-}\log(P)] \\ \qquad\qquad -(1+0.3178Sk)\log(1/2)\}, \; P < 1/2, \\ -(\sqrt{3}/\pi)(\sigma)(1+0.3178Sk)\log(P), \qquad\qquad P \geq 1/2, \end{cases} \qquad (23.7)$$

where P= F(u), and F(.) is the distribution function of Y.

For many stochastic optimization models, both the quantile function and the loss function are integral components of the objective function, which needs optimization (refer to the numerical examples below). From the quantile function (23.2), with Z given by (19.8), and the LF given by (23.7), we obtain formulae expressed explicitly in terms of P. The optimal value of P may now be easily found, either numerically, by implementing a minimization routine, or analytically, by differentiation with respect to P, and equating to zero. The latter will be demonstrated in the numerical example of Section 23.4.

The above expressions for the quantile function and the loss function are based on a piece-wise linear approximation that preserves (exactly or approximately, depending on how the parameters have been calculated) the first three moments of the underlying distribution. For example, if Y represents lead-time demand, the first three moments of this demand distribution are preserved. As related recently by Lau and Lau (2003), the third moment seems to be important in determining optimal solutions for inventory problems. The fact that the piece-wise linear approximations preserve that moment (either exactly or approximately) adds to the relevance of these approximations to the general area of inventory analysis.

Consider now the general case, when Z can be any standardized symmetrically distributed r.v. (like the various "candidates" addressed in Chapter 21). Further assume that Z has simple expressions for the quantile function in terms of P (as just demonstrated for the standardized logistic).

It is easy to show that for Y, given by the quantile function (23.1), the LF in terms of the LF of Z, $L_Z(P)$, is

$$L_Y(P) = \begin{cases} A_1 L_Z(P) + (A_2 - A_1)L_Z(0.50), & P < 1/2, \\ A_2\, L_Z(P), & P \geq 1/2. \end{cases} \qquad (23.8)$$

[Derivation of this relationship follows the same course that led to the derivation of (23.7).]

Let us use one of Shore's five-moment approximations for Z (Shore, 1982), given in Section 21.3 and reintroduced here for convenience:

$$z_p = -0.4115\{(1-P)/P + \log[(1-P)/P] - 1\}, \quad P \geq 1/2. \qquad (23.9)$$

The associated loss function is

$$L_Z(P) = \begin{cases} 0.4115\{1 - \log[P/(1-P)]\}, \ P < 1/2, \\ 0.4115(1-P)/P, \ \ P \geq 1/2. \end{cases} \qquad (23.10)$$

Introducing into (23.8) we have

$$L_Y(P) = \begin{cases} A_1(0.4115)\{1 - \log[P/(1-P)]\} + (A_2 - A_1)(0.4115), \ P < 1/2, \\ A_2(0.4115)(1-P)/P, \ \ P \geq 1/2. \end{cases} \qquad (23.11)$$

Since Z used is an approximation for the standard normal variate, the parameters are (refer to Section 21.4)

$$A_1 = \sigma(1 - 0.4177Sk), \ B_1 = \mu - (1/3)\sigma Sk,$$

$$A_2 = \sigma(1 + 0.4177Sk), \ B_2 = \mu - (1/3)\sigma Sk. \qquad (23.12)$$

The two loss functions, given by (23.7) and (23.11), are comparable in terms of accuracy (refer to the afore-cited references for some numerical comparisons). It should be born in mind, however, that when a certain loss function is used in a stochastic optimization problem, the same Z should be used both in the expressions for the quantile function and for the loss function. Adhering to this principle tends to result in simpler expressions because some terms may cancel out in the process of deriving an analytical optimal solution for the problem on hand.

23.3. Second Approach - The Quantile Function and Loss Function

The piece-wise linear approximations, on which the first approach is based (Section 23.2), describe the quantile function of a random variable that is (at least theoretically) unbounded in both directions (since Z used is unbounded). By contrast, stochastic quantities in inventory modeling are often non-negative. Lead-time demand, or lead-time, are good examples. This characterization calls for developing an approach that preserves the simplicity of the first approach, is capable of accommodating a large spectrum of differently-shaped distributions, and at the same time reflects the actual range of variation of the random variable being modeled.

In this section, we use for inventory analysis a new four-parameter family of distributions (Shore, 1998), which satisfies these requirements. As was done in the previous section, the quantile function and the loss function are first developed. A maximum likelihood estimation procedure is then derived to estimate the parameters of the new distribution. In Section 23.5 the new family of distributions is applied to obtain an optimal solution for a certain inventory problem.

Let Y be the response (a r.v.), and assume that Y is bounded by zero from below, and is unbounded, otherwise. Let F(y) be the cumulative distribution function (CDF), and denote by μ and σ the mean and the standard deviation, respectively. Let y be the P-th quantile, namely, F(y)=P.

Consider the following inverse distribution function (the quantile function):

$$y(P) = \begin{cases} A_1[P/(1-P)]^{B_1}, & P < 1/2 \\ A_2\{[P/(1-P)]^{B_2} - 1\} + A_1, & P \geq 1/2, \end{cases} \qquad (23.13)$$

where $\{A_i, B_i\}$ (i=1,2) are parameters that need to be determined. The motivation for introducing this quantile function, and some of its characteristics, are treated in Shore (1998).

The density function (d.f.) of Y is

$$f[y(P)] = \begin{cases} [1/(A_1 B_1)]P^{1-B_1}(1-P)^{1+B_1}, & P<1/2, \\ [1/(A_1 B_1)]P^{1-B_2}(1-P)^{1+B_2}, & P \geq 1/2. \end{cases} \qquad (23.14)$$

Denote by $M_i(Y)$ the i-th upper partial moment of Y, namely,

$$M_i(Y) = \int_{P=1/2}^{1} [y(P)]^i\, dP, \qquad (23.15)$$

with y(P) given by (23.13), and dP= f(y)dy. Note that in (23.15) integration is with respect to P (not with respect to y). This practice will often be repeated where P, rather than y, is used as the variable of integration or differentiation.

Pursuing the two-moment (partial and complete) matching approach (refer to Chapter 21 for motivation and details), an estimation procedure may be developed to identify the parameters in (23.13) (refer to Shore, 1995b, 1998). However, the procedure requires a numeric search for the values of the parameters. This renders the estimation procedure somewhat cumbersome for practitioners. A modified procedure is highly desirable, and it will be developed later on.

Next, consider Y that is symmetrically distributed and unbounded in *both* directions. Denote the quantile of the *standardized* variable by $u=(y-\mu)/\sigma$. The following quantile function replaces (23.13) for this characterization of the range of variation of Y:

$$u(P) = \begin{cases} A\{[P/(1-P)]^B - 1\}, & P < 1/2, \\ -A\{[P/(1-P)]^{-B} - 1\}, & P \geq 1/2. \end{cases} \qquad (23.16)$$

This may easily be recognized as a generalization of the five-moment approximation to the quantile function (inverse distribution function) of the standard normal distribution, given in Shore (1982). Therein we have obtained, for the standard normal: A= 5.531, B= 0.1193.

Finally, if Y is bounded by LB from below and by UB from above (for example, for the standard beta distribution LB=0, UB=1), we have for the inverse distribution function of Y:

$$y(P) = \begin{cases} A_1[P/(1-P)]^{B_1} + LB, & P < 1/2, \\ A_2[(1-P)/P]^{B_2} + UB, & P \geq 1/2. \end{cases} \qquad (23.17)$$

Henceforth, we will address only (23.13), which adequately represents the range of variation typical to stochastic r.v.s in inventory analysis (like lead-time and lead-time demand).

When (23.13) is used to represent some commonly encountered distributions (using a quantile matching procedure), the accuracy obtained is satisfactory. If a two-moment procedure is employed to fit (23.13) to a specified skewed distribution, the fitted (23.13) has skewness measure that is well preserved. The kurtosis measure also does not depart appreciably from the exact value.

Some accuracy comparisons may be found in the afore-cited references.

The allied loss function can now be derived. Define

$$L_i(P) = \int_P^1 [A_iB_i][F/(1-F)]^{B_i}(1/F)dF,$$

$$(i=1 \text{ for } P<1/2, i=2, \text{ otherwise}). \qquad (23.18)$$

It may easily be shown that the loss function, associated with (23.13), is

$$L_Y(P) = \begin{cases} L_1(P) + [L_2(1/2) - L_1(1/2)], & P < 1/2, \\ L_2(P), & P \geq 1/2. \end{cases} \qquad (23.19)$$

This expression will be used in subsequent sections.

Suppose now that the moments, needed to determine the parameters of (23.13) via moment matching, are unknown, and have to be estimated from sample observations. It has been previously noted (Shore, 1996, 1998) that estimates of the parameters of (23.13) tend to have large sampling variability due to the fact that B_1 and B_2 appear as exponents. In particular, B_2 tends to be close to zero and is particularly susceptible to sampling deviations. An alternative approach that circumvents this difficulty is needed.

To develop the modified procedure, let us impose a requirement for continuity of the d.f. at $P=1/2$. This implies that [refer to (23.14)]: $A_1B_1 = A_2B_2 = K$, say, and (23.13) becomes

$$y(P) = \begin{cases} (K/B_1)[P/(1-P)]^{B_1}, & P < 1/2, \\ (K/B_2)\{[P/(1-P)]^{B_2} - 1\} + A_1, & P \geq 1/2. \end{cases} \qquad (23.20)$$

The parameter "B_2" is expected to be close to zero (observe representative values in Shore, 1995b). Since, for non-negative u, $\text{Lim}_{\lambda \to 0} [(u^\lambda - 1)/\lambda] = \log(u)$, we may rewrite (23.20) as

$$y(P) = \begin{cases} A_1[P/(1-P)]^{B_1}, & P < 1/2, \\ A_2\log[P/(1-P)] + B_2, & P \geq 1/2. \end{cases} \qquad (23.21)$$

Two new parameters are introduced for $P \geq 1/2$, to allow for a two-moment fitting. The "penalty" to be paid is the loss of continuity of the quantile function at $P=1/2$.

For a two-moment fitting procedure, using partial and complete moments, the expressions for the parameters in (23.21) are (Shore, 1998):

$$B_1 = 1.70993\{0.5\mu_2(V) - [\mu_1(V)]^2\}^{0.5},$$

$$A_1 = \exp\{2[\mu_1(V) + 0.69314B_1]\},$$

$$A_2 = 1.70993\{0.5M_2(Y) - [M_1(Y)]^2\}^{0.5},$$

$$B_2 = 2[M_1(Y) - 0.69314A_2], \qquad\qquad (23.22)$$

where V is the log transformation

$$V = \begin{cases} \log(Y), & Y < M, \\ 0, & Y \geq M, \end{cases} \qquad\qquad (23.23)$$

$\mu_i(V) = E[V^i]$ and M is the median of Y. It is interesting that all the parameters in (23.21) may be calculated directly from (23.22), and no numeric search routine is required. Also note that in calculating sample estimates for the moments of V, the sum in the nominator should be divided by n, the complete sample size, even though only observations smaller than the median participate in estimating the moment.

For a set of distributions, approximated by (23.21) with parameters given by (23.22), Table 23.1 displays the resulting parameters' values and the skewness and kurtosis values. High accuracy is obtained for the skewness measure even though only moments of second degree at most were used in the fitting procedure. Better accuracy is obtained in fitting the original (23.13), however the allied fitting procedure requires a numerical search for the parameters. Furthermore, when only sample data are available, there is a sampling accuracy problem associated with estimating the parameters, as explained earlier.

When moments are unknown, but sample estimates are available, all moments in (23.22) are replaced by their estimates. It is numerically demonstrated in Shore (1998) that for small to medium sized samples (say n<100), the skewness and kurtosis estimates, numerically calculated from (23.21) [with parameters estimated from (23.22)], have mean-squared-errors smaller than those of corresponding direct sample statistics. This is typical to moment-matching based on the first two moments (partial and complete). (Refer also to Shore, 2004b.)

Table 23.1. Parameters, skewness (Sk.ap) and kurtosis (Ku.ap), of the fitted (23.21). Parameters calculated from (23.22). The exact values are Sk and Ku, respectively.

Distribution	A1	B1	A2	B2	Sk.ap	Ku.ap
Gamma:						
$\alpha=3$, $\beta=2$,	5.337	.3972	2.507	5.162	1.142	1.950
Sk= 1.15, Ku= 2.00						
$\alpha=7$, $\beta=2$,	13.15	.2311	3.464	13.34	.8003	1.157
Sk=.756, Ku=.857						
Lognormal:						
$\mu_0=1.0$, $\sigma_0=0.3686$	2.636	.1900	0.8065	2.628	1.134	2.052
Sk= 1.20, Ku=2.66						
$\mu_0=1.0$, $\sigma_0=0.5106$	2.605	.2631	1.3602	2.420	1.506	3.245
Sk= 1.80, Ku=6.26						
Weibull:						
$\alpha=2$, $\beta=5$	4.378	.4615	1.450	4.271	.6841	.7547
Sk= 0.631, Ku=.245						
$\alpha=1.375$, $\beta=5$	4.1212	.6712	2.488	3.679	1.227	2.143
Sk= 1.23, Ku=1.97						
$\alpha=1.125$, $\beta=5$	3.948	.8204	3.459	3.076	1.560	3.314
Sk= 1.676, Ku=4.04						

The loss function of (23.21) is

$$L_Y(P) = \begin{cases} L_1(P) + [L_2(1/2) - L_1(1/2)], \ P < 1/2, \\ -A_2 \ \log(P), \ P \ge 1/2, \end{cases} \qquad (23.24)$$

with $L_1(P)$ given by (23.18), and $L_2(P)= -A_2 \log(P)$ ($P \ge 1/2$). Refer also to (23.19).

Until this point, the parameters of the quantile function of the response, Y, have been estimated via moment matching. However, the parameters in (23.13) may be easily estimated via maximum-likelihood.

Assume that there are n available observations $\{y_{(k)}\}$ (k=1,..,n), where $y_{(k)}$ is the k-th order-statistic (there are k-1 smaller observations). The likelihood function (L) is, from (23.14),

$$L = [1/(A_1B_1)]^{(n/2)} \prod_{k=1}^{n/2} [P_{(k)}{}^{1-B1}(1-P_{(k)})^{1+B1}]$$

$$x\ [1/(A_2B_2)]^{(n/2)} \prod_{k=n/2+1}^{n} [P_{(k)}{}^{1-B2}(1-P_{(k)})^{1+B2}], \qquad (23.25)$$

where $P_{(k)} = F[y_{(k)}]$, and n is assumed to be an even number. Introducing from (23.13) for $P_{(k)}$ in terms of $y_{(k)}$, (23.25) is expressed in terms of the given observations. The maximum-likelihood estimates for $\{A_i, B_i\}$ (i=1,2) may now be identified by maximizing log(L) either via differentiation or by a numerical search. A similar procedure can be developed for (23.21).

Other fitting and estimation procedures, including procedures for right- and left-censored observations, are developed in Shore (1996). The case of truncated observations may be particularly interesting for models with lost sales. The reader is referred to the above reference for the appropriate fitting procedures.

23.4. First Approach- Newsboy Problem with Order-up-to Policy

A general solution for the classical newsboy problem is derived, pursuing the first approach (Section 23.2). We assume that only the first three moments of the demand in the period, D, are known. Let S be the order-up-to level, and IC(S) be the expected inventory cost (ordering, holding and penalty) for the period. It is well known that IC(S) is given by

$$IC(S) = cS + h \int_{t=0}^{S} (S-t)f_D(t)dt + p \int_{t=S}^{\infty} (t-S)f_D(t)dt, \quad (23.26)$$

where $f_D(d)$ is the d.f. of D, and $\{c, h, p\}$ are the unit ordering cost, unit holding cost and unit penalty cost, respectively. It is assumed that penalty cost, p, is applied to excessive demand (that which exceeds available stock and thus create a shortage).

It may be easily shown that

$$\int_{t=0}^{S} (S-t)f_D(t)dt = L_D(P) + S - E(D), \qquad (23.27)$$

where $L_D(P)$ is the value of the loss function of D at its CDF value $F_D(S)=P$, and E(D) is the expected value of D. Introducing from (23.27) into (23.26), and then for S (the P-th quantile of D) from (23.1), we obtain for (23.26)

$$IC(S) = (c+h)(A_2 z_p + B_2) + (h+p) L_D(P) - (h)E(D), \qquad (23.28)$$

where z_p is the P-th quantile of Z, and $L_D(P)$ may be given by any of the formulae in Section 23.2. Note, that it is assumed that for the optimal solution P*>1/2. This explains why the parameters $\{A_2, B_2\}$ appear in (23.28).

Consider using for z_p the approximation for the standard normal variable given in (23.9). The corresponding loss function is given by (23.10), and the parameters $\{A_2, B_2)$ by (23.12). Note that both z_p and $L_D(P)$ are now expressed in terms of a single term, namely (1-P)/P. This requires differentiating of (23.28) with respect to (1-P)/P (assuming that P>1/2), rather than with respect to P.

We obtain the well-known formula (exact)

$$P^* = (p\text{-}c) / (p+h), \qquad (23.29)$$

or

$$(1\text{-}P^*)/P^* = (h+c) / (p\text{-}c), \qquad (23.30)$$

where P* is the optimal P.

Introducing from (23.30) into (23.9) and then into (23.1), with the parameters given by (23.12), the optimal S is

$$S^* = A_2 z^* + B_2$$

$$= A_2 \, (\text{-}0.4115)\{ \, (1\text{-}P^*)/P^* + \log[(1\text{-}P^*)/P^*] - 1\} + B_2$$

$$= \sigma_D \, [1 + (0.4177)Sk(D)](\text{-}0.4115)\{ \, (h+c) / (p\text{-}c)$$

$$+ \log[(h+c) / (p\text{-}c)] - 1\} + E(D) - (1/3)[Sk(D)] \, \sigma(D), \qquad (23.31)$$

where z* is the optimal quantile of Z (that corresponds to P*), and σ_D and Sk(D) are the standard deviation and the skewness measure of the demand, D. Equation (23.31) is an *explicit* expression for S*, the optimal value of the decision variable, in terms of the cost parameters and of the first three moments of D. If S* is assumed to be discrete, a value of 1/2 should be subtracted from the RHS of (23.31). Similar expressions may be developed for P*<1/2.

Introducing for $S^* = A_2 z^* + B_2$ back into (23.28), and similarly for $L_D(P^*)$, IC(S*) is obtained in terms of the costs and the first three moments of the demand, D. This general solution is not based on the normal assumption [$Sk(D) \neq 0$], and is valid while the first three moments are specified.

The reader may wish to examine the accuracy of this solution for some numerical data and specified non-normal distributions.

Numerical comparisons for this model and others, where the first approach was used to obtain solutions for a demand distribution that is only partially specified (via the first three moments), can be found in the afore-cited references.

23.5. Second Approach - Two Examples

23.5.1. The Continuous-Review (Q,R) Model

This is the most common continuous-review model. Given its "old" age (refer, for example, to Hadley and Whitin, 1963), it still attracts relatively large research effort. A recent example is Platt, Robinson and Freund (1997). We address here the approximate backorder version, as treated by Lau and Lau (1993).

The continuous review (Q,R) model assumes that when inventory level (which is continuously monitored) goes down to R, an order of size Q is issued. Lead-time demand is a r.v., Y, with CDF F(y). The long-run average cost per period is

$$C(Q,R) = KE(D)/Q + h[Q/2 + R - \mu] + \pi E(D)L_Y(P)/Q, \quad (23.32)$$

where Q (the lot size) and R (the reorder-point) are decision variables, $L_Y(P)$ is the loss function at $P=F(R)$, K is the (fixed) order cost, h and π are the unit carrying cost per period and the unit shortage cost, respectively, E(D) is the average demand *per period* and μ is the average lead-time demand (average of Y).

Introducing for R from (23.13) and for $L_Y(P)$ from (23.19), and differentiating with respect to Q and to P, we obtain the optimality conditions

$$(Q^*)^2 = [2E(D)K]/h + \{[2E(D)\pi]/h\}L_Y(P^*),$$

$$P^* = 1 - (Q^*h)/[E(D)\pi],$$

$$L_Y(P^*) = \begin{cases} L_1(P^*) + [L_2(1/2) - L_1(1/2)], & P^* < 1/2, \\ L_2(P^*), & P^* \geq 1/2. \end{cases} \qquad (23.33)$$

From these expressions, the optimal (Q^*, P^*) may be identified, and thence R^* from (23.13). Since in most inventory models it is expected that $P^* > 1/2$, this interval should be searched first for the optimal solution. Also note, that in (23.33) Q^* and P^* are the exact expressions, and only $L_Y(P^*)$ is particular to the new solution procedure.

An alternative procedure may be derived by using the modified expressions (23.21)- (23.24). For the common case of $P^* \geq 1/2$, one obtains the simpler optimality conditions

$$(Q^*)^2 = [2E(D)K]/h + \{[2E(D)\pi]/h\}[-A_2 \log(P^*)]$$

$$= [2E(D)K]/h + \{[2E(D)\pi]/h\}\{-A_2 \log[1-(Q^*h)/(E(D)\pi)]\}, \ P^* \geq 1/2$$

$$P^* = 1 - (Q^*h)/[E(D)\pi] \qquad (23.34)$$

Only a routine root-finding routine is now required to identify Q^*. Further details about the derivation of this model may be found in Shore (1999).

A numerical example

For the cost and demand parameters (taken from Lau and Lau's examples, 1993):

$$E(D) = 1000, \ K = 10, 200, \ h = 1, \ \pi = 5, 30.$$

Table 23.2 shows exact and approximate solutions, derived for some cases under various distributional scenarios. The three distributions selected for Table 23.2 represent mild skewness (Weibull), medium skewness (gamma) and extreme skewness (lognormal). We observe that for all three cases, the new approach yields highly accurate solutions. For the normal case, a solution procedure that employs approximations for the standard normal quantile function is presented and numerically demonstrated in Shore (1982).

To better appreciate the effectiveness of the new approach, suppose that the actual lead-time demand distribution is either lognormal or

Table 23.2. Comparison of exact solution (Q*, P*) and approximate solution (Qa*,Pa*), for the (Q,R) model. Approximate solutions using either (23.33) (upper entry) or (23.34) [lower entry, with parameters given in Table 23.1). The lowest two entries for the lognormal and Weibull cases display solutions derived under the gamma assumption.

Distribution	K=10				K= 200			
	π=5		π=30		π=5		π=30	
	Q*	Qa*	Q*	Qa*	Q*	Qa*	Q*	Qa*
	P*	Pa*	P*	Pa*	P*	Pa*	P*	Pa*
Gamma:								
α=3, β= 2,	144.0	144.2	143.9	143.9	635.2	635.2	635.0	635.2
Sk= 1.15, M=5.3		144.0		144.0		635.1		635.0
	.9704	.9712	.9952	.9952	.8729	.8729	.9788	.9788
		.9712		.9952		.8730		.9788
Lognormal:								
μ_0=1.0, σ_0=0.5106	143.1	143.2	143.4	143.2	634.0	634.0	634.2	634.3
Sk= 1.8, M=2.72		142.8		142.8		633.9		633.8
	.9714	.9713	.9952	.9952	.8732	.8732	.9789	.9789
		.9714		.9952		.8732		.9789
		142.6		142.6		633.8		633.6
		.9683		.9914		.8859		.9749
Weibull:								
α=2, β= 5	142.6	142.7	143.4	142.4	633.9	633.9	633.6	633.7
Sk= 0.631, M=4.2		142.9		142.9		634.0		633.9
	.9715	.9714	.9952	.9952	.8732	.8732	.9789	.9789
		.9714		.9952		.8732		.9789
		143.1		142.9		634.3		634.1
		.9779		.9984		.8656		.9853

Weibull, with parameters as given in Table 23.2. However, we have mistakenly identified the underlying lead-time distribution as gamma, with parameters that preserve the actual mean and standard deviation, namely, $\alpha= (\mu/\sigma)^2$, $\beta= \sigma^2/\mu$, where μ and σ are the actual mean and standard deviation of the lead-time demand distribution.

The lowest two entries of Table 23.2 for the lognormal and the

Weibull cases present the exact optimal solutions under this scenario. Note, that P_a^* stands for the *actual* value of P associated with the optimal R^*, where the latter had been identified under the gamma assumption.

Examining Table 23.2, one realizes that although both the mean and the variance are preserved, the wrongly assumed gamma distribution consistently results in solutions that are worse than those obtained under the new approach (using approximations). Although the differences may seem negligible in this example, cumulative experience with the new approach suggests that the results obtained here are representative of the more general case, namely, that a wrongly selected distribution (that preserves the *exact* first two moments) generally yields worse results than those derived by the new approach, where skewness is preserved. Numerical evidence may be found in Shore (1996, 1997).

Alternatively, we may wish to use some other four-parameter families of distributions (like Pearson's), which would be fitted by four-moment matching. When the exact moments are unknown, and only sample estimates are available, this approach would also provide less accurate solutions (relative to the new approach) due to the large sampling errors associated with estimates of skewness and kurtosis (relate to a discussion of this problem in Section 20.3).

To sum up, relative to the two other options available to the practitioner, namely, identifying the lead-time demand distribution (and risking mis-identification), or using some flexibly-shaped four-parameter distribution (to be estimated by four-moment matching), the new approach seems to consistently outperform for cases where sample data are needed in the solution procedure.

23.5.2. Safety lead-times for purchased components

This problem was initially formulated in Hopp and Spearman (1993, henceforth HS). Consider the problem of setting safety lead-times for purchased components, where the only manufacturing operation is final assembly. There are n components, and delivery times (for each component) are independent random variables with CDF (for component i) of $F_i(y)$ and mean μ_i (i=1,2,..,n). The optimization problem is to determine the lead-times used to order the components (n decision variable), given relevant costs (that will be detailed shortly).

The authors (HS) developed an approximate solution procedure, assuming normality for lead times. An alternative solution procedure, extended for non-normal cases and numerically demonstrated for the normal distribution, is given in Shore (1995a). In the latter reference, the solution procedure is based on the piece-wise linear approximations (Chapter 21). However, as alluded to at the beginning of Section 23.3, these approximations are unbounded in both directions and may deliver unsatisfactory accuracy, for lower-tail quantiles, if the approximated random variable is bounded from below (as lead-time demand is). Additionally, being linear, these approximations fail to deliver satisfactory accuracy if the optimal quantile value resides in the far right-tail of the distribution.

Here, the family of distributions, introduced in Section 23.3, is employed to derive a general solution for this problem. The accuracy obtained for non-normal populations is demonstrated. Only the first of the two formulations in HS will be addressed, however extension to the other formulation is straightforward.

Denote by l_i the lead-time used to order component i (a decision variable, i=1,2,..,n). Assume that the holding cost per day, for a component that arrives too early, is $h_i = rp_i/365$, where r is the annual inventory carrying cost rate, and p_i is the cost of component i. It is assumed that assembly cannot be performed early, namely, assembly does not start when the last component has arrived, but according to a pre-planned schedule. However, if *at least* one component is late, assembly is delayed at a time-independent cost of C. Denote by $W(l_i)$ the average waiting time for component i:

$$W(l_i) = \int_{y=0}^{li} (l_i - y)dF_i(y). \qquad (23.35)$$

The cost equation is (find further details in HS)

$$Z = \sum_{i=1}^{n} h_i W(l_i) + C[1 - \prod_{i=1}^{n} P_i], \qquad (23.36)$$

where $P_i = F_i(l_i)$, namely, P_i is the probability that component i will arrive on time (not be late). It is easy to verify that $W(l_i) = L_Y(P_i) + l_i - \mu_i$.

Introducing into (23.36), and then for $L_Y(P_i)$ from (23.19) and l_i from (23.13), one obtains, on differentiating with respect to P_i and equating to zero:

$$(A_k B_k) [P_i*/(1-P_i*)]^{Bk+1} (h_i) - CH = 0,$$

$$(i=1,2,..,n; k=1 \text{ or } 2) \tag{23.37}$$

where $P_i* = F_i(l_i*)$, l_i* is the optimal lead-time for component i, and

$$H = \prod_{j=1}^{n} P_j*. \tag{23.38}$$

From (23.37) and the expression for H (23.38), the values $\{P_i*\}$ may be computed.

A numerical example

Suppose that there are two components, and delivery times are independently and identically gamma distributed with parameters $\alpha=7$, $\beta=2$ (second case in Table 23.1). Further assume that $F(l_i*)>1/2$ [k=2 in (23.37)]. The cost data, taken from Case 1 in HS, are

$$p_1 = 979, p_2 = 298, r = 0.25.$$

These yield $h_1 = 0.6705$, $h_2 = 0.2041$. For C we take C=150 (instead of C=500 in HS; The latter's numerical example comprises 20 components, and using C=500 for our case would yield a trivial solution of $P_1* = P_2* \cong 1$). For the above gamma distribution: $A_2 = 780.1$, $B_2 = 0.00442$ [obtained from a two-moment numerical fitting of (23.13) for k=2].

Introducing into (23.37), three equations need to be solved:

$$(3.4480) [P_1*/(1-P_1*)]^{0.00442}(0.6705)[-1+1/(1-P_1*)]-150H = 0$$

$$\text{(for the first component)}$$

$$(3.4480) [P_2*/(1-P_2*)]^{0.00442} (0.2041)[-1+1/(1-P_2*)]-150H = 0$$

$$\text{(for the second component)}$$

$$P_1*P_2* = H \tag{23.39}$$

Solving these equations, the approximate optimal solution is

$$P_1^* = 0.9842, \ P_2^* = 0.9951.$$

This solution yields total cost of $Z^* = 15.83$.

The exact solution, searched numerically with the exact statistical expressions (including repeated numerical integration), is

$$P_1^* = 0.9855, \ P_2^* = 0.9958.$$

This solution yields total cost of $Z^* = 15.82$.

A highly accurate solution is obtained. By deriving the optimal solution using an incorrect lead-time distribution whilst preserving the actual mean and standard deviation, the reader may wish to examine the sensitivity of this solution, as was implemented for the previous example.

References

[1] Hadley, G., Whitin, T. (1963). *Analysis of Inventory Systems*. Prentice-Hall, Englewood Cliffs, NJ.

[2] Hopp, W. J., Spearman, M. L. (1993). Setting safety lead-times for purchased components in assembly systems. *IIE Transactions, 25*, 2-11.

[3] Lau, A. H., Lau, H. S. (1993). A simple cost minimization procedure for the (Q,R) inventory model: Development and evaluation. *IIE Transactions, 25*, 45-53.

[4] Lau, H. S., Lau, A. H. (2003). Nonrobustness of the normal approximation of lead-time demand in a (Q,R) system. *Naval Research Logistics, 50*(2), 149-166.

[5] Moon, I., Silver, E. A. (2000). The multi-item newsvendor problem with a budget constraint and fixed ordering costs. *Journal of the Operational Research Society, 51*(5), 602-608.

[6] Platt, D. E., Robinson, L. W., Freund, R. B. (1997). Tractable (Q,R) heuristic models for constrained service levels. *Management Science, 43*, 951-965.

[7] Shore, H. (1982). Simple approximations for the inverse cumulative function, the density function and the loss integral of the normal distribution. *Applied Statistic, 31*(2), 108-114.

[8] Shore, H. (1986a). Simple general approximations for a random variable and its inverse distribution function based on linear transformations of a non skewed variate. *SIAM J. Sci. Stat. Comput., 7*(1), 1-23.

[9] Shore, H. (1986b). General approximate solutions for some common inventory models. *Journal of the Operational Research Society, 37*(6), 619-629.

[10] Shore, H. (1995a). Setting safety lead-times for purchased components in assembly systems- a general solution procedure. *IIE Transactions, 27*, 634-637.

[11] Shore, H. (1995b). Fitting a distribution by the first two moments (partial and complete). *Journal of Computational Statistics and Data Analysis, 19*, 563-577.

[12] Shore, H. (1996). Optimum schedule for preventive maintenance: A general solution for a partially specified time-to-failure distribution. *Journal of Production and Operations Management, 5*, 148-162.

[13] Shore, H. (1997). A general formula for the failure-rate function when distribution information is partially specified. *IEEE Transactions on Reliability, 46,* 116-121.

[14] Shore, H. (1998). Approximating an unknown distribution when distribution information is extremely limited. *Communication in Statistics (Simulation and Computation), 27* (2), 501-523.

[15] Shore, H. (1999). Optimal solutions for stochastic inventory models when the lead-time demand distribution is partially specified. *International Journal of Production Economics,* 59, 477-485.

[16] Shore, H. (2004a). A general solution for the newsboy model with random order size and possibly a cutoff transaction size. *Journal of the Operational Research Society, 55* (11), 1218-1228.

[17] Shore, H. (2004b). Non-normal populations in quality applications- A revisited perspective. *Quality and Reliability Engineering International,* 20(4), 375-382.

Review Questions

Comment. The review questions are intended to address RMM-related issues that the reader may find beneficial to elaborate on in order to clarify various aspects of the RMM methodology. No attempt is made here to demonstrate numerical solutions via RMM other than the numerical examples, already given in the text. The reader, however, is encouraged to address real-life problems, from her/his own experience or from the literature (peer-reviewed papers or textbooks), and compare solutions derived by existing methodologies to RMM-based solutions.

The author welcomes such comparisons, to be possibly published (with proper attribution) in future editions of the book.

Question A. In Chapter 1, we have made a two-way distinction between various modes of modeling a response variation. This may be summarized in a tabular form, as shown in the table below:

Modeling:	Systematic Variation	Random Variation
Theory-based		
Empirically		

Discuss how modeling is currently carried out for each entry in the table. Relating to the two-way partition given, what drawbacks do you find in the various methodologies currently used to model a response variation (refer also to Chapters 4-6).

Question B. For each of the ten requirements of empirical modeling, expounded in Chapter 6, provide an example, from your own experience, where a modeling effort failed to meet a requirement. Discuss the actual consequences and implications (managerial, economical or otherwise) of failure to meet the requirement. To what extent does RMM provide a remedy (if at all).

Question C. In Part II of the book the RMM approach is developed. It is emphasized that there are two possible ways to define each error. Additionally, two equivalent forms were introduced of representing one of the two correlated standard normal variables in terms of the other *plus* an independent standard normal variable. This two-way classification of how the two errors can be presented generates altogether 8 different variations of the RMM model.

Introduce the eight variations in full, and provide, from your own professional know-how, instances where one version would deliver better representation than the others.

Question D. In Part II of the book, fitting and estimation procedures were developed for the basic RMM model. Select two of the RMM other seven variations, and likewise develop fitting and estimation procedures.

Question E. Explain and demonstrate why modeling a log-transformed response via RMM ensures that when the linear predictor increases, the variance stabilizes and thus is decoupled from the mean. Is this effect equally effective for all 8 variations of the RMM model? Demonstrate with numerical data.

Question F. Select any of the numerical problems in the text (or in any other source), and model the response variation, using three different versions of the RMM model. Assess, according to the goodness-of-fit statistics obtained, which is the most adequate to model the process that has generated the data.

Question G. For any of the numerical problems given in the book, prepare a complete statistical report about the parameters' estimates for $\{\alpha, \lambda, \mu_2\}$, obtained as a result of implementing non-linear least-squares (NL-LS). What conclusions do you derive from this report? (For example, sample size adequate? Parameters' estimates too highly correlated which requires model re-definition? *etc.*)

Question H. In Chapter 12, the Johnson families of distributions were presented as special cases of the RMM model (except for a modified definition of the errors). For a data set of your selection, use standard procedures to estimate the parameters of the Johnson family (be reminded that estimation comprises both selecting the proper family and estimating the parameters). Likewise, estimate parameters of the RMM model.

Based on goodness-of-fit statistics, compare the results of the two modeling efforts.

Question I. For the models in Chapter 2, not addressed in Chapter 12, show that they are indeed special cases of the RMM model.

Question J. Employing Monte-Carlo simulation, generate a thousand samples (with sample size $n=25$) from a highly-skewed known distribution (for example, gamma). For each sample:

- *Estimate* the skewness and kurtosis measures (calculate the sample's skewness and kurtosis)
- Fit an RMM error distribution (an INT may be used), applying any of the two-moment estimation procedures detailed in Chapter 20
- *Calculate* skewness and kurtosis from the fitted quantile function (use numerical integration).

Compare the mean-squared-errors (MSEs) obtained by the two approaches. How would this comparison fair if the sample size is increased to, say, $n=100$?

Question K. Fatigue life is commonly modeled by the Weibull distribution. Alternatively, one may model life-time by the error distribution associated with a relational model, for example, the RFL model of Pascual and Meeker (refer for details to Chapter 15).

Using data from any formerly published source, model the process that has generated the given fatigue-life data with the Weibull model, the RFL model and the RMM model. Evaluate and compare the results.

Can you think of reasons (historic or otherwise), why error distributions, associated with relational models, are *rarely* used in practice to model random variation?

Question L. In Chapter 18, two numerical examples are given for forecasting S-shaped diffusion processes. It was suggested that the correct way to assess forecasting capability of a given model is to fit the model with some of the data (for example, two thirds of the observations), and then extrapolate (prepare forecasts) to the rest of the data (observations excluded from the fitting routine).

This suggestion was implemented for the first numerical example, given in Section 18.4. Extend it to the second example.

Question M. In Section 20.6, two new normalizing transformations (NTs) are introduced, based on the inverse of the corresponding INTs. Use sample data that have been formerly published (in the literature or in textbooks) to estimate the parameter of the Box-Cox transformation. Likewise estimate the parameters of the two new NTs (for example, by moment matching). Evaluate the goodness-of-fit obtained from the three transformations. Also evaluate how normalized the data are after applying each of the estimated transformations.

Question N. For the two NTs introduced in Section 20.6, develop maximum likelihood procedures, similarly to the procedures developed for the Box-Cox transformation (Box and Cox, 1964).

Question O. Relating to the numerical examples of Chapter 23, demonstrate, with some skewed distributions of your choice, that it is preferable to use in an optimization model approximations (like those in Chapters 20 and 21), that had been fitted by a two-moment (partial and complete) procedure, than using a *mis-specified* distribution that preserves the actual (true) first two moments of the data-generating random variable.

Author Index

Note: Indented names relate to authors' full names as they appear in "References" sections. Non-indented names relate to authors' names in the text (for authors whose names appear in a "References" section).

Subject Index